The Biology of Mediterranean-Type Ecosystems

THE BIOLOGY OF HABITATS SERIES

This attractive series of concise, affordable texts provides an integrated overview of the design, physiology, and ecology of the biota in a given habitat, set in the context of the physical environment. Each book describes practical aspects of working within the habitat, detailing the sorts of studies which are possible. Management and conservation issues are also included. The series is intended for naturalists, students studying biological or environmental science, those beginning independent research, and professional biologists embarking on research in a new habitat.

The Biology of Streams and Rivers
Paul S. Giller and Björn Malmqvist

The Biology of Soft Shores and Estuaries
Colin Little

The Biology of the Deep Ocean
Peter Herring

The Biology of Soil
Richard D. Bardgett

The Biology of Polar Regions, 2nd Edition
David N. Thomas et al.

The Biology of Caves and Other Subterranean Habitats
David C. Culver and Tanja Pipan

The Biology of Alpine Habitats
Laszlo Nagy and Georg Grabherr

The Biology of Rocky Shores, 2nd Edition
Colin Little, Gray A. Williams, and Cynthia D. Trowbridge

The Biology of Disturbed Habitats
Lawrence R. Walker

The Biology of Freshwater Wetlands, 2nd Edition
Arnold G. van der Valk

The Biology of Peatlands, 2nd Edition
Håkan Rydin and John K. Jeglum

The Biology of African Savannahs, 2nd Edition
Bryan Shorrocks and William Bates

The Biology of Mangroves and Seagrasses, 3rd Edition
Peter J. Hogarth

The Biology of Deserts, 2nd Edition
David Ward

The Biology of Lakes and Ponds, 3rd Edition
Christer Brönmark and Lars-Anders Hansson

The Biology of Coral Reefs, 2nd Edition
Charles R.C. Sheppard, Simon K. Davy, Graham M. Pilling, and Nicholas A.J. Graham

The Biology of Mediterranean-Type Ecosystems
Karen J. Esler, Anna L. Jacobsen, and R. Brandon Pratt

The Biology of Mediterranean-Type Ecosystems

Karen J. Esler
Department of Conservation Ecology & Entomology, Stellenbosch University, South Africa

Anna L. Jacobsen
Department of Biology, California State University, Bakersfield, USA

R. Brandon Pratt
Department of Biology, California State University, Bakersfield, USA

OXFORD
UNIVERSITY PRESS

OXFORD
UNIVERSITY PRESS

Great Clarendon Street, Oxford, OX2 6DP,
United Kingdom

Oxford University Press is a department of the University of Oxford.
It furthers the University's objective of excellence in research, scholarship,
and education by publishing worldwide. Oxford is a registered trade mark of
Oxford University Press in the UK and in certain other countries

© Karen J. Esler, Anna L. Jacobsen & R. Brandon Pratt 2018

The moral rights of the authors have been asserted

First Edition published in 2018
Impression: 1

Published in the United States of America by Oxford University Press
198 Madison Avenue, New York, NY 10016, United States of America

British Library Cataloguing in Publication Data
Data available

Library of Congress Control Number: 2017951487

ISBN 978–0–19–873913–5 (hbk.)
ISBN 978–0–19–873914–2 (pbk.)

DOI 10.1093/oso/9780198739135.001.0001

Printed in Great Britain by
Bell & Bain Ltd., Glasgow

Links to third party websites are provided by Oxford in good faith and
for information only. Oxford disclaims any responsibility for the materials
contained in any third party website referenced in this work.

To all mediterranean-type ecosystem researchers, past and present, who have contributed to understanding and conserving these unique and beautiful regions.

Preface and Acknowledgements

The mediterranean-type climate (MTC) regions of the world are special places. We have been enchanted and inspired by the beautiful and interesting landscapes and species that are found in MTC regions. They are the birthplaces of humans (South Africa) and human civilization (the Mediterranean). The mild MTC is attractive to the millions of people that make their homes in these densely populated regions, as well as to the millions of visitors that take holidays in these regions. MTC region biodiversity is notably high and rivals that of the wet tropics in some taxonomic comparisons. Critically, and like many areas of the globe, these regions are threatened by a wide range of factors that include habitat loss owing to development and agriculture, disturbance, invasive species, and climate change.

Comparisons among MTC regions have yielded critical insights into important questions that are at the cutting edge of ecological and evolutionary research. These regions continue to be a model system for comparisons that attract scholars and students from many areas of the globe. By writing this book, as part of Oxford's Biology of Habitat Series, we hoped to share some of what we find so fascinating about these regions, particularly to younger scientists, who will be the future of MTC region research. But, beyond the enchanting and inspiring, we were also strongly motivated to shine a spotlight on the threats that are despoiling these special places at alarming rates. Many challenges lie ahead and it will require energetic and talented scholars, land managers, and policy-makers to ensure that these regions remain special places.

The structure and format of this book was designed to appeal to general readers as well as students, researchers, and land managers working within MTC regions. The book is divided into four sections. The first section introduces mediterranean-type ecosystems (MTEs) and the key characteristics that define them. The second section introduces the organisms and communities of MTC regions and the traits that uniquely suit them to the characteristics of these regions. The third section focuses on the processes that lead to current community compositions within the MTC regions and the functions of organisms within them. Finally, the fourth section addresses the challenges faced by the natural communities within MTC regions and how these challenges may be addressed in the future. We have tried to highlight areas of interest and future research questions. To aid the reader, we have included in **bold** important terms within each chapter. These terms are defined in the text where they occur and, for those wishing to expand their reading to primary resources, these terms are important in finding and understanding relevant literature. Students may find these particularly useful as they build their scientific vocabulary in diverse areas of research.

Within many of the chapters, we have included contributed 'Case Studies'. Our intent with these contributions was to enrich the text with in-depth examples on highlighted topics, draw attention to debate and provocative questions and topics, highlight areas of future research, and illustrate the personal experiences of researchers in these regions. In keeping with the book theme, we endeavoured to ensure that each Case Study makes comparisons between at least two of the five MTEs.

We acknowledge the many inspired studies of colleagues past and present that have illuminated fascinating aspects of MTC region biology, which have engendered the interest of so many scientists, including ourselves. With the enormous body of literature on MTC region biology, we have had to omit specific mention of countless important studies. We would like to specially acknowledge and thank all of our colleagues who contributed Case Studies. Many individuals assisted in the development of the outline, content, and details of this book. This includes attendees of the 2014 MEDECOS, held in Olmué, Chile, who attended a workshop held on this book. They suggested many topics and themes that we incorporated into our book outline and many of these attendees also volunteered to contribute Case Studies. We would like to thank our colleagues who graciously answered email inquiries from us as we developed the text, including Mary Arroyo, Stephen Cousins, Richard Cowling, Stephen Davis, Stuart Hall, Joey Hulbert, Jon Keeley, Byron Lamont, Curtis Marean, Guy Midgley, Kenneth Oberlander, Fernando Ojeda, Phil Rundel, Mlungele Nsikani, Petr Pyšek, Hugh Safford, and Martina Treurnicht. We would also like to thank Martin D. Venturas, Tessa Oliver, and Stuart Hall for contributing photos, and Stephen Myburgh and Mymoena Londt for tracking down references.

We hope that this book provides an overview of key issues and concepts relevant to MTEs and brings together the diverse and sometimes fragmented literature specific to these systems. We have thoroughly enjoyed our collaborations with international researchers, and the many friends that we have made along the way. We hope that this book serves as a catalyst and inspiration for the next generation of MTEs researchers.

Karen J. Esler, Anna L. Jacobsen, and R. Brandon Pratt
18 July 2017

Contents

List of Plates   xiii

Case Study Contributors and Affiliations   xv

SECTION 1 **Setting the Scene: Mediterranean Landscapes**

1 **Introduction**   3

1.1 Ecosystems   3

1.2 Mediterranean-type ecosystems   5

1.3 Defining mediterranean-type ecosystems   7

1.4 A brief history of early comparative mediterranean research   11

CASE STUDY 1: Early mediterranean ecosystem comparisons.
   A personal history   14
   Contributed by H.A. Mooney

1.5 Hotspots of biodiversity   17

2 **Characteristics of Mediterranean-Type Ecosystems**   23

2.1 Origins and biogeography   23

CASE STUDY 2: Assembly of mediterranean-type floras—
   convergence, exaptation, and evolutionary predisposition   27
   Contributed by David D. Ackerly and Renske E. Onstein

CASE STUDY 3: The role of environmental stability in explaining
   variation in plant diversity in mediterranean-type ecosystems   34
   Contributed by Richard M. Cowling

2.2 Introduction to ecosystem drivers and processes   35

   2.2.1 Climate   35

CASE STUDY 4: Freezing as an understudied driver of
   plant distribution within mediterranean-type ecosystems   39
   Contributed by Stephen D. Davis and George Matusick

   2.2.2 Fire   43

   2.2.3 Topography and geology   47

   2.2.4 Soils and mineral nutrition   51

2.3 Evolutionary convergence   53
2.4 An overview of mediterranean-type ecosystem characteristics   56

SECTION 2   Painting the Picture: The Living Template

3   Organisms and their Interactions   69

3.1 Organismal adaptations   69
    3.1.1 Plants   70
      3.1.1.1 Phenology and life form   70
      3.1.1.2 Roots, soils, and symbioses   77
CASE STUDY 5: Pioneering research on proteoid root clusters in three mediterranean regions   80
Contributed by Byron B. Lamont and María Pérez-Fernández
      3.1.1.3 Fire   84
CASE STUDY 6: Central Chile compared to California: the enigma of the Chilean mediterranean-type climate flora   85
Contributed by Mary T.K. Arroyo
    3.1.2 Animals   87
    3.1.3 Microbes   88
3.2 Diversity and endemism   89
3.3 Ecological and evolutionary context   91
    3.3.1 Edaphic communities   92
    3.3.2 Animal and plant interactions   92
      3.3.2.1 Plants and herbivores   93
      3.3.2.2 Plants and pollinators   94
      3.3.2.3 Plants and seed dispersers   95

4   Diversity and Community Structure   109

4.1 Assembling plant communities   109
4.2 Shrublands   118
    4.2.1 Evergreen sclerophyll shrublands   118
CASE STUDY 7: Mediterranean-type vegetation outside of mediterranean-type climate regions   119
Contributed by Philip W. Rundel
    4.2.2 Heathlands—a component of evergreen sclerophyll shrublands   123
    4.2.3 Drought-deciduous soft-leaved shrublands   124
    4.2.4 Desert shrublands along arid margins   125
4.3 Woodlands and forests   128
CASE STUDY 8: Large old (venerable) trees of fire-prone mediterranean-type climate regions   129
Contributed by Grant W. Wardell-Johnson
4.4 Grasslands   133

4.5 Riparian communities   136
4.6 Vegetation dynamics: patchiness in space and time   138

SECTION 3 **Choreography: Life in Motion**

5 **Evolution and Diversity**   149

5.1 Origins   149
5.2 Macroevolutionary patterns   155
    5.2.1 Divergence   155
    5.2.2 Convergence   160
CASE STUDY 9: Mediterranean heathland and fynbos:
    a neglected example of convergence between
    mediterranean climate regions   160
    Contributed by Fernando Ojeda
    5.2.3 Extinction   165
5.3 Diversity   165
5.4 Drivers of diversity   166
CASE STUDY 10: The origin of the mediterranean biome   167
    Contributed by Philip W. Rundel

6 **Form and Function of Mediterranean Shrublands**   177

6.1 Structure and physiology   177
    6.1.1 Sclerophyllous leaves   177
    6.1.2 Leaf traits other than sclerophylly   182
6.2 Photosynthesis and growth   186
6.3 Responding to limited water   196
6.4 Demography and population dynamics: the key role of fire   200
    6.4.1 Fire regime   200
CASE STUDY 11: Linking fire traits and historical fire regimes
    in the mediterranean-type environment   202
    Contributed by Juli G. Pausas
    6.4.2 Shrub response to fire: from individuals
        to populations   204
    6.4.3 Evolution of traits in response to fire   206

7 **Ecosystems processes**   219

7.1 Ecosystem structure and primary productivity   219
7.2 Mineral nutrients   225
    7.2.1 Plant adaptations to low nutrients   227
    7.2.2 Nutrients, fire, and decomposition   229
7.3 Hydrology   231

SECTION 4 **The Modern Stage: Transformation**

8 **Transformation**   239

8.1 Threats to mediterranean-type climate regions   239

8.2 Human interactions with mediterranean-type climate
region landscapes   241

8.3 Land-use, land-cover change   244

CASE STUDY 12: Australian acacias—super invaders
of mediterranean-type ecosystems   247
Contributed by David M. Richardson

CASE STUDY 13: Land-use changes in an urbanizing world: a
comparison between the city of Cape Town, South Africa
and Los Angeles County, USA   251
Contributed by Patricia M. Holmes and Alexandra D. Syphard

8.4 Habitat fragmentation   258

8.5 Fire regime changes and habitat type conversion   259

8.6 Invasive species   263

    8.6.1 Invasive plants   263

    8.6.2 Invasive animals   267

    8.6.3 Pathogens   268

8.7 Nutrient enrichment   270

8.8 Climate change   271

8.9 Conclusion   276

9 **Planning for the future**   291

9.1 Introduction   291

9.2 Conservation approaches   293

    9.2.1 Protected area expansion   296

    9.2.2 Conservation stewardship   297

    9.2.3 Biodiversity and spatial planning   300

9.3 Ecological restoration and related activities   302

9.4 Climate change, altered disturbance regimes, and fire
management   307

CASE STUDY 14: Fire and climate change in
mediterranean-type ecosystems   310
Contributed by Max A. Moritz and Enric Batllori

9.5 Conclusion   312

Index   323

List of Plates

Plate 1 Landscape photos from each of the five mediterranean-type ecosystem (MTE) regions (a–e) (see p. 5).

Plate 2 Convergent evolution of a succulent plant form in different deserts (a–d) (see p. 6).

Plate 3 Conspicuous common northern hemisphere and southern hemisphere taxa (see p. 26).

Plate 4 Examples of plants that re-establish through resprouting following fire (see p. 45).

Plate 5 Examples of plants that re-establish through seeding following fire (see p. 46).

Plate 6 Plant physical defence structures related to herbivory (see p. 94).

Plate 7 Flowers may be suited for generalist or specialist pollinators (see p. 96).

Plate 8 Arid shrub communities (see p. 126).

Plate 9 Grassland in a mediterranean-type climate region (see p. 134).

Plate 10 Relict species currently have limited distributions (see p. 156).

Plate 11 Resprouting and non-resprouting species from the same genus (see p. 159).

Plate 12 Adaptive leaf traits (see p. 184).

Plate 13 Carnivorous and parasitic plants (see p. 230).

Plate 14 Mediterranean region traditional land use (a, b) (see p. 246).

Plate 15 A degraded Chilean landscape (see p. 257).

Plate 16 Invasive alien plants (see p. 264).

Plate 17 Pathogen impacts on mediterranean-type climate region plants (see p. 269).

Plate 18 Drought-induced mortality (see p. 274).

Case Study Contributors and Affiliations

David D. Ackerly
Department of Integrative Biology, University of California, Berkeley (UC Berkeley), Berkeley, California, USA

Mary T.K. Arroyo
Departamento de Ciencias Ecológicas, Instituto de Ecología y Biodiversidad (IEB), Universidad de Chile, Santiago, Chile

Enric Batllori
CEMFOR-CTFC, InForest Joint Research Unit, CSIC-CTFC-CREAF, Solsona, Spain; CREAF, Cerdanyola del Vallès, Spain

Richard M. Cowling
Centre for Coastal Palaeosciences; Nelson Mandela University, Port Elizabeth, South Africa

Stephen D. Davis
Natural Science Division, Pepperdine University, Malibu, California, USA

Patricia M. Holmes
Biodiversity Management Branch, Environmental Management Department, City of Cape Town, South Africa

Byron B. Lamont
Department of Environment and Agriculture, Curtin University, Perth, Australia

George Matusick
State Centre of Excellence for Climate Change Woodland and Forest Health, School of Veterinary and Life Sciences, Murdoch University, Murdoch, Australia

H.A. Mooney
Department of Biology, Stanford University, Stanford, California, USA

Max A. Moritz
Department of Environmental Science, Policy, and Management, University of California, Berkeley (UC Berkeley), Berkeley, California, USA

Fernando Ojeda
Departamento de Biología-IVAGRO, Universidad de Cádiz, Campus Río San Pedro, Spain

Renske E. Onstein
Institute for Biodiversity and Ecosystem Dynamics (IBED), University of Amsterdam, Amsterdam, The Netherlands

Juli G. Pausas
Centro de Investigaciones sobre Desertificacion, Consejo Superior de Investigaciones Científicas (CIDE-CSIC), Valencia, Spain

María Pérez-Fernández
Ecology Área, University Pablo de Olavide, Sevilla, Spain

David M. Richardson
Centre for Invasion Biology, Department of Botany and Zoology, Stellenbosch University, Stellenbosch, South Africa

Philip W. Rundel
Department of Ecology and Evolutionary Biology, University of California, Los Angeles (UCLA), Los Angeles, California, USA

Alexandra D. Syphard
Conservation Biology Institute, Corvallis, Oregon, USA

Grant W. Wardell-Johnson
Department of Environment and Agriculture, Curtin University, Perth, Australia

Setting the Scene: Mediterranean Landscapes

1 **Introduction**

Abstract

Mediterranean-type climate (MTC) regions have long been of interest to scientists and they formed the basis for many early ecological studies. This has included comparisons of the vegetation within these regions (mediterranean-type vegetation) as well as other functional, climatic, and historical studies and comparisons. Comparing MTC regions and the species that occur within them has been used to test the evolutionary convergence hypothesis. Continuing scientific interest in MTC regions is linked to their unusually high levels of species richness and biodiversity. These regions have the highest species richness outside of the tropics, particularly in vascular plant diversity, as well as high levels of endemism. International research activities and meetings have provided the opportunity for scholars to collaborate across MTC regions and have fostered an active comparative research environment from the 1960s to the present.

1.1 Ecosystems

Early humans had a keen sense of how species distributions varied in space and time, as their very existence depended on the resources obtained from wild plants and animals. Even though modern societies now rely primarily on domesticated species, we are still dependent on wild species for the **ecosystem services** they provide. Ecosystem services include the myriad ways that functioning natural systems benefit humankind, including clean air and water. Moreover, these wild species and the natural habitats in which they occur continue to hold great fascination for us; the cultural, spiritual, and recreational benefits of ecosystems are important components of ecosystem services. Ecosystems and the species within them teach us critical details

The Biology of Mediterranean-Type Ecosystems. Karen J. Esler, Anna L. Jacobsen, and R. Brandon Pratt,
Oxford University Press (2018). © Karen J. Esler, Anna L. Jacobsen, and R. Brandon Pratt 2018.
DOI 10.1093/oso/9780198739135.001.0001

about our planet's past, present, and potential future as well as enriching our lives with their beauty.

Modern understanding of species distributions has its roots in the era of scientific discovery during the eighteenth and nineteenth centuries. At this time, scientific explorers travelled the world collecting and describing species that were previously unknown to western science. In addition to cataloguing new specimens, mapping new areas, and looking for valuable resources, some of these scientists were formulating grand ideas that forever changed science. From these early scientists, like Alexander von Humboldt (1769–1859) and Charles Darwin (1809–1882), the understanding of factors determining species distributions came into focus and the study of these factors lay the foundation for the science of ecology (Lomolino et al. 2004).

Species distributions vary in space. The dominant drivers of these distributions are well established, and chief among them is climate (Holdridge 1947; Walter 1973). Distant parts of the globe may have similar climates and these climatically similar regions often contain assemblages of plants and animals that appear to be superficially similar in the structure of their vegetation. The term **biome** is used to refer to regions of similar climate and vegetation structure, and examples include the tropical rainforests, boreal forests, tundra, grasslands, and mediterranean-type shrublands.

Not only are regions that share a common biome structurally similar but, according to ecological theory, they should also function similarly. In other words, biomes are predicted to identify and describe **ecosystems** (Chapter 7). An ecosystem includes the organisms in an area and how they interact with their **environment**, including both the living and non-living components of their surroundings. The flow of energy and material that occurs through the interactions between many different species and their shared environment gives rise to a particular **trophic structure** (how organisms are organized into different feeding relationships) and cycling of materials that characterize an ecosystem. The vegetation of terrestrial biomes fills the role of 'producer' within these systems, capturing solar energy that is then converted to the biomass that is food to consumer organisms (Chapter 7). Plants are most often used to define biomes and ecosystems, and biomes are often named after the type of vegetation they contain.

The precise location of biomes and the functioning of ecosystems are determined by several factors. As mentioned above, one of the most important determinants is climate; however, there are other **ecosystem drivers** that affect species distributions and ecosystem function. The most important of these drivers are soils, physiography, disturbance regime, and history. These drivers are discussed more fully in Chapter 2 as they relate to our focal ecosystem type, mediterranean-type ecosystems (MTEs).

1.2 Mediterranean-type ecosystems

There are five regions of the world that are commonly referred to as MTEs: southwestern Australia, California, central Chile, the Mediterranean Basin, and the Cape Region of South Africa.* All five of these regions are dominated by vegetation that forms a structurally similar shrubland (Figure 1.1). The dominant species within these shrublands are evergreen shrub species with

Figure 1.1   **Landscape photos from each of the five mediterranean-type ecosystem (MTE) regions.** Each of the mediterranean-type ecosystem regions of the world is typified by an evergreen shrub-dominated vegetation. This similarity gives the regions a very similar appearance at the landscape level. The images included here show mountain views from each of the five regions: (a) Stirling Range, Western Australia, Australia, (b) Santa Ynez Mountains, California, USA, (c) Cordillera de la Costa, Valparaíso Region, Chile, (d) Corsica, France, and (e) Western Cape, South Africa. Each region and country uses different names for this vegetation type and, for the shown images, the regional vegetation type names are: (a) mallee, (b) chaparral, (c) matorral, (d) maquis, and (e) fynbos. (See Plate 1) *Source:* Photos c and d from Anna L. Jacobsen and a, b, and e from R. Brandon Pratt.

* Note that a lower case 'm' is used when referring to these mediterranean regions generally and a capital 'M' is used when referring to the specific region of the Mediterranean.

sclerophyllous, thick and leathery, leaves (Chapter 6). It is these evergreen shrublands that are recognized as a biome, the mediterranean-type shrubland biome, and also as a unique vegetation type, mediterranean-type vegetation (MTV). It is not generally established that the five MTEs function similarly at the level of an ecosystem, and indeed the differences between these systems are in many ways more illuminating than their similarities. For this reason, we generally avoid referring to these regions as MTEs and instead refer to them as **mediterranean-type climate** (MTC) regions (Section 1.3).

Many of the taxa that dominate these regions are distantly related evolutionarily, particularly between the northern and southern hemispheres. This has led to the hypothesis that features unique to MTEs have selected for this growth form and that these shrublands represent an example of **convergent evolution** (Schimper 1903). Convergent evolution describes the pattern of distantly related species independently evolving similar features as a result of similar selective pressures. One of the most commonly cited examples of convergent evolution is the separate appearance of fleshy and thickened

Figure 1.2   **Convergent evolution of a succulent plant form in different deserts.** Within the winter-rainfall deserts that occur adjacent to regions dominated by mediterranean-type vegetation (MTV), distantly related plant families have evolved a common fleshy body form that allows them to store water. In California, USA, an example of this growth form is found within the Cactaceae (a) and it is also found within the Euphorbiaceae, as shown in an example plant from South Africa (b). Although these plants have a superficially similar form, their distant evolutionary relatedness is evident in the extreme divergence in floral structure between these Cactaceae (c) and Euphorbiaceae (d) species. (See Plate 2)
Source: Photos from Anna L. Jacobsen.

tissues specialized for storage of water (succulence) in species from the plant families Cactaceae (new world) and Euphorbiaceae (old world) within different arid regions in the new and old worlds (Figure 1.2).

One of the chief reasons scientists have been interested in comparing MTC regions and the species that occur within them is to test the evolutionary convergence hypothesis. This idea will form a dominant theme of this book. Importantly, convergent evolution may be apparent across many different levels, ranging from ecosystem functioning, community form and function (as seen in Figure 1.1), the structure and physiology of individual species, such as through the common presence of sclerophyllous leaves or stem succulence (as seen in Figure 1.2), or in the types of interactions found among species. The topic of convergence and the scales at which it may occur are discussed further in Chapter 2.

1.3 Defining mediterranean-type ecosystems

Ecosystems vary in space and time, are generally complex, and are spatially extensive; therefore, it is often impractical or impossible to study a whole ecosystem. Instead, scientists typically focus on one or a few key drivers when studying a particular ecosystem (for instance, climate or soil type). Many studies have compared various organismal traits that relate to resource acquisition and use. A wide range of variables has been used in studies and as part of different approaches and focuses in the comparison of MTC regions (Table 1.1).

Table 1.1 Systems for identifying mediterranean-type ecosystems (MTEs). MTEs may be identified using one or more proxies or indicators for ecosystem function. Within each of the broad categories used for identification (geography, climate, physiography, phytogeography, plant morphology and function, and ecosystem), researchers have used differing broad or narrow definitions, and a sample of such definitions is included for each category. The most commonly used definition for MTEs within a category is included in italics.

Geography: based on global location

West-facing coasts of continents

Extending between latitudes of 25° and 49° (broadest definitions) to 32°–35° (narrowest definitions)

Commonly: west coasts of continents between 32° and 40° latitude

Climate: based on regional temperature, precipitation regimes, and rainfall reliability

Precipitation: from no requirement to falling within a narrow range of 350–650 mm annual precipitation

Seasonality: predominantly winter rainfall (November–April northern hemisphere, May–October southern hemisphere)

Continued

Table 1.1 Continued

Seasonality: from >50% to >90% winter precipitation

Seasonality: climate diagrams that show summer water deficit and winter surplus

Water stress: from no requirement to a minimum of 1 warm month with no precipitation

Winters: from no specific requirement to 1 month with an average temperature <15°C

Freezing: from no requirement to no severe frosts with no more than 3% of hours below 0°C

Commonly: warm dry summers, cool moist winters, with a pronounced summer dry period

Category used to identify 'mediterranean-type climate regions'

Physiography: based on terrain, drainage, and soil types within a broad climate region

Complexes of eroded mountains and plains

Thin soils in upland areas, with deeper soils in valleys and plains

Mosaics of different substrates and older and newer soils

Mediterranean, California, and Chile predominantly younger soils, compared to South Africa and Australia

River systems have strong seasonal pulses and flow variation

Commonly: landscapes of eroded mountains, heterogeneous low-nutrient soils, and pulse-driven river systems

Phytogeography (i.e. floristic provinces): based on the origin and distribution of plant taxa

For each region, there are related plant species or species compositions that identify a geographic area

Identified by species evolutionary histories, distributions, and patterns of endemism

Not generally useful in making connections among different MTEs, but useful within a region

Evolutionary relatedness may identify 'mediterranean-type vegetation' outside of mediterranean-type climate (MTC)

Commonly: floristic affinities and species relations may be used to identify 'mediterranean-type vegetation' (MTV) within regions

Plant morphology and function: based on species structure, physiology, and other organismal features

Relies on the assumption that ecosystem drivers have shaped common organismal features

Usually focuses on dominant plant species as having the largest influence on biotic communities

May include traits such as leaf structure and function, nutrient acquisition, and post-fire recovery

The basis for identification of the archetypal vegetation communities of different MTEs

Commonly: communities dominated by evergreen sclerophyllous shrubs with many species displaying fire-cued seed germination

Ecosystem: based on the functional interactions between species and their environment

Driven by many processes, including climate, soils, disturbance, and the biogeographic history of species

The level at which succession, diversity, resilience, and stability of a system are evaluated

Many species and communities linked through energy, nutrient, and water cycling

Commonly: encompasses all of the above categories and the processes that link them

Climate has been the most commonly used driver to define MTEs. Taking climate as an example, MTEs have been generally characterized as regions having hot dry summers and cool moist winters. Although different authors have set varying limits on the seasonality of precipitation, the total precipitation, and minimum temperature, there is broad agreement on the core areas that represent MTC regions. Depending on how strictly they are delimited, the five global MTC regions closely correspond to areas dominated by MTV, but may also include areas that are drier, wetter, or colder than often considered MTC regions (Figure 1.3), and which may be dominated by different vegetation communities (Chapter 4) (Blumler 2005). The same patterns and complexities in how limits are applied are also relevant to other key drivers that have been used to define MTEs.

Although the five MTC regions share much in common, they also have some notable differences in the importance of different ecosystem drivers, with implications for ecosystem function. For example, South Africa and Australia generally have more extensive areas with soils that are impoverished in nutrients than the other regions. Another difference is the fire regimes of the different regions, both historical and current. While all regions currently experience crown fires during the late summer or autumn, the amount of time that typically passes between successive fires, the fire return interval, varies between some regions. In Chile, extreme topography likely limited natural ignition sources, resulting in a much lower historical frequency of fires than other MTEs prior to the modern occurrence of anthropogenic ignitions. Much of the rest of this book will discuss these and other similarities and differences that have shaped the five global MTEs, with a special focus on the processes and drivers that apply to all regions.

For our purposes, we have been very careful to identify explicitly how we are defining MTC regions, especially when we are using a more specific definition for certain topics. In general, we have purposefully used the broadest and most inclusive criteria in our identification of MTEs. Species do not generally turn over at strict borders and communities and instead gradually transition from one to another. Thus, for the purposes of studying an ecosystem, where one of the most important factors is the interaction of species, it seems appropriate to be more inclusive. This is not a critique of studies that have been narrower in their definitions in the past—this was often necessary for the purposes of making the comparisons that were required as part of individual studies—but is rather a way to ensure that we are including all of the varied communities, processes, and drivers that compose and impact MTC regions.

As a point of practicality, we will generally be using two different identifying criteria throughout the remainder of this book. First, we will discuss **MTC regions**, broadly defined as those that receive primarily winter rainfall and experience cool, moist winters and warm, dry summers. We are not including

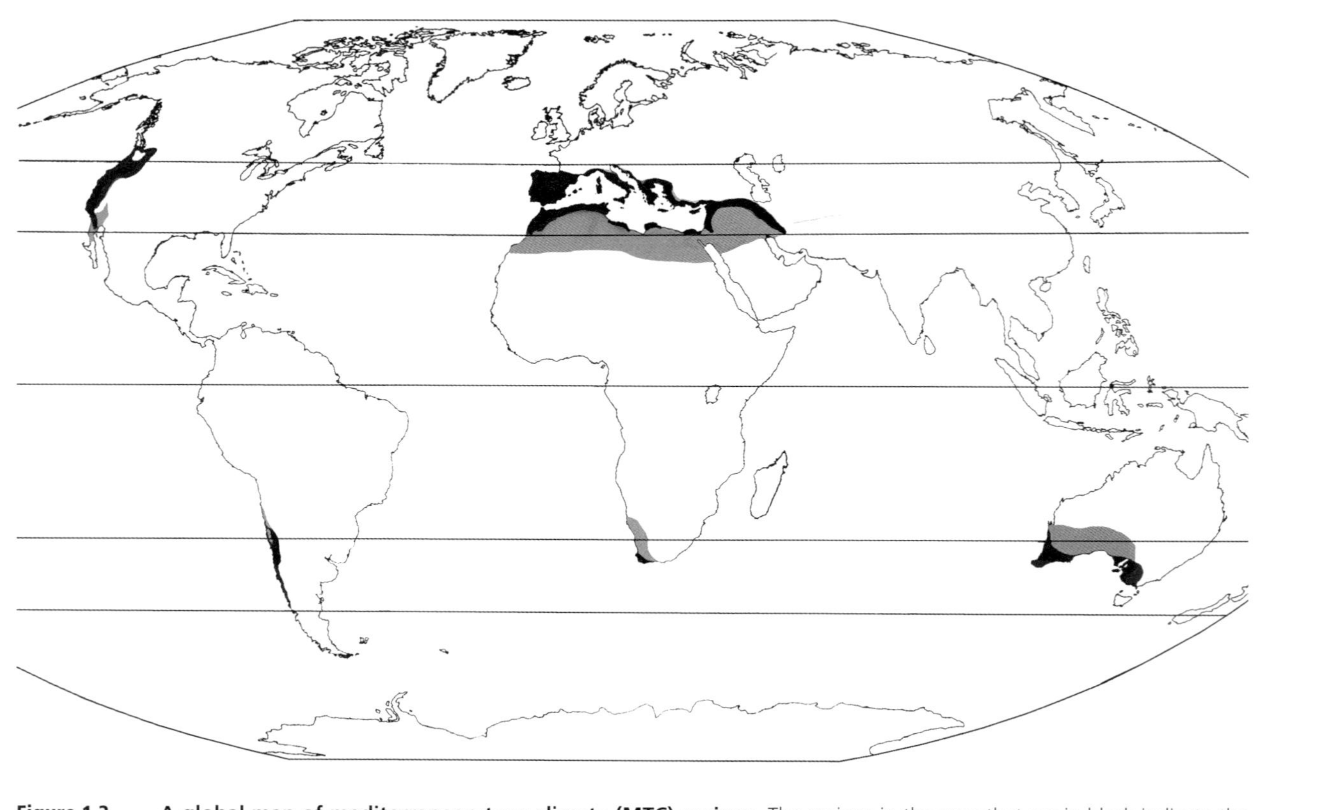

Figure 1.3   **A global map of mediterranean-type climate (MTC) regions.** The regions in the map that are in black indicate the location of MTC regions as most narrowly defined. The grey regions represent areas that may be included within MTC regions depending on how limits are set on the seasonality of precipitation, the total precipitation, and minimum temperature. The MTC regions as included in this figure are based on Aschmann (1973). Horizontal lines indicate 30° and 45° latitude in both hemispheres as well as the equator.

any further limitations on these criteria and we will include both high and low rainfall areas and areas that experience severe frosts and freezing as well as warmer areas within our definition of MTC. Second, we will discuss **mediterranean-type vegetation (MTV)**, using a floristic definition, in large part because comparisons of related and similar plant species that occur both inside and outside of MTC regions allows for interesting insights into the species traits that are specific to MTC regions and therefore MTEs.

Finally, it is worth specifically mentioning that we will not be adhering to strict plant morphological or function-based criteria for identification of regions. Although we will necessarily focus on many of the species and traits of the archetypal sclerophyllous shrub communities of MTEs, because they have been the focus of the vast majority of MTEs research and comparisons, we are also interested in the many other communities that occur within MTEs, some of which have often been ignored. This includes the communities dominated by drought-deciduous 'soft-leaved' (malacophyllous) shrubs, forests inhabiting the wetter areas of MTC regions, arid shrub and succulent communities, and grasslands. Many of these communities fall within even the most narrowly geographically, climatically, or physiographically defined MTC regions and have been excluded from prior comparisons based on plant morphology alone.

1.4 A brief history of early comparative mediterranean research

Comparative MTC region research has been of particular interest to scientists for many years and is a focus of the present book. The earliest comparisons were often based on the synthesis of written accounts from different regions, the analysis of preserved species, or the examination of living specimens housed within botanical gardens. With the advent of more efficient means of transport and communication between regions, comparisons across multiple regions could be conducted by individuals or teams working themselves in multiple regions. Many of these studies and collaborations have provided unique insights that could only be gained from study and comparison of the systems in situ and we have highlighted these types of activities and research projects in many 'case studies' presented throughout this book.

Naveh (1967) wrote that 'comparisons of mediterranean vegetation types are as old as ecology itself', highlighting the early interest of scientists in understanding these environments. Early comparative research has been the subject of numerous excellent reviews (Naveh 1967; Specht 1969a; Johnson 1973). Many of the earliest studies focused on comparing the most common vegetative elements of MTV and postulating likely reasons for their similarity in structure, with most studies focused on climate. Grisebach (1872) was one of the first to suggest that vegetative similarity between different

Figure 1.4   **Species from each mediterranean-type ecosystem (MTE) region representing sclerophyllous genera that were discussed by Schimper (1903).** Schimper (1903) included example photos and line drawings of the 'sclerophyllous flora' of each of the five MTC regions. The photos included here are modern, but they correspond to figures that were originally presented in Schimper (1903), and we have included his original figure numbers parenthetically below to indicate the original example to which our modern photos correspond. Shown are *Pistacia* from the Mediterranean (a) (Figure 286), *Leucadendron* at Table Mountain in South Africa (b) (Figure 290), *Banksia* from Australia (c) (Figure 297), *Arctostaphylos* from California, USA (d) (Figure 300), and *Lithraea* (formerly *Rhus*) from Chile (e) (Figure 305).

Source: Photos from Anna L. Jacobsen.

mediterranean-type regions was a result of their similarities in climate, or, in modern terms, that the similar structure of MTV was a result of evolutionary convergence in response to climatic factors.

Schimper (1903) also emphasized climate as a driving factor in the structure of mediterranean plants (identified as 'sclerophyllous woodland' in 'districts of the warm temperate belts with moist winters'). He identified evergreen sclerophyllous leaves and an absence of succulence as key features of the semi-arid plants from these regions (Figure 1.4). Considerable research focus on sclerophyllous leaves has since found that these types of leaves are also adaptive in other water-limited and nutrient-poor environments (Loveless 1962), thus they are not unique to MTC regions (Chapter 6). However, the original assessment by Schimper (1903) that sclerophylly is important in MTC environments has been supported by later studies (Mooney and Dunn 1970a).

Later work continued to examine patterns of plant **physiognomy** (the general appearance of vegetation) and biogeography, but expanded to also include physiology, community types, climate ranges, soils, successional patterns, and disturbance factors. These studies concluded that there are many similarities between the vegetation in these regions, but they also highlighted notable differences, particularly for soils (Specht 1969a, b) and history of disturbance (Naveh 1967). Some of these studies were fully modern in the sense that they addressed questions that are still quite relevant today and employed approaches and methods that are still widely used. Naveh (1967) compared the climate, soils, and ecological histories of MTC regions in California and Israel. Specht (1969a, b) compared stands in Australia, California, and France, including their succession patterns and nutrient relations, and an analysis of biomass production along a fire chronosequence. Mooney and Dunn (1970a, b) examined photosynthetic patterns of the dominant vegetation in California and Chile. They also examined the distribution of different leaf habits and growth forms along environmental gradients and found that evergreen species were most abundant at the wetter end of the MTC and were replaced by drought-deciduous and succulent species at drier sites (Mooney and Dunn 1970a, b; Mooney et al. 1970). Based on their study, Mooney and Dunn (1970a) proposed an important model regarding the relative advantages of evergreen sclerophyllous leaves in the context of environment that provides a framework for why such leaves are favoured in an MTC (Chapter 6).

The largest effort to compare MTEs was ushered in by the International Biological Programme (IBP) with support from the National Science Foundation, USA (Case Study 1). At the same time that a mediterranean component of this program was being initiated, in 1971, the International Society for Mediterranean Ecology (ISOMED) was founded. Both of these milestones brought researchers together from the different regions and spawned many studies comparing different MTC regions. These efforts resulted in a spurt of

Case Study 1 Early mediterranean ecosystem comparisons. A personal history

H.A. Mooney, Department of Biology, Stanford University, Stanford, California, USA

When it came time for my first sabbatical leave from the University of California at Los Angeles I came up with the idea to modestly build on the legacy of Alexander von Humboldt, the founder of modern biogeography. The particular issue that I wanted to explore was von Humboldt's conclusions relating climate and vegetation and the similar plant forms, not closely related, that were present in similar climatic regions. The question I posed was 'do these structural analogs in form translate into analogs in physiological function?' I received a grant to pursue this study and went to Santiago, Chile to initiate the work in 1966.

I was very fortunate in meeting three individuals in Santiago who were critical in the subsequent studies on ecological convergence. One was Jochen Kummerow, a plant physiologist at the University of Chile. We collaborated in the first physiological comparative studies and he became a colleague and major contributor to subsequent expanded efforts. The second person who I was extremely fortunate to meet early on, and who subsequently played a big role in my own early career development, was Francesco di Castri, who at that time was the head of the Veterinary School at the University of Chile. At this time, and by coincidence, the University of California and the University of Chile were just initiating a 10-year collaborative program across all disciplines funded by the Ford Foundation (Philip Rundel was supported by this program and that resulted in his long-term association with MEDECOS). The program involved exchanging faculty and students between universities. Di Castri had written a proposal to become engaged in this effort.

When I met with di Castri he said he wanted to make a climate-biotic communities study comparing Chile and California. He had already done such a study along climatic gradients in Chile with his colleague, Ernesto Hajek (di Castri and Hajek 1961). So naturally, I became very interested in interacting with him while I was in Chile. The opportunity to do so came with the development of the US International Biological Program (IBP) that became operational at the end of 1967. Most nations focused all of their efforts into the main theme of the program, which was to study the productive capacity of ecosystems on Earth. The US program did this, innovating systems ecology in the process, but also added a number of other elements, one of which was on the convergence of the structure and function of ecosystems. This opportunity played directly into the interests of di Castri and me. An IBP planning meeting was held in Caracas, Venezuela in the Fall of 1967. I attended and presented a program structure that encompassed the convergence theme for Chile and California while Otto Solbrig was promoting a comparison of the deserts of North and South America.

Subsequently, funding for these proposals from the National Science Foundation was obtained. Di Castri and I led the Chile-California effort. We wanted to take a whole system approach to the problem that looked at the drivers of the form and function as well as the biotic components. Our first task was to find out what the knowledge base was as we initiated our new effort. Before we started new studies we brought scientists from around the world to Valdivia, Chile in 1971 to share their knowledge on various aspects of mediterranean-climate systems (di Castri and Mooney 1973). This gathering has been

Case Study 1 Figure. A picture of di Castri and Mooney, conference co-organizers, at the first MEDECOS meeting in Valdivia, Chile in 1971. This photo was taken while on a cruise ship fieldtrip as part of meeting activities. (Photo from H.A. Mooney.)

considered as the initial formal meeting of mediterranean climate scientists (Case Study 1 Figure). Subsequent meetings in later years were termed MEDECOS, an acronym derived from 'International Mediterranean Ecosystems Conference' (Hobbs et al. 1995).

The initial planning for the new research effort took place while di Castri was located in Santiago. We had decided that for the new comparative research effort we would establish carefully matched (climatically and physiographically) sites in both Chile and California. The principal matched sites were to be located initially in Santiago, Chile and San Diego, California, but then di Castri moved to the University of Valdivia so our planning shifted to a comparison centred in Valdivia and the San Francisco area. However, di Castri then moved once again, this time to Paris to head up the new Man and Biosphere Program, so we shifted our planning back again to the regions of Santiago and San Diego. The experimental design was centred on mediterranean-type scrublands in both of these areas. The design also included, on each continent, matched sites that were both moister and drier than the central sites (forests to arid scrub). The design thus gave us the opportunity to ask whether sites along the climatic gradient were more similar to each other on a given continent than they were between continents. Detailed descriptions

Continued

Case Study 1 (*Continued*)

of these sites are given in a *Chile-California Mediterranean Scrub Atlas* (Thrower and Bradbury 1977). The synthesis of the scientific findings of the overall project was reported by Cody and Mooney (1978) and by Mooney (1977).

Subsequent MEDECOS meetings initially focused on comparisons of single environment drivers and the related responses of structure and function across mediterranean-type ecosystems (MTEs); Stanford 1977 (Fire), South Africa 1980 (Nutrients), San Diego 1981 (Management), Australia 1984 (Resilience), France 1987 (Water), Greece 1991 (Plant–Animal Interactions), and Chile 1994 (Biogeography). This model revealed important differences among these systems, especially driven by geological history and plant nutrition (e.g. phosphorus limitation). Subsequent meetings that occurred have been open meetings or meetings with cross-cutting themes that could include all researchers in these areas, no matter what their research focus. My most recently attended meeting, the Thirteenth MEDECOS, was held in Olmué, Chile during October, 2014 with a theme of 'Crossing Boundaries across Disciplines and Scales'.

The joy of these meetings has always been interacting with investigators who lived on different continents but who were essentially working on the same problem and thus facing the same research challenges. For me personally, I was able to meet and become long-term colleagues with Richard Hobbs, Richard Groves, Mary Arroyo, Zev Naveh, David Richardson, Ray Specht, Eduardo Fuentes, Jacque Roy, Fred Kruger, Brian Huntley, Francesco di Castri, and many others from California. I have been publishing together with Hobbs for decades and never ceased to be amazed by his energy and vision. Brian Huntley was one of the key organizers of the Third MEDECOS meeting in South Africa. He was interested in the IBP ecosystem level approach and was just setting up programs in the fynbos. I subsequently collaborated with him through the years on many biodiversity issues. Fred Kruger was particularly important in my subsequent career. I had met him in Australia while we both were on leave. I then visited his research site at Stellenbosch during the Third MEDECOS. It was at that time that we both decided to embark on a global program on invasive species sponsored by the Scientific Committee on Problems of the Environment. The program occupied us for many years and grew beyond our focus on mediterranean-type systems that was our original proposal. My early interactions with di Castri continued and in later years we launched a program on Biodiversity Science called DIVERSITAS.

Thus, research on the comparisons of mediterranean-climate systems through the years has been professionally rewarding to me in ways that I would never have originally imagined. I am sure that this is also true for many other mediterranean-climate scientists and will be true for many future scientists as well.

research activities, meetings, conferences, and collaborations, many of which still continue, and produced many edited volumes addressing various aspects of MTEs biology and ecology (di Castri and Mooney 1973; Mooney 1977; Thrower and Bradbury 1977; di Castri et al. 1981; Miller 1981; Kruger et al. 1983; Dell et al. 1986; Tenhunen et al. 1987; Groves and di Castri 1991; Arianoutsou and Groves 1994; Arroyo et al. 1994; Moreno and Oechel 1994; Davis and

Richardson 1995; Moreno and Oechel 1995; Roy et al. 1995; Rundel et al. 1998; Richardson et al. 2001; Arianoutsou and Papanastasis 2004).

While the IBP ended in the mid-1970s, ISOMED is still in operation (http://www.incomme.org/isomed.html). This society continues to organize mediterranean-type ecosystem (MEDECOS) meetings every 3–4 years, with these meetings always occurring in one of the five regions (Case Study 1 Figure). These meetings continue to be incredibly important for bringing together scientists and ideas regarding the ecology, biology, and conservation of MTEs. Many of the scientists who have worked across MTC regions to conduct the comparative studies that will be the focus of the rest of this book first connected through attendance at one or more of these meetings. Through MEDECOS meetings, ISOMED has fostered an extensive network of shared educational experiences, collaborations, and connections between many of the researchers who have been instrumental in elucidating the biology of MTEs and who continue to work in these regions. The legacy of these efforts is that MTC regions are the most extensively compared of any terrestrial system and these areas continue to attract great scientific interest.

1.5 Hotspots of biodiversity

Continuing scientific interest in MTC regions is linked to their unusually high levels of species richness and biodiversity. These regions have the highest species richness outside of the tropics, particularly in vascular plant diversity (Figure 1.5 and Figure 1.6). They also have very high levels of **endemism**, which refers to species that do not occur anywhere else. Both tropical and MTC regions were among the first to be recognized as **biodiversity hotspots** (Myers 1990), and all five of them, including Central Chile, the California Floristic Province, the Cape Floristic Province and the Succulent Karoo (South Africa), the Mediterranean Basin, and the Southwest Australia Region, are repeatedly included in lists of the most biodiverse regions on Earth (reviewed in Mittermeier et al. 2011).

An example of just how diverse the MTC regions are was the inclusion of all five MTC regions within a study that identified 25 global hotspots. These 25 identified regions covered a mere 1.4 per cent of global land area, yet included 44 per cent of all vascular plant species (Myers et al. 2000). Recent revisions of this analysis have continued to affirm the inclusion of MTC regions as unique global biodiversity hotspots (Mittermeier et al. 2011; Williams et al. 2011). Analysed more broadly, MTC regions cover less than 5 per cent of the global land area yet contain almost 20 per cent of vascular plant species (Cowling et al. 1996). The Cape Region of South Africa, in particular, has been identified as being uniquely diverse when compared to predicted levels of global biodiversity (Kreft and Jetz 2007).

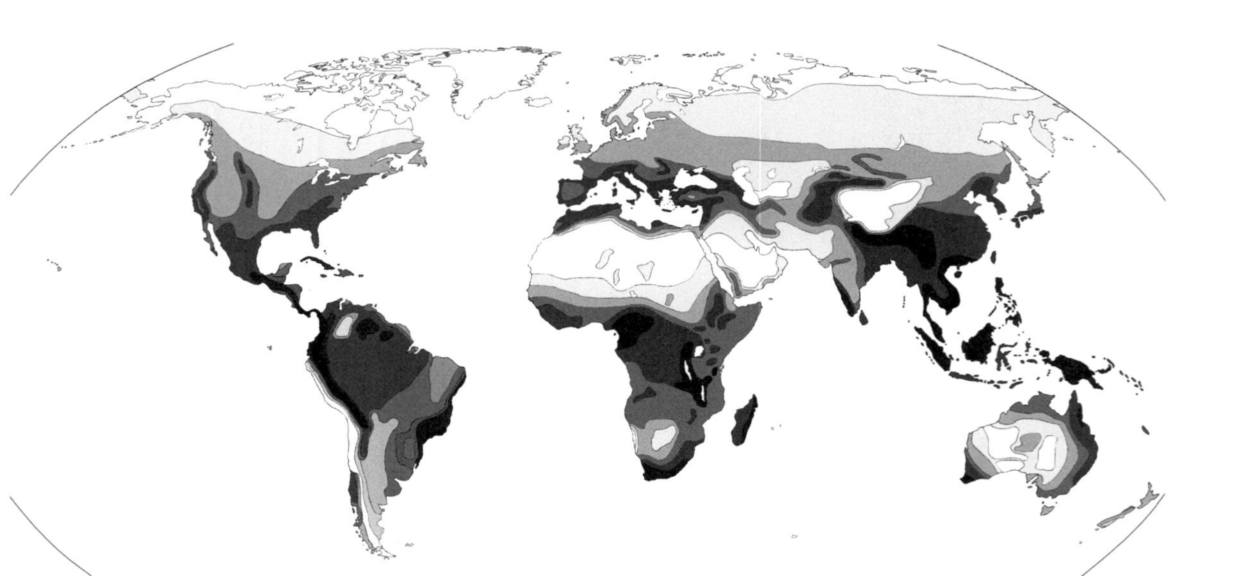

Figure 1.5 **Global species richness of vascular plants.** The number of vascular plant species per land area, a measure of species richness, for generalized biotic zones of the globe vary with region. Darker shades of grey indicate areas that contain high numbers of vascular plant species and light grey areas indicate regions of low vascular plant diversity. Plant diversity is greatest in tropical regions and in MTC regions. (Modified from Barthlott et al. 1996 and Mutke and Barthlott 2005.)

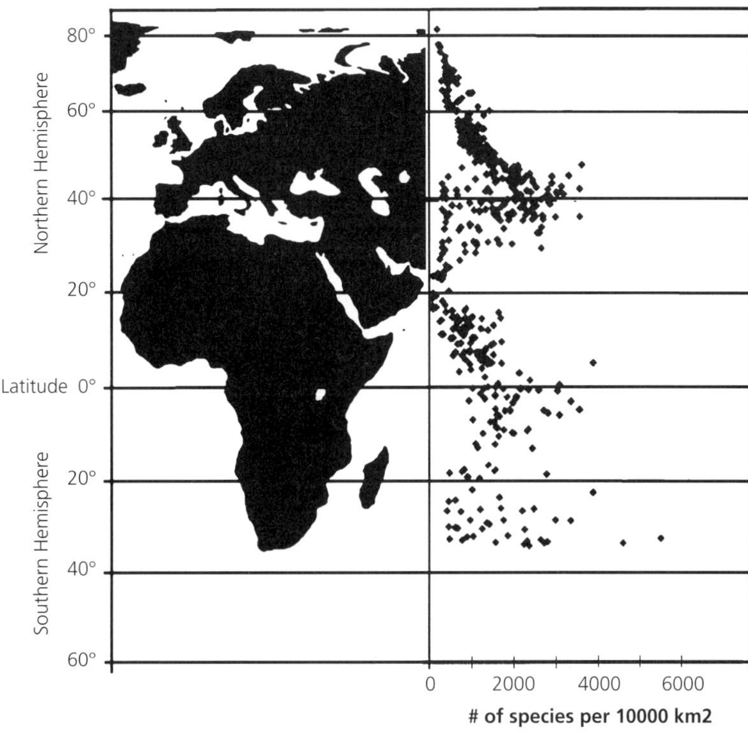

Figure 1.6 **Species richness of vascular plants with latitude.** The number of vascular plant species per land area, a measure of species richness, for sites occurring along a cross-continental north–south transect through Europe and Africa. Species richness is high in tropical zones located around the equator (0° latitude) and in MTC regions, including a peak in the northern hemisphere at the latitudes corresponding to the Mediterranean Basin and the highest measured species richness for the transect in the southern hemisphere at the latitudes corresponding to the Cape Region of South Africa. (Data from Mutke and Barthlott 2005.) A map is included in the panel on the left that aligns with the indicated latitudes to assist with interpretation of the latitude-based data presented in the panel on the right.

Concomitant with high levels of biodiversity, biodiversity hotspots are identified in part because of unique threats to the biodiversity of these regions (Myers 1990). The criteria for inclusion as a hotspot in Myers (1990) required that a region have at least 1500 endemic vascular plant species and have less than 30 per cent of the original extent of the native vegetation extent remaining. Thus, the inclusion of all MTC regions in these is due in part to the large-scale human transformation of these regions (Chapter 8). This includes land-use changes, climate change, the introduction of non-native species, and changes in the key drivers of these systems, especially through changes in fire regimes and disturbance.

Literature cited

Arianoutsou M, Groves RH, eds. 1994. Plant-animal interactions in mediterranean-type ecosystems. Springer Science+Business Media, Dordrecht.

Arianoutsou M, Papanastasis VP, eds. 2004. Ecology, conservation and management of mediterranean climate ecosystems. Millpress, Rotterdam.

Arroyo MTK, Zedler PH, Fox MD, eds. 1994. Ecology and biogeography of mediterranean ecosystems in Chile, California, and Australia. Springer-Verlag, Berlin.

Aschmann H. 1973. Distribution and peculiarity of mediterranean ecosystems. Pages 11–20 in di Castri F, Mooney HA, eds. Mediterranean type ecosystems: origin and structure. Springer-Verlag, Berlin.

Barthlott W, Lauer W, Placke A. 1996. Global distribution of species diversity in vascular plants: towards a world map of phytodiversity. Erdkunde 50: 317–27.

Blumler MA. 2005. Three conflated definitions of Mediterranean climates. Middle States Geographer 38: 52–60.

Cody ML, Mooney HA. 1978. Convergence versus nonconvergence in mediterranean-climate ecosystems. Annual Review of Ecology, Evolution and Systematics 9: 265–321.

Cowling RM, Rundel PW, Lamont BB, Arroyo MK, Arianoutsou M. 1996. Plant diversity in mediterranean-climate regions. Trends in Ecology & Evolution 11: 362–6.

Davis GW, Richardson DM, eds. 1995. Mediterranean-type ecosystems: the function of biodiversity. Springer-Verlag, Berlin.

Dell B, Hopkins AJM, Lamont BB, eds. 1986. Resilience in mediterranean-type ecosystems. Dr W. Junk Publishers, Dordrecht.

di Castri F, Hajek ER. 1961. Indices pluviotérmicos como base para una clasificación del país en zonas bioclimáticas. IV Convención de Médicos Veterinarios. Santiago (Chile), Agosto: 19–23.

di Castri F, Mooney HA, eds. 1973. Mediterranean type ecosystems: origin and structure. Springer-Verlag, Berlin.

di Castri F, Goodall DW, Specht RL, eds. 1981. Mediterranean-type shrublands. Elsevier, Amsterdam.

Grisebach AHR. 1872. Die vegetation der erde nach ihrer klimatischen anordnung. (The vegetation of the earth according to their climatic arrangement.) Leipzig, Engelmann.

Groves RH, di Castri. 1991. Biogeography of mediterranean invasions. Cambridge University Press, Cambridge.

Hobbs RJ, Richardson DM, Davis GW. 1995. Mediterranean-type ecosystems: opportunities and constraints for studying the function of biodiversity. Pages 1–42 in Davis GW, Richardson DM, eds. Mediterranean-type ecosystems: the function of biodiversity. Springer-Verlag, Berlin.

Holdridge LR. 1947. Determination of world plant formations from simple climatic data. Science 105: 367–8.

Johnson AW. 1973. Historical view of the concept of ecosystem convergence. Pages 3–7 in di Castri F, Mooney HA, eds. 1973. Mediterranean type ecosystems: origin and structure. Springer-Verlag, Berlin.

Kreft H, Jetz W. 2007. Global patterns and determinants of vascular plant diversity. Proceedings of the National Academy of Sciences 104: 5925–30.

Kruger FJ, Mitchell DT, Jarvis JUM. 1983. Mediterranean-type ecosystems: the role of nutrients. Springer-Verlag, Berlin.

Lomolino MV, Sax DF, Brown JH, eds. 2004. Foundations of biogeography: classic papers with commentaries. University of Chicago Press, Chicago, Illinois.

Loveless AR. 1962. Further evidence to support a nutritional interpretation of sclerophylly. Annals of Botany 26: 551–61.

Miller PC. 1981. Resource use by chaparral and matorral: a comparison of vegetation function in two mediterranean type ecosystems. Springer-Verlag, Berlin.

Mittermeier RA, Turner WR, Larsen FW, Brooks TM, Gascon C. 2011. Global biodiversity conservation: the critical role of hotspots. Pages 3–22 in Zachos FE, Habel JC, eds. Biodiversity hotspots: distribution and protection of conservation priority areas. Springer-Verlag, Berlin.

Mooney HA. 1977. Convergent evolution in Chile and California: mediterranean climate ecosystems. Dowden, Hutchinson & Ross, Inc., Stroudsburg, Pennsylvania.

Mooney HA, Dunn EL. 1970a. Photosynthetic systems of mediterreanean-climate shrubs and trees of California and Chile. American Naturalist 104: 447–53.

Mooney HA, Dunn EL. 1970b. Convergent evolution in mediterranean-climate evergreen sclerophyll shrubs. Evolution 24: 292–303.

Mooney HA, Dunn EL, Shropshire F, Song L. 1970. Vegetation comparisons between the mediterranean climatic areas of California and Chile. Flora 159: 480–96.

Moreno JM, Oechel WC. 1994. The role of fire in mediterranean-type ecosystems. Springer-Verlag, New York.

Moreno JM, Oechel WC. 1995. Global change and mediterranean-type ecosystems. Springer-Verlag, New York.

Mutke J, Barthlott W. 2005. Patterns of vascular plant diversity at continental to global scales. Biologiske Skrifter 55: 521–31.

Myers N. 1990. The biodiversity challenge: expanded hot-spots analysis. Environmentalist 10: 243–56.

Myers N, Mittermeier RA, Mittermeier CG, de Fonseca GAB, Kent J. 2000. Biodiversity hotspots for conservation priorities. Nature 403: 853–8.

Naveh Z. 1967. Mediterranean ecosystems and vegetation types in California and Israel. Ecology 48: 445–59.

Richardson DM, Esler KJ, Cowling RM. 2001 Mediterranean-type ecosystems—past, present and future. An introduction. Journal of Mediterranean Ecology 2: 123–5.

Roy J, Aronson J, di Castri F. 1995. Time scales of biological response to water constraints: the case of the mediterranean biota. SPB Academic Publ., Amsterdam.

Rundel PW, Montenegro G, Jaksic F, eds. 1998. Landscape disturbance and biodiversity in mediterranean-type ecosystems. Springer-Verlag, Berlin.

Schimper AFW. 1903. Plant-geography upon a physiological basis. (Transl. by WR Fisher). Oxford, Clarendon Press.

Specht RL. 1969a. A comparison of the sclerophyllous vegetation characteristic of mediterranean type climate in France, California, and southern Australia. I. Structure, morphology, and succession. Australian Journal of Botany 17: 277–92.

Specht RL. 1969b. A comparison of the sclerophyllous vegetation characteristic of mediterranean type climate in France, California, and southern Australia. II. Dry matter energy, and nutrient accumulation. Australian Journal of Botany 17: 293–308.

Tenhunen JD, Catarino FM, Lange OL, Oechel WC, eds. 1987. Plant response to stress: functional analysis in mediterranean ecosystems. Springer-Verlag, Berlin.

Thrower NJW, Bradbury DE, eds. 1977. Chile-California mediterranean scrub atlas. Dowden, Hutchinson & Ross, Inc., Stroudsburg, Pennsylvania.

Walter H. 1973. Vegetation of the earth in relation to climate and the eco-physiological conditions. Springer-Verlag, New York.

Williams KJ, Ford A, Rosauer DF, De Silva N, Mittermeier R, Bruce C, Larsen FW, Margules C. 2011. Forests of East Australia: the 35th biodiversity hotspot. Pages 295–310 in Zachos FE, Habel JC, eds. Biodiversity hotspots: distribution and protection of conservation priority areas. Springer-Verlag, Berlin.

2 Characteristics of Mediterranean-Type Ecosystems

Abstract

Modern mediterranean-type ecosystems (MTEs) are shaped by key ecosystem drivers that affect their function. The most important of these drivers are climate, topography, soils, and fire. There are important geographical, climatic, and fire histories that are crucial to understanding these systems. Mediterranean-type climate (MTC) is defined as a cool wet winter (winter-wet) and a warm dry summer, which is a unique pattern of seasonality and one that is rare globally. All of the MTC regions have nutrient-poor soils, particularly as related to nitrogen (N), and some also have extensive phosphorus-poor soils. There is considerable variation both within and between regions in their degree of nutrient impoverishment of soils. Through these shared ecosystem drivers, selection has operated within each ecosystem to shape the communities and the organisms within them. This has resulted in the communities and organisms displaying similar structures and processes.

2.1 Origins and biogeography

To understand and compare mediterranean-type ecosystems (MTEs) as they occur today, one must first appreciate the origins of modern landmasses, the **biogeographical history** of the floras, and the timing and emergence of mediterranean-type climate (MTC) and other important ecosystem drivers, such as fire. Many of these large-scale patterns continue to shape MTEs today and contribute to some of the broad similarities and differences across these regions.

Of particular importance is the timing of the arrival of the MTC. Palaeoclimate reconstructions and fossil assemblages show that climate has dramatically changed over the course of Earth's history. A range of techniques is used to infer paleoclimates (Graham 1999), but key variables in these

The Biology of Mediterranean-Type Ecosystems. Karen J. Esler, Anna L. Jacobsen, and R. Brandon Pratt,
Oxford University Press (2018). © Karen J. Esler, Anna L. Jacobsen, and R. Brandon Pratt 2018.
DOI 10.1093/oso/9780198739135.001.0001

reconstructions are temperatures and seasonality of the climate. Temperature reconstructions indicate how hot or cool conditions were and characterize glacial and interglacial periods. Seasonality characterizes the annual variability in temperature and precipitation. Both of these factors are important because current MTC is both dry as well as strongly and uniquely seasonal, and these

Era	Sub-era	Period		Phyticera	Ma
Cenozoic	Quaternary	Neogene	Holocene	Cenophytic	
			Pleistocene		2.56
	Tertiary		Pliocene		5.3
			Miocene		23
		Palaeogene	Oligocene		34
			Eocene		56
			Palaeocene		66
Mesozoic		Cretaceous		Mesophytic	145
		Jurassic			201
		Triassic			252
Palaeozoic		Permian		Palaeophytic	299
		Carboniferous			323
					359
		Devonian			419
		Silurian		Proterophytic	443
		Ordovician			485
		Cambrian			541

Figure 2.1 **The geological timescale.** The major divisions within the geological timescale include eras and periods. The approximate time in millions of years before present (Ma) that transitions between times occurred are shown. These eras and periods are based largely on when major changes occurred within the fossils of multicellular animals. The phyticera are eras classified based on major changes in the plant fossil record. (Based on Willis and McElwain 2014.)

are key drivers of these systems. In addition, more seasonal environments are prone to fire, also a key driver of these systems (Keeley et al. 2012).

Some of the iconic taxa that comprise modern mediterranean-type vegetation (MTV) were present by the late Cretaceous and early Cenozoic (Figure 2.1). This period constituted the beginning of the Cenophytic, approximately 140 million years ago (Ma), during the early evolution of flowering plants (angiosperms). During this time, the Earth's plates were arranged in two large landmasses, Gondwana in the southern hemisphere and Laurasia in the northern hemisphere (Figure 2.2). The floristic affinities of different MTV still

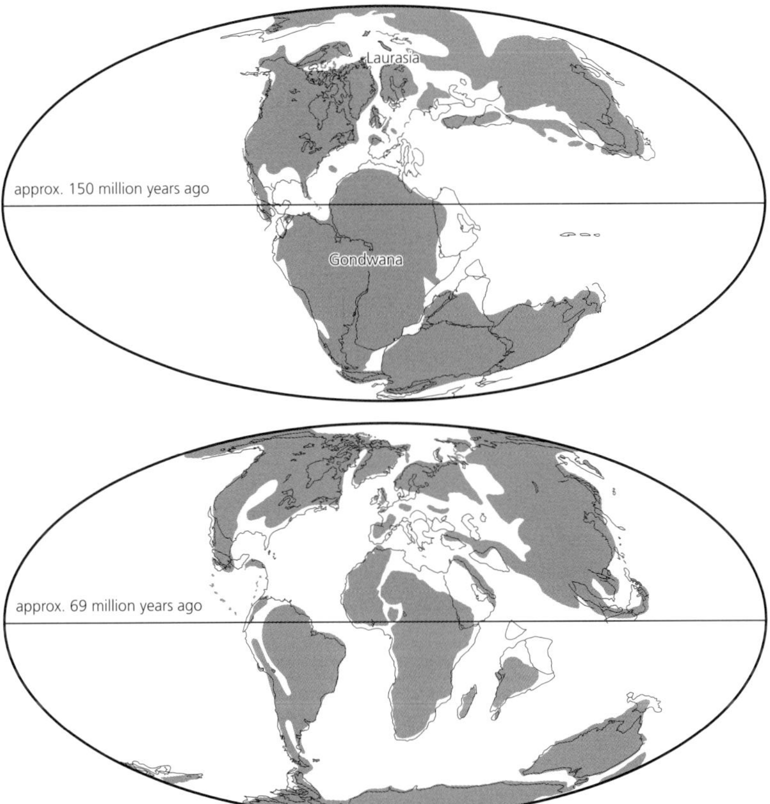

Figure 2.2   **A global reconstruction of past continental geography.** The locations of continental landmass during the late Mesophytic (approximately 150 million years ago) (top image) and during the late Cretaceous (approximately 69 million years ago) (bottom image). Modern continents and landmasses are outlined. Grey regions represent the areas that would have likely been land. During the early evolution of angiosperms, the Earth contained two large landmasses, Gondwana, sometimes referred to as Gondwanaland, compromised of the modern southern hemisphere continents, and Laurasia, comprised of the modern northern hemisphere continents (top image). These supercontinents soon broke apart and moved toward their modern locations (bottom image). (Modified from McLoughlin 2001.)

Figure 2.3 **Conspicuous common northern hemisphere and southern hemisphere taxa.** In the northern hemisphere mediterranean-type ecosystems, there are many common families and genera, but the most conspicuous shared taxon is oak (*Quercus*). From California (panel a), a large live oak (*Quercus agrifolia*) is shown. From the Mediterranean (panel b), a Spanish cork oak (*Quercus suber*) is shown following the harvest of the trunk bark for cork. In the southern hemisphere MTEs, the most conspicuous shared taxon is the Protea family (Proteaceae). There are numerous genera in this family. From Australia, a *Grevillea* sp. is shown (panel c), and other common Proteaceae genera include *Banksia* and *Hakea*. From South Africa, a Protea species is shown (panel d), and other common Proteaceae genera include *Leucadendron*, *Leucospermum*, and *Serruria*. Proteaceae are also present in Chile, including the genera *Embothrium* and *Lomatia*, but these genera contain few species and represent only a minor element within the Chilean flora. (See Plate 3)

Source: Photos c and d from Anna L. Jacobsen and photos a and b from R. Brandon Pratt.

bears the legacy of these past connections (Raven 1973). Among the northern hemisphere MTEs, there are many common families and genera, the most conspicuous taxon being the oaks (*Quercus* spp.) (Figure 2.3; see Plate 2). Many of the shared northern sclerophyllous genera are derived from Madro-Tertiary lineages (Axelrod 1958; Palamarev 1989); that is, lineages that first evolved during the Cretaceous and spread across the warmer regions of the Laurasian landmasses during the Tertiary. In the southern hemisphere, the separation of Gondwanan land elements occurred during the time of early angiosperm evolution. There are some common elements across these regions that pre-date the angiosperms and as well as angiosperm taxa whose origins occurred before or during the period when the seas between continents were relatively small and the continents were only a small distance apart (McLoughlin 2001). The most conspicuous taxon that occurs across the southern hemisphere MTV is the Protea family (Proteaceae) (White 1986) (Figure 2.3; see Plate 2).

Importantly, during these early time periods MTC had not yet established, and so these early common taxa were not 'MTV' but rather the early ancestors of taxa that would later persist and diversify within areas of newly emergent MTC. Many iconic modern MTV lineages have early origins and many traits, including sclerophyllous leaves, that pre-date the establishment of an MTC (Chapter 6). The question of what traits 'predisposed' certain lineages for later establishment and evolutionary change within MTC areas and what traits were selected for following the establishment of an MTC has been an interesting area of evolutionary research, particularly as new tools emerge to study plant evolution and to reconstruct past climatic history (Case Study 2). From shared floristic origins, each newly separated continent experienced different climate histories and geologic events through the

---

Case Study 2 Assembly of mediterranean-type floras—convergence, exaptation, and evolutionary predisposition

David D. Ackerly, Department of Integrative Biology, University of California, Berkeley (UC Berkeley), Berkeley, California, USA

Renske E. Onstein, Institute for Biodiversity and Ecosystem Dynamics (IBED), University of Amsterdam, Amsterdam, The Netherlands

It is well documented that the mediterranean-type ecosystems (MTEs) of the world are, on geological timescales, quite young. As discussed in Chapter 2, freezing of the Antarctic ice sheet played a key role in altering ocean currents and contributing to the summer high-pressure systems at mid-latitudes off the western coasts of the five major landmasses. While cooling and aridification had been under way through most of the Cenozoic, the distinctive mediterranean-type climate (MTC) and associated floras seem to emerge starting about 10 million years ago (Ma), with rapid diversification of many lineages in just the last 3 million years.

Given the recent appearance of MTC, the large geographic distances among regions, and the absence of similar climates in the intervening areas, it is not surprising that MTC region floras have been assembled from floristic elements of their respective regions.

Continued

Case Study 2 (*Continued*)

Independent assembly, combined with distinctive functional traits—particularly evergreen, sclerophyllous leaves—made MTC region floras prime candidates (and textbook examples) of convergent evolution (Cody and Mooney 1978). Subsequent research has confirmed that many of the distinctive plants of these regions, such as chamise (*Adenostoma* sp.) that dominates California chaparral, and several lineages of Proteaceae (e.g. in genera *Grevillea* and *Banksia*) in Australian kwongan, evolved from mesophytic ancestors and did indeed undergo dramatic shifts in leaf form as they adapted to arid, MTCs (Ackerly 2004; Onstein et al. 2016).

However, two observations have always lurked in the background of this story of convergence. First, if convergence arises from strong selection in response to MTC, then the traits should have evolved in sync with, or after, the origin of the climates; alternatively, as long proposed by Axelrod (1975, 1989), it is possible that many of these lineages evolved sclerophyllous leaves, and other traits suitable for MTC, either in response to earlier exposure to arid climates or in response to other factors entirely, particularly low nutrient soils (which are prevalent in South Africa and Australia). If the traits evolved first, predisposing these lineages to colonize, expand, and diversify in MTC regions, then they would be considered exaptations (traits that evolved under one selective regime, and then became adaptive in a different context; Gould and Vrba 1982) with respect to MTC, and the functional similarity of these floras would reflect the importance of ecological sorting and assembly processes, with convergent, adaptive evolution operating in parallel or perhaps in the background. There is considerable evidence supporting the view that many of the critical functional traits of woody mediterranean-type lineages do in fact pre-date the appearance of MTC, based on comparative phylogenetic analyses (Onstein and Linder 2016; Ackerly 2004; Kadereit and Baldwin 2012) and fossil evidence (Sniderman et al. 2013; Hill 1998; Axelrod 1989). The second observation is that there are in fact a number of closely related lineages that occur in two or more of the MTC regions. Several hypotheses can be advanced to explain the occurrence of close relatives across MTC regions, and each in turn presents a twist that sheds light on the processes underlying assembly and evolution of the world's MTC region floras, and of regional floras more generally.

Based on a preliminary tabulation of woody, angiosperm genera across the five MTC region floras, we found at least forty genera that occur in two or more regions, with nine occurring in three regions (a full analysis awaits resolution of taxonomic and nomenclatural issues across all regions). There are no genera occurring naturally in four or five regions, although there are four families in common across all five (Fabaceae, Malvaceae, Rhamnaceae, Sapindaceae). In pairwise comparisons, the number of shared genera ranges from one to sixteen. The greatest number is between the MTC regions of South Africa and Australia, reflecting both their shared biogeographic history and the greater overall diversity (Verboom et al. 2014). The next highest is between the MTC regions of California and the Mediterranean (thirteen genera), which is especially striking as the total diversity of genera is much lower. Some of these taxa (e.g. *Arbutus*) were key to Axelrod's Tethyan hypothesis of mid-latitude stepping stone dispersal across the then smaller Atlantic (Axelrod 1975), see (Wen and Ickert-Bond 2009; Kadereit and Baldwin 2012).

While the number of shared genera is fairly low (relative to total diversity), it is instructive to consider alternative hypotheses of why we see these closely related taxa occupying

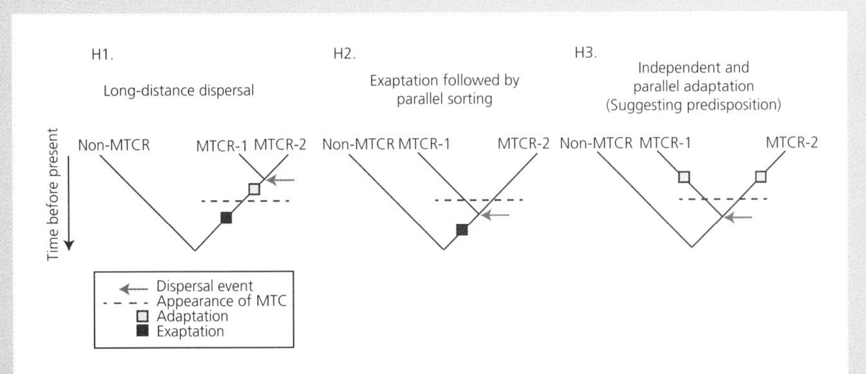

Case Study 2 Figure. Hypotheses for the sharing of traits by species in mediterranean-type climate (MTC) regions. Three hypotheses (H1–H3) for the occurrence of close relatives sharing functional traits among MTC regions, illustrated on a phylogenetic tree. Adaptation refers to the evolution of traits in response to MTC, whereas exaptation refers to the evolution of traits under another selective regime, which then become adaptive under MTC. H1: Exaptive or adaptive evolution of traits to MTC in one region, followed by long-distance dispersal to another region. H2: Exaptive evolution of traits and long-distance dispersal prior to appearance of MTC and independent sorting of taxa from each continent into the respective regions. H3: Independent and parallel adaptive evolution of traits from common ancestors in response to MTC after dispersal to the respective regions. MTCR, Mediterranean-type climate regions.

geographically disparate, but climatically similar, regions. Three hypotheses may be advanced that would explain the occurrence of close relatives sharing functional adaptations to MTC, such as sclerophylly (Case Study 2 Figure; Kadereit and Baldwin 2012): 1) evolution in one region, followed by long-distance dispersal to the other; 2) evolution of sclerophylly prior to appearance of MTC, and independent sorting of taxa from each continent into the respective regions; and 3) independent and parallel evolution of sclerophylly from common ancestors in response to MTC. We find the last of these three hypotheses particularly intriguing as it suggests shared predispositions in the related lineages that facilitate the evolution of sclerophylly or related traits. These could involve genetic factors making it 'easier' to evolve the trait, or ecological preferences that allow the two lineages, in parallel, to succeed and expand in the drying climates of the emerging MTC regions, thus putting them under parallel section pressures.

Kadereit and Baldwin (2012) re-evaluated twenty-five disjunctions between California and the Mediterranean flora, including four woody groups (Arbutoideae, *Cercis*, *Platanus*, *Styrax*). Based on phylogenetic relationships, divergence ages, and ecological characteristics of each group, they concluded that at least nine of these (including three of the woody groups) represent parallel evolution of arid adapted traits in the two regions from more mesic-adapted common ancestors. However, they also emphasized the difficulties of drawing strong historical inferences from the limited biogeographic and ecological evidence available in modern taxa. The identified examples of parallel evolution offer excellent model systems to examine ecological and genetic mechanisms that may predispose closely related lineages to undergo parallel adaptive change.

Continued

Case Study 2 (*Continued*)

At the family level, the Rhamnaceae (Rosales) stand out for their ecological importance across all five MTC regions, and potentially illustrate this pattern of predisposition to evolution of sclerophylly. Rhamnaceae comprises ca. 1055 species of predominantly warm temperate shrubs, with approximately one-third of these species occurring in MTC regions. Rhamnaceae is one of the few angiosperm families with taxa occurring across the five MTC regions. However, Rhamnaceae is not restricted to the wet-winter, summer-dry MTC, as the remaining two-thirds of the species within the family occur from tropical rain forests to deserts on all continents except Antarctica. Consequently, Rhamnaceae shows spectacular morphological variation, from spiny shrubs to large tropical forest trees or lianas, and leaves can be entire, toothed or revolute, deciduous or evergreen, and in several species leaves are even absent (aphyllous) and photosynthesis is performed by the stems. The evolution of these traits happened during the last ca. 84–101 Ma, when crown Rhamnaceae originated (Onstein et al. 2015).

The typical sclerophyllous MTE traits are found within several clades of the Rhamnaceae: the Phyliceae tribe in South Africa, the Pomaderreae tribe in Australia and genus *Ceanothus* (subclade *Cerastes*) in California. Onstein and Linder (2016) estimated the possible age of these traits by performing ancestral state reconstructions on the dated phylogeny of Rhamnaceae. They compared these trait ages to the age of the onset of MTC in the respective regions. They concluded that these traits must have been pre-adaptations or exaptations to the MTC in South Africa and Australia, but may have been adaptive in California. To illustrate, the colonization of the MTC region of South Africa by the Phyliceae occurred ca. 22.6–41.7 Ma (intervals reflect uncertainty in the estimated age of a node) (Onstein et al. 2015) and this shift co-occurred with the evolution of sclerophyllous leaf traits (Onstein et al. 2014; Onstein and Linder 2016). However, the age of the MTC in South Africa is probably 10–15 Ma, and therefore much younger than the age of sclerophyllous leaf traits. In California, *Ceanothus* subgenus *Cerastes* and its sclerophyllous leaf traits originated 4.2–49 Ma, although most probably between 24–49 Ma, and the MTC 2–5 Ma. The adaptation hypothesis can therefore not be rejected for *Ceanothus* subgenus *Cerastes*. Onstein and Linder (2016) hypothesized that the selective regime that has led to sclerophyllous Rhamnaceae in South Africa and Australia may have been infertile, low nutrient soils. Nevertheless, the 'independent' evolution of sclerophyllous leaf traits in Rhamnaceae does suggest that lineages may have been predisposed for MTC, perhaps through deep-phylogenetic cryptic genetic and developmental precursors.

Although convergence may have been important in explaining the similarity in morphology among MTC region floras, in-situ radiations of clades possessing sclerophyllous traits may also have contributed to the functional and taxonomic diversity found in MTC regions today. The match between morphology and environment may, for example, increase speciation rates, or, more likely, decrease extinction rates on macroevolutionary timescales (Lancaster and Kay 2013; Onstein and Linder 2016; Onstein et al. 2016). If this diversification rate (speciation rate—extinction rate) is driven by the possession of sclerophyllous leaf traits, as was shown for the Rhamnaceae occurring in the MTC regions of South Africa and Australia, this may generate a flora dominated by sclerophyllous species, even if the deeper phylogenetic diversity is non-sclerophyllous. Convergent and parallel evolution, as well as trait-dependent diversification, are thus all-important processes to consider if we aim to understand species diversity and functional similarities in MTC region floras and in regional floras more generally.

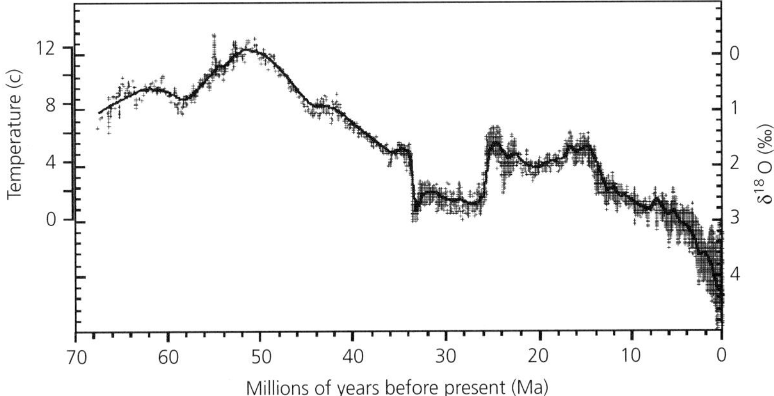

Figure 2.4 **Global oxygen isotope records from the deep sea.** Oxygen isotope levels derived from samples representative of deep ocean waters and gathered from deep-sea sediment cores are indicative of changes in global climate. The relationship between changes in temperature and changes in oxygen isotopes is included on the left axis. The temperature scale does not extend to lower oxygen isotope levels because the formation of ice sheets below the shown level complicates this relationship. (Modified from Zachos et al. 2001.)

Cenozoic that contributed to the evolution of different floras with a high number of endemic species unique to each region.

From the late Cretaceous to the present, the Earth's climate changed dramatically. These changes were driven by a range of factors such as alterations in Earth's orbit, plate tectonics, and atmospheric gas concentrations (Zachos et al. 2001). The long-term temperature trend, with considerable periods of variation, was a sustained cooling of the earth following a peak about 50 Ma (Figure 2.4). Seasonality increased during the late Cretaceous and into the Palaeogene (Davies-Vollum 1997), and there were likely regions of winter-wet climate at this time (Horrell 1991; Willis and McElwain 2014). However, this changed in the mid-Palaeogene (the Eocene; 60–50 Ma) when the Earth became generally hotter and wetter and there was limited seasonality in the regions with future MTC.

In the late Palaeogene (the Oligocene; approximately 34 Ma), an abrupt cooling occurred that was associated with plate tectonics. At this time, global ocean circulation patterns changed owing to the opening of the Tasmanian Gateway between Australia and Antarctica and the later opening of Drake Passage between South America and Antarctica. These openings allowed the Antarctic circumpolar current to develop (Kennett 1977). This was followed by the formation of a connection between North and South America, which also changed ocean circulation patterns. The late Palaeogene and Neogene were also periods of ice sheet development, both in the Antarctic and Arctic (Zachos et al. 2001). These changes led to a spread of drier climates and increased seasonality. Drying, cooling, and increased seasonality were all important in the development of MTC, with changing global ocean currents providing the conditions necessary for winter rainfall precipitation. All five modern MTC regions occur on the western sides of continents, where cool ocean currents flow from the poles toward the equator: for California the California current system, for Chile the Humboldt current, for South Africa the Benguela current, in Australia the West Australian current, and the Canary current for Western Europe and North Africa.

Communities of arid and semi-arid vegetation types emerged coincident with polar glaciation around 25 Ma (Singh 1988), with the modern MTC regions likely developing sometime around 10 Ma (Janis 1993), and a period of recent drying around 5–3 Ma. The timing of MTC region emergence in their modern locations was generally similar across regions, especially in the context of the large scale of geological time. There is considerable variation in region-specific and between-region estimates of when MTC regions were established, and these estimates are complicated by the reconstruction of past climates and differences in how MTC regions may be defined or identified. The generality of the timing of emergence of MTC regions is indicative of the broad global patterns in changing aridity and seasonality that drove their development.

Amidst these broad scale patterns during the Cenozoic, shorter term climate oscillations, such as decadal patterns, created variability that would have affected the available habitat for drought-adapted vegetation as well as the prevalence and extent of fire (Keeley et al. 2012). These factors would also have been impacted by changing topographical relief, especially in Chile and California, and latitudinal gradients. This would have created heterogeneous environments that would potentially allow vegetation-types to persist in small relictual pockets or to expand to cover broad extents. Thus, the key drivers for MTEs as well as the adaptations important for MTV likely arose throughout this timeframe, from around 25 Ma to the present.

In brief, for each region, a modern MTC region developed according to broadly similar timelines:

- In the Mediterranean, climate drying and seasonality developed during the Miocene, approximately 12–8 Ma (Thompson 2005; van Dam 2006), with a pronounced shift to a drier climate around 5–3 Ma (Suc 1984; Lo Presti and Oberprieler 2009). The evolutionary divergence of many MTV lineages and vegetation reconstructions support this timeline (Fauquette et al. 1999; van Dam 2006; Kovar-Eder et al. 2006; Lo Presti and Oberprieler 2009; Salvo et al. 2010).
- In Australia, an MTC was established around 5 Ma and was coincident with diversification of sclerophyllous taxa there (White 1986); however, it may have existed for some time prior to this. Some authors have suggested that an MTC was in place as early as 20 Ma, with a subsequent period of drying 5–2 Ma (Hopper and Gioia 2004; Martin 2006) or 15–12 Ma with

drying 4–1 Ma (Byrne et al. 2008; Rix et al. 2015). Phylogenetic data indicate periods of diversification for Australian MTV between 25–15 Ma and 5–3 Ma (Crisp et al. 2004).

- In South Africa, an MTC was likely established around 10 Ma. Climate reconstructions suggest an onset of drying between 15 and 10 Ma (Hoetzel et al. 2015), a shift toward MTV plant types between 10 and 6 Ma (Dupont et al. 2011), and increasing aridification and an expansion of shrublands 5–3 Ma (Hoetzel et al. 2015; Diekmann et al. 2003). Some have suggested that an MTC may have developed earlier, perhaps as early as 19.5 Ma, based on phylogenetic analyses of plant diversification (Bytebier et al. 2010).
- In California, sclerophyllous shrublands were established by 10 Ma (Raven and Axelrod 1978), with a winter-wet climate pattern likely having developed between 15 and 10 Ma (Axelrod 1973).
- In Chile, an MTC was likely established during a period of geological uplift that isolated the west coast of South America and blocked weather systems from the east, with MTC development likely occurring between 15 and 8 Ma (Armesto et al. 2007). Andean uplift was initiated around 20 Ma, with the Central Andes reaching half of their current elevation approximately 10 Ma followed by a rapid period of uplift between 2 and 5 Ma (Gregory-Wodzicki 2000). As early as 15 Ma there is evidence of the establishment of drying conditions, with modern conditions established by 3–4 Ma (Hartley 2003). Evolutionary studies suggest that an MTC may have developed later in Chile, perhaps only during the past 6–4 Ma during the final period of uplift and coincident with evolutionary radiations in several lineages (Bleiweiss 1998; Givnish et al. 2011).

All of the MTC regions were impacted by the most recent glacial period, the Last Glacial Maximum (LGM), which was the culmination of a period of cooling during the last 100,000 years. This period of cooling caused the spread of ice sheets across the northern hemisphere high latitudes, changed global temperature and weather patterns, and impacted the distributions of MTC areas. Two regions, Chile and California, may have been particularly impacted by these changes owing to their more recent geological uplift and consequently greater topographical variation compared to the other MTC regions. These two regions have experienced the most variation in climate and topography since the end of the LGM (Cowling et al. 2015), and this may have limited the ability of taxa to persist through this period of change and therefore reduced their current levels of biodiversity relative to other MTC regions (Case Study 3).

Thus, we arrive at the present-day MTEs and the key **ecosystem drivers** that affect ecosystem function. The most important of these drivers are climate (MTC), topography, geology, and soils of each region, which are determined by their geographical and climatic histories, and their disturbance regimes, particularly as related to fire.

Case Study 3 The role of environmental stability in explaining variation in plant diversity in mediterranean-type ecosystems

Richard M. Cowling, Centre for Coastal Palaeosciences; Nelson Mandela University, Port Elizabeth, South Africa

All five mediterranean-type ecosystems (MTEs) are recognized as hotpots of plant diversity and endemism (Mittermeier et al. 2005) and two of them—the South African Cape and southwestern Australia—have the world's richest extratropical vascular plant floras (Cowling et al. 1996; Kreft and Jetz 2007). What explains the variation in plant diversity among MTEs? Decades of research on ecological convergence of MTEs provided some of the best documented cases of this phenomenon globally; however, it also revealed striking cases of divergence. Furthermore, regional-scale richness patterns are not well explained by contemporary ecological factors such as climate (including energy regimes), soil nutrients, and topographical heterogeneity (a surrogate for habitat diversity) (Valente and Vargas 2013; Bradshaw and Cowling 2014). Instead, diversity appears to be under the control of differences in the environmental histories of MTEs and their impacts on patterns of diversification (speciation minus extinction) (Linder 2008; Valente et al. 2011).

In a recent paper, Cowling et al. (2015) hypothesized that older, more climatically stable, MTEs would support more species, because they have had more time for species to accumulate than MTEs that were historically subject to greater topographic upheavals and fluctuating climates (see also Hopper 2009). They tested this hypothesis by investigating the relationship between plant diversity (measured as the intercepts of species–area curves that are homogeneous in slope across all regions) and measures of Pleistocene climatic stability and Cenozoic topographical stability. The results were consistent with their hypothesis: the most diverse MTEs—South Africa and Australia—had the highest environmental stability, and the least diverse—Chile and California—had the lowest stability (Case Study 3 Figure). Spain, representative of the western Mediterranean Basin, an area of low topographic stability and high climatic stability, occupied an intermediate position in terms of diversity. Greece, representative of the eastern Mediterranean Basin, had a lower diversity than expected from its stability, possibly owing to an island effect (all diversity data were from islands).

Data from numerous dated molecular phylogenies suggested that variation in plant diversity in MTEs is likely to be a consequence not of differences in diversification rates, but rather the persistence of numerous pre-Pliocene clades in the more stable South African and Australian regions (e.g. Sauquet et al. 2009; Valente and Vargas 2013; Onstein et al. 2015). The extraordinary plant diversity of the South African Cape is a consequence of the combined effects of both mature and recent radiations, the latter associated with increased habitat heterogeneity produced by mild tectonic uplift in the Neogene (Cowling et al. 2009; Verboom et al. 2014). These results are consistent with research for many different biomes and taxa: regions of high environmental stability during the Cenozoic are associated with high species (and genetic) diversity (e.g. Graham et al. 2006; Araújo et al. 2007; Carnaval et al. 2009; Werneck et al. 2012).

These results are consistent with the notion that, given sufficient stability, plant hyperdiversity can develop outside the humid tropics (Cowling et al. 2009; Sniderman et al. 2013), implying that water and energy variables are not consistent predictors of high regional-scale

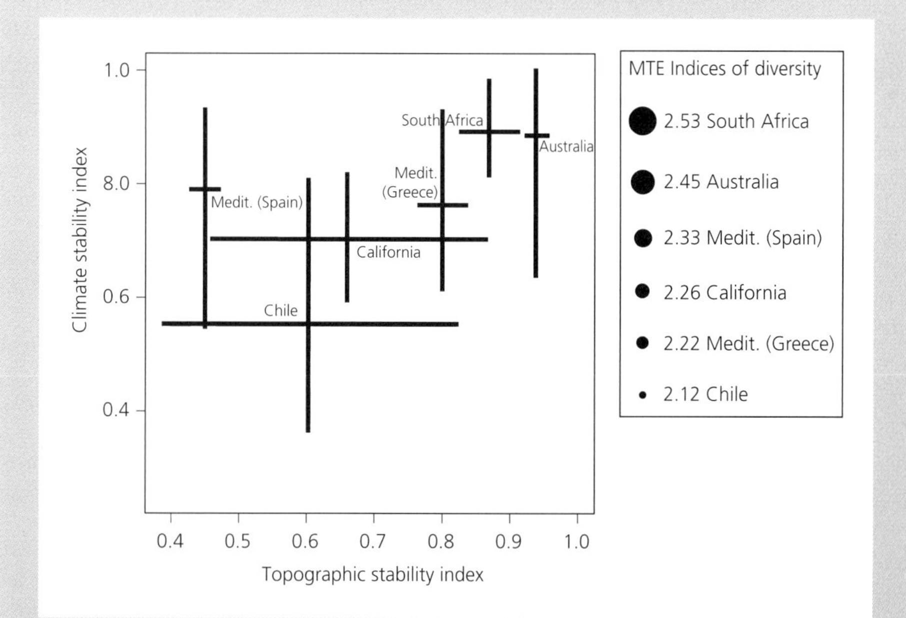

Case Study 3 Figure. Relationships between diversity indices and mean values of topographic and climate stability across mediterranean-type ecosystems (MTEs). Diversity indices were calculated as the slope of species–area regressions for each region (Cowling et al. 2015) and they are shown on the right; a higher diversity index indicates more species per area within that region. Climate and topographic stability indices were also calculated. A stability index of 1 indicates complete stability. The black lines for each region represent the topographical stability range over the time period from the late Oligocene to the present day and the climate stability range from mediterranean climate models (Köppen, Maxent, GAM) and downscaled global climate simulations (MIROC and CCSM). In general, diversity indices increase as both climate and topographic stability increase. (Figure modified from Cowling et al. 2015.)

plant richness (Davies et al. 2005; Kreft and Jetz 2007). The concentration of plant species in the humid tropics of the world is likely to be a consequence of Cenozoic environmental stability at these latitudes (Ricklefs 2004) rather than characteristics of contemporary environments.

2.2 Introduction to ecosystem drivers and processes

2.2.1 Climate

MTC is typified by a cool wet winter (winter-wet) and a warm dry summer. This unique pattern of **seasonality** is relatively rare globally, with MTC being reported as occurring in less than 1 per cent to approximately 5 per cent of the Earth's land surface depending on how it is defined (Aschmann 1973; Cowling et al. 1996).

Seasonality is often evaluated using climate diagrams ('climadiagrams'; Walter 1973) and we present a simplified version of this diagram type in Figure 2.5. A key feature of this type of diagram is that left axis ticks, which represent temperature (°C), are equal to ½ the value of the equivalent right axis ticks, which represent precipitation (mm) (e.g. 20°C would be aligned

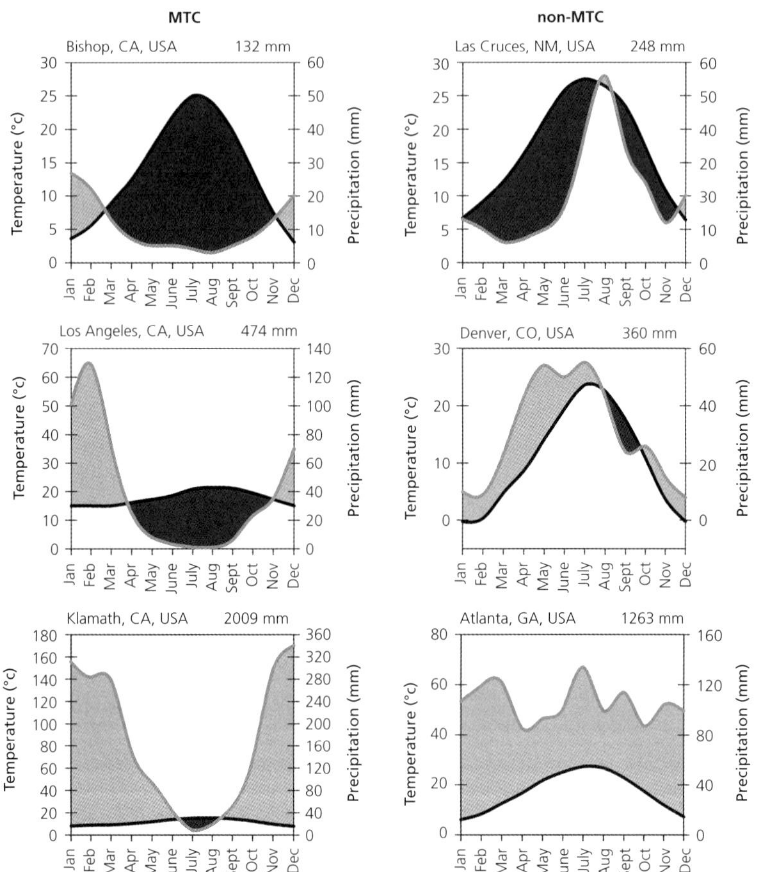

Figure 2.5   **Sample climadiagrams from different cities within the USA.** The climadiagrams in the left column are indicative of mediterranean-type climate (MTC), with only two seasons (wet and cool, dry and warm). In MTCs, precipitation occurs predominantly in the winter months to create a period of winter moisture surplus (light grey fill) and a period of summer water deficit (dark grey fill). This pattern is consistent across MTCs, even if they have very low mean annual precipitation (Bishop, CA) or very high precipitation (Klamath, CA). In contrast, non-MTC regions (right column), even if they have similar levels of mean annual precipitation to MTCs, have very different climate patterns, including having two short periods of moisture availability, one each in the summer and winter (Las Cruces, NM), favourable moisture conditions for most of the year, including much of the summer (Denver, CO), or no periods of regular water scarcity (Atlanta, GA). For each panel, the location is in the upper left and the mean annual precipitation is on the upper right. Data are from US Climate Data (http://www.usclimatedata.com). Note that axes scaling varies for each panel.

with a value of 40 mm precipitation). Precipitation that is greater than twice the average temperature within a month is considered to represent a period of water availability, while precipitation that is less than twice the average temperature is considered to represent a period of water scarcity. Ignoring potential limitations on maximum or minimum temperatures or mean annual precipitation, MTC climadiagrams show two distinct seasons annually, a period of water availability during the winter months and a period of summer water deficit (Figure 2.5, left panels). This is not a pattern that is present in other climate regions, even ones that occur near MTC or which have similar levels of precipitation (Figure 2.5, right panels).

All five MTC regions have the characteristic MTC pattern of seasonality, as would be expected for this key ecosystem driver (Figure 2.6). Within each region, MTC seasonality patterns occur in regions of low to high mean annual precipitation, similar to the range shown in Figure 2.5 (left column) for

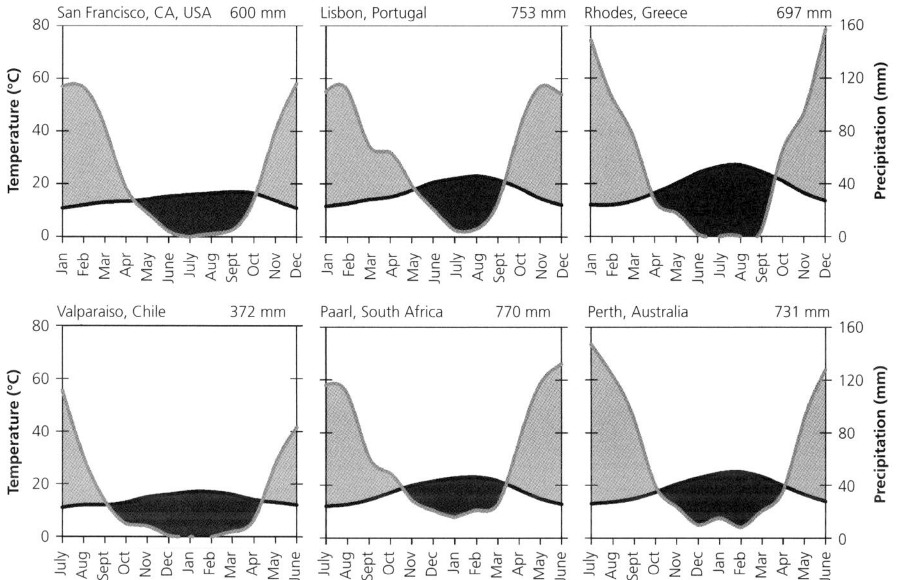

Figure 2.6   **Climadiagrams for selected cities from each of the five global mediterranean-type climate (MTC) regions.** Across the global MTC regions, the seasonality of the climate is very similar. All five regions have a pronounced winter moist period and a dry period of summer water deficit. Note that the water year (Jan–Dec for the northern hemisphere, top row; July–June for the southern hemisphere, bottom row) has been placed as the *x* axis in all figures rather than the months in order of the calendar year. Periods of water availability are indicated with light grey fill and periods of water scarcity are indicated with dark grey fill. For each panel, the location is in the upper left and the mean annual precipitation is on the upper right. The data are from http://www.usclimatedata.com (San Francisco), http://www.climatedata.eu (Lisbon and Rhodes), http://www.valparaiso. climatemps.com (Valparaiso), http://en.climate-data.org (Paarl), and http://www. weatherzone.com.au (Perth). All panels are shown with the same scaling.

California. Thus, a general pattern of MTC seasonality may exist across a range of absolute levels of mean annual precipitation (MAP). For example, many areas in South Africa with an MTC have high MAP, of over 1000 mm, but they still have seasonality and MTV. The reason that areas of such high MAP in South Africa are not forested, as they are in all other MTC regions at such high MAP, is an interesting question.

MTC regions are often viewed as being semi-arid to arid, although authors vary in how they have set a precipitation limit defining an MTC. For instance, at the low end, MTC regions have been suggested to require at least 100 mm in MAP (Joffre and Rambal 2001), at least 200 mm (Miller et al. 1977), at least 250 mm (Cowling et al. 2005), at least 275 mm (Aschmann 1973), or at least 350 mm (Miller and Hajek 1981). There has also been variability at the upper end, limiting MTC to semi-arid regions and excluding moister uplands or forests, including MTC regions being defined as typically receiving less than 650 mm MAP (Miller and Hajek 1981), less than 800 mm (Miller 1981), or less than 900 mm (Aschmann 1973). Some have also suggested requirements associated with the number of months of moisture availability (i.e. the grey regions in Figures 2.5 and 2.6), such as 5–10 months of moisture availability (Joffre and Rambal 2001). Others have also suggested that MTC must receive not just a majority of precipitation during the winter months (>50 per cent), but a larger proportion, such as 65 per cent (Aschmann 1973; Chapter 4 Case Study 8), 70 per cent (Cowling et al. 2005), or 90 per cent (Rundel 1999). Even though limitations such as those described above are often mentioned, most authors recognize that MTC regions in 'the broad sense' also include adjacent areas, including many winter-rainfall deserts such as the Mojave (California), Succulent Karoo (South Africa), and Atacama (Chile) (Rundel 1999), moister adjacent forest communities (Sander and Wardell-Johnson 2012), and low to mid-elevation mountain tops within MTC areas that may themselves have less seasonal precipitation due to their elevation (Agenbag et al. 2008).

Temperatures, particularly at the cold end, have also been subject to variable definitions for MTC regions. The often-used Köppen classification system sets a temperature limit for 'Cs' climates, which coincide with MTCs. For these climates, temperature must be greater than –3°C average during the coldest month of the year and no more than 6 per cent of the hours of the year may be below freezing (Köppen and Geiger 1936). Others have suggested that winter average temperatures are typically less than 15°C with rare frosts that are never severe and no more than 3 per cent of the hours per year below 0° C for MTCs (Aschmann 1973) or, with ranges for both the seasonal highs and lows, such as temperatures of 6–10°C average during winter and 20–24°C average during summer (Miller et al. 1977).

As stated in Chapter 1, we use an inclusive definition for MTC (MTC sensu lato), except for where certain discussions necessitate narrowing of this definition. Where we are discussing MTC regions as more narrowly defined, such as by some of the limits mentioned above (MTC sensu stricto), we will indicate it in the text. Using a broad definition of MTC, especially since narrower definitions were often formulated to match MTV distributions specifically, includes many other plant communities, including deserts and moist forests that occur within the globe's winter-rainfall regions. Additionally, particularly in the case of freezing, narrow definitions for MTC may exclude regions that occur within the heart of even the strictest MTC-defined areas. Freezing can be a strong determinant of plant distributions within MTC, and a reduction in freezing extent or occurrence may have important impacts on these systems. In part because of a tradition of excluding freezing from MTC, this has been an understudied driver of species distributions within MTC regions (Case Study 4).

Case Study 4 Freezing as an understudied driver of plant distribution within mediterranean-type ecosystems

Stephen D. Davis, Natural Science Division, Pepperdine University, Malibu, California, USA

George Matusick, State Centre of Excellence for Climate Change Woodland and Forest Health, School of Veterinary and Life Sciences, Murdoch University, Murdoch, Australia

Not only is freezing a neglected topic in climate descriptions of mediterranean-type ecosystems (MTEs; Larcher 1981); it evokes a long list of provocative implications, especially in regard to the advent of Anthropocene global warming (Hannah et al. 2014) and the concomitant assumption that freezing will become less important. However, freezing can have profound impacts on the distribution of evergreen sclerophyllous plants that comprise mediterranean-type tree and shrub communities, and altering this barrier to species distributions may have large impacts. It is well understood that cold temperatures play an important role in limiting the distributional range of vegetation at high latitudes and high elevations, but less well appreciated are landscape factors that lead to important thermal transects from hilltop to valley bottoms.

The mediterranean-type climate (MTC) regions of coastal southern California and southwestern Australia have winters that are mild and relatively warm and there are microclimate effects, such as radiation freezes on calm, clear nights that regulate overall community structure and species distribution across short distances. There have been documented freezing gradients along thermal transects that are more than threefold steeper, and in the opposite direction to the adiabatic lapse rate calculated for decreasing temperature with altitude (~5°C decrease per 1000 m; Nobel 1999). Inland valleys of coastal mountains (California) and plateaus (Australia) are sheltered from regional advective influences of the ocean, thus forming a primary gradient in decreasing temperature from coastal exposures inland (e.g. gradient from 0 to −12°C). Superimposed on this primary gradient is a secondary gradient in decreasing temperature from hilltop to valley floor (e.g. gradient from −6 to −12°C). Such patterns are commonplace in areas of complex terrain and are caused by radiative cooling on calm clear nights in combination with the pooling of cold-air down slope into valley bottoms (Dobrowski 2011).

Continued

Case Study 4 (*Continued*)

In both California and Australia, microclimatic freezing occurs because of: 1) strong thermal gradients from coastal to inland sites resulting from the ameliorating influence of ocean waters on air temperature and localized advection; 2) strong radiation freezes on calm, clear nights intensified by cold air drainage and pooling into valleys; and 3) the ghost of Pleistocene past that has left residual refugia on cold hilltops and even colder valley bottoms (Case Study 4 Figure).

For example, in the Santa Monica Mountains of southern California, the three dominant chaparral species in coastal exposures experience minimum temperatures of only 0°C, persist at –6°C on hilltops, but are absent 5 km inland in valley bottoms, where minimum temperature approaches –12°C (Case Study 4 Figure, left panel). Along with these distribution shifts are also shifts in tissue freezing resistance. The lethal temperature for 50 per cent cell death of leaves (LT50) for each of these adult species matched their distribution pattern and was –6°C (Ml), –8°C (Cs), and –9°C (Cm) (Boorse et al. 1998). At –12°C inland sites, cold tolerant congeneric species displace the coastal thermophilic species and display much lower LT50 values of –16°C (Ro), –18°C (Cc), and –19°C (Co) (Case Study 4 Figure, left panel). LT50s of seedlings were found to be even more sensitive to freezing injury than adults, precluding initial establishment (Pratt et al. 2005).

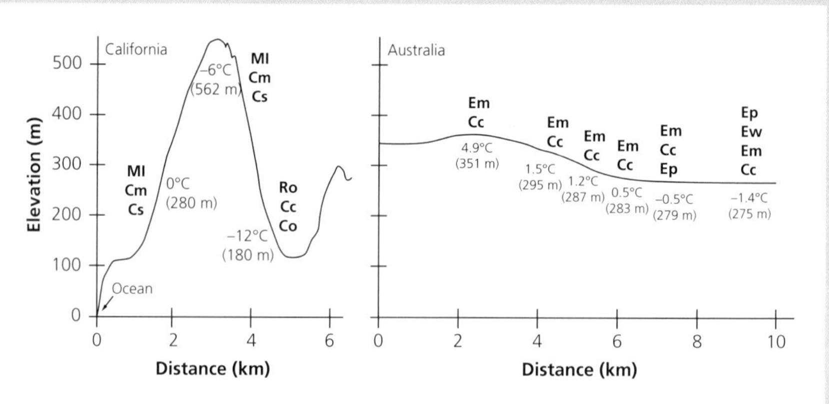

Case Study 4 Figure. Diagrams of freezing gradients with changes in topography in California and Australia. In California (left panel), a primary freezing gradient occurs from ocean to inland, 0 to –12°C, and secondary freezing gradients from hilltop to valley floor, –6 to –12°C. Data is from the coastal Santa Monica Mountains of southern California (Davis et al. 2007a, 2007b). Species that dominate chaparral shrub communities at warm coastal sites and hilltops (Ml, Cm, and Cs) are displaced by cold-tolerant, congeneric species that form microrefugia in cold inland valleys (Ro, Cc, and Co). Species listed are: Ml = *Malosma (Rhus) laurina*, Cm = *Ceanothus megacarpus*, Cs = *Ceanothus spinosus*, Ro = *Rhus ovata*, Cc = *Ceanothus crassifolius*, and Co = *Ceanothus oliganthus*. In Australia (right panel), a freezing gradient is shown as was measured in a prominent valley of the Darling Plateau near Perth, Australia during a severe cold event in May 2012. The overstory Eucalypts that dominate the uplands (listed in order of dominance, Em and Cc) are susceptible to frost dieback at lower slopes and decrease in proportional composition. More frost-tolerant species (Ep and Ew) dominate the colder valley bottoms. Species listed are: Cc = *Corymbia calophylla*, Em = *Eucalyptus marginata*, Ep = *Eucalyptus patens*, and Ew = *Eucalyptus wandoo*.

The role of frost in determining species distributions in southern Australia is much less clear for a variety of reasons. The absence of strong topographic relief in the region contributes to relatively moderate extreme low temperatures at inland sites (Case Study 4 Figure, right panel). None the less, strong temperature gradients leading to frost do form periodically on the local and microscales, which causes differential damage and mortality patterns within and between species in tree communities (Matusick et al. 2014). Since most species distribution studies have focused on larger scales, the role of microsite temperatures has been largely overlooked, with the exception of Ball et al. (1991, 1997). Distributional studies investigating climate as a driver have been limited to overstory trees (mainly *Eucalyptus* and allied species), which are generally less impacted by cold-air pooling than sclerophyllous shrubs owing to their elevated canopy height. However, juvenile trees are susceptible and repeated frosts lead to successive dieback and mortality (McChesney et al. 1995), which has the potential to alter microsite composition. Indeed, tree species are distributed along a minimum temperature gradient in prominent valleys of the Northern Jarrah Forest (Case Study 4 Figure, right panel). *Eucalyptus marginata*, the dominant upland tree species throughout the Darling plateau, is much less tolerant of frost than species that dominate lower slopes and valley bottoms, including *E. patens* and *E. wandoo*. While these distributions have been historically attributed to soil and moisture gradients (Bell and Heddle 1989), current work is also considering the role of frost.

The microrefugia at cold sites represent habitats with an unfavourable microclimate (cold) within a favorable regional climate (warm). This has allowed freeze-tolerant species to persist outside their normal distributional range (cold mountain elevations or polar latitudes). The formation and persistence of cold microrefugia in valley bottoms has two possible explanations, a historical one and a physiographic one. During the Quaternary, 2.5 million years ago to present, there were multiple oscillations between glacial and interglacial periods, averaging 80,000 years per cycle for glaciation and 10,000 years per cycle for warm interglacials (Gavin et al. 2014). This suggests greater opportunity for the formation of freezing microrefugia than thermophile microrefugia. It is also possible that freezing mortality and associated community effects may occur rapidly compared to gradual warming where species with different thermal tolerances may persist in mixed communities for longer. Gene flow over the short distances considered here (Case Study 4 Figure) may preclude adaptation of cold-adapted species to the warmer conditions and vice versa for thermophilic species during cold glacial periods. Presumably, during glaciation, thermophile microrefugia persisted at warm microclimates benefiting from the ameliorating influence of the ocean (coastal exposures). In contrast, during short duration interglacials, cold-tolerant microrefugia remained in cold-air drainages, allowing persistence of cold-adapted species.

Importantly, climate conditions leading to radiative cooling on the local landscape scale, including topography, are typically not incorporated into regional and global climate models. Furthermore, conditions leading to temperature inversion are disconnected from synoptic weather patterns (Daly et al. 2010), undermining the predictive power of global and regional scale climate models to predict radiation frost. While climate change is altering frost patterns, the impact on radiation frosts may be amplified, because nighttime sky temperatures, that drive radiative cooling, are predicted to increase more rapidly than mean air temperature. Indeed, regional warming over the past century has already been observed as asymmetrical, with average minimum temperatures increasing nearly

Continued

Case Study 4 (*Continued*)

twice as rapidly as maximum temperature (IPCC 2007). Cloud coverage changes, which are currently unpredictable, could also affect night-time frosts in the future.

The disconnect between local scale and regional climate models is evident in southwestern Australia, where, despite a strong increasing trend in mean minimum temperatures, extreme cold events have increased (Zheng et al. 2012). Also, increased variability during spring and autumn has caused a lengthening of the frost window, increasing the potential for damage to vegetation (Alexander et al. 2007). To enable more accurate predictions of future impacts of frost on vegetation in MTEs, radiation frost must be specifically addressed by incorporating topography and other local governing factors into regional climate models.

Further complicating modelling efforts, freezing does not work alone and interacts with other factors. When drought was found to extend into winter months, there was a strong interaction between freezing and tissue dehydration that inflated freeze-thaw induced embolism of stem xylem, exacerbating overall cold-damage to shoots (Davis et al. 2007a, 2007b). In addition, significant interactions with wildfire have been found that inflate post-fire seedling mortality (100 per cent seedling mortality in cold valley bottoms; Pratt et al. 2005) and caused stunted growth because of lost vigour owing to frequent post-freeze resprouting (every 3–5 years) compared to typical post-fire resprouting every 32 years (Witter et al. 2007). Interactions among freezing events and these secondary factors will likely intensify with predicted climate change, such as increased incidence of wildfire and drought in California and Australia (Diffenbaugh et al. 2015).

Freezing is a fruitful topic for future research in MTEs and has been a neglected topic. Freezing can interact with other factors, such as drought, which is expected in increase owing to climate change. Freezing can act beyond broad latitudinal and elevation scales, where at local scales it can impact species diversity near distributional centres and can drive shifts in plant community structure (Loarie et al. 2009; Klausmeyer and Shaw 2009). Recent models predict that anthropogenic warming in MTEs is now unidirectional, not cyclical, and thus will not return to present climate conditions (Hannah et al. 2014). Indeed, warming over the past century has already been observed as asymmetrical, with minimum temperatures increasing nearly twice as rapidly as maximum temperatures (IPCC 2007).

A final element of MTC is the presence of seasonal winds. All five MTC regions experience predictable wind patterns that coincide with the period of summer water deficit (Keeley et al. 2012). These may be **foehn winds**, which are downslope winds that flow from high elevations to low valleys or coasts, most often as part of a prevailing weather pattern. There may also be **katabatic winds**, which occur because of high interior pressure cells. Winds may be both and these are the types of winds (katabatic foehn winds) that are present for most MTC regions. Each region, and country within the region for the Mediterranean Basin, has a different name for these summer and autumn winds. These names include, but are not limited to, Berg winds

(South Africa), Puelche (Chile), Brickfielder (Australia), Santa Ana (California), Mistral (France), and Poniente (Spain) winds (Keeley et al. 2012). These dry season winds contribute to weather patterns that dry vegetation and convert it from potential fuels to readily combustible fuels that are an important part of the periodic crown fire regimes within MTC regions (Chapter 6).

2.2.2 Fire

Fire had likely been present within the contemporary MTC regions since the middle of the Tertiary and became more widespread through the later Tertiary (Singh 1988; Dodson and Ramrath 2001, reviewed in Keeley et al. 2012). Sclerophyllous MTV lineages likely originated in dry (shallow soils) marginal habitats and, in the northern hemisphere, these taxa would have occurred to a limited extent in a matrix of more mesic forests. During dry periods, fire would have burned much of the globe and fire-adapted sclerophyllous species would expand. In contrast, forests and other fire-sensitive taxa would have likely been dominant during moister periods and fire-adapted taxa limited to dry relictual pockets. It appears that in the South African and Australian MTC regions the conditions were different. In these areas, nutrient-poor soils led to an earlier development of MTV (middle Tertiary) and the predictable fire associated with these communities. Chile likely experienced a pattern similar to the northern hemisphere MTC regions, where many MTV lineages evolved on dry sites by the late Tertiary. Unique to Chile, during the late Miocene, the uplift of the Andes blocked summer storms and lightning ignitions, which severely limited fires.

The early evolution of fire-related traits suggests a long history of fire across most MTC regions. This is especially the case for the traits related to the release (serotiny) and germination of seeds in response to fire-cues, which is selected for and maintained only if fires are recurrent (Keeley and Bond 1997; Keeley et al. 2012, Pausas and Keeley 2014). There are a great number of lineages and species in present-day South African and Australian MTV that are obligate seeders after fire, which is a life history almost certainly associated with fire (Keeley et al. 2012; Chapter 6). Interestingly, although Chile has a more limited history of fire than the other regions, many Chilean taxa have lignotubers, which is most common among present-day MTV species and probably linked to fire. Thus, some Chilean species may harbour traits that were adaptive to earlier fire regimes.

Natural fires across all of the regions are caused predominantly by lightning. In South Africa, fires have also been linked to rock falls (Moll et al. 1980; Kruger and Bigalke 1984). Volcanic activity may have been an important past ignition source in Chile (Fuentes and Espinoza 1986) and recent evidence supports the presence of wildfires associated with volcanic activity during the Neogene in this region (Abarzúa et al. 2016) (see also Figure 2.11). Within modern times, fire, both from natural ignitions and associated with human activities, has been a prevalent occurrence across all MTC regions. The first written observations of the Los Angeles area of California are associated with fire. In 1542, Juan Rodriguez Cabrillo described smoke filling the Los Angeles bay as they arrived from sea and it was accordingly originally named the Baya de los Fumos, or Bay of Smoke (Kelsey 1998). Similarly, in 1827, during European exploration of the Perth area of Australia for potential settlement, James Stirling found that the ground around the Swan River had been cleared by a fire shortly before their arrival (Hallam 2014). Indeed, across all five MTC regions, there is a long history of anthropogenic fire associated with indigenous, settler, and modern populations (Hall 1984; Mensing et al. 1999; Vanniere et al. 2008; Contreras et al. 2011; Hallam 2014; see also Chapter 8).

MTC regions experience periodic wildfires that occur primarily as **crown fires**, especially within the communities dominated by MTV. In a crown fire, the above ground portion of plants, including the crown of over-story plants, burns. Plants that persist following a fire may do so by epicormically resprouting from buds protected by thick bark along the trunks and large stems of plants, basally sprouting from roots or other specialized below-ground structures such as lignotubers (Figure 2.7; see Plate 3), and re-establishing through the germination of seeds that are able to survive the fire by being protected by soil or specialized structures such as serotinous woody cones or fruit (Figure 2.8; see Plate 4). Fires may occur as surface fires in other MTE communities such as moister forests, burning along the ground but leaving the woody plant canopy intact, or may be rare to absent in low-fuel arid communities such as the winter-rain deserts. Plant adaptations to fire are discussed in Chapter 6.

Fires occur as part of a **fire regime** (Krebs et al. 2010). That is, communities within MTC areas experience fires with a certain frequency, intensity, severity, seasonality, extent, and pattern of spread (Chapter 6). Fire regimes may differ between communities occurring within one region, for instance between grasslands, shrublands, and forests (Gibson et al. 2015). They also differ between comparable communities across different MTEs owing to location-specific differences in ignition sources, plant growth, and climate and weather patterns.

Different communities and regions have experienced different past and current fire frequencies and this is often discussed as differences in their fire return interval (FRI). FRI is a measure of the amount of time that typically passes between recurring fires at a site. Within the sclerophyllous shrubland communities of MTEs, the modern FRI is currently relatively short owing to increased anthropogenic ignitions and potentially getting shorter (Wilson et al. 2010; see Chapter 8). In South African fynbos the typical modern FRI is 10–13 years and it is relatively rare for a stand to be older than 45 years (van Wilgen et al. 1992; van Wilgen et al. 2010). In Australian kwongan, a similar range has been reported, with reports varying from 7 to 16 years depending on the site and vegetation characteristics (Enright et al. 2005).

Figure 2.7   **Examples of plants that re-establish through resprouting following fire.** A common recovery strategy post-fire is for plants to resprout from specialized structures. This includes species such as *Protea nitida* in South Africa (a) which form epicormic sprouts along the burnt trunk post-fire (b). The white arrow in panel b points to a newly emerging epicormic sprout in this species. In Australia, *Eucalyptus diversicolor* is shown epicormically resprouting following fire (c). Epicormic resprouting also occurs in many *Quercus* tree species (d) in northern hemisphere mediterranean-type ecosystems. Other woody species form specialized woody root structures, termed lignotubers, that enable them to resprout from their root crown post-fire, including in some *Banksia* species (e) in Australia and species such as *Malosma laurina* (f) in California. The white arrow in panel g points to a newly emerging resprout from a bud within the lignotuber of *Adenostoma fasciculatum* in California. Many non-woody species also survive fire via underground structures, including many geophyte species such as the post-fire flowering bulb from South Africa shown in panel h. (See Plate 4)

Source: Photos from Anna L. Jacobsen.

California chaparral has a longer FRI, with approximately 30 years between fires, but this is an increase over a natural interval of as long as 50–100 years (Keeley et al. 1999), and some sites experience very short FRIs (Jacobsen et al. 2004). Similar to California, FRIs have been reported as 50–100 years in Spain (Vázquez and Moreno 1998), and 16 to more than 50 years in southeast France (Ganteaume and Jappiot 2015). Fire has historically been less frequent in Chile compared to other MTEs owing to the reduced occurrence of lightning during the summer (Aravena et al. 2003), but anthropogenic fire is currently relatively common (Segura et al. 1998; Lara et al. 1999; Montenegro et al. 2003).

Figure 2.8 **Examples of plants that re-establish through seeding following fire.** Many mediterranean-type ecosystem species re-establish following fire through the germination of seeds that are cued to release after fire or are triggered to germinate by cues associated with fire. These examples include some South African Protea species that release seeds from protective canopy fruit structures following fire (a). Seeds are released in large numbers and can be seen as the loose orange and brown piles on the ground around the bases of burnt stems of the same species (Plate 4). Some pine species, such as *Pinus coulteri*, which occurs in the Coast Ranges of California, have serotinous cones that protect seeds from fire and then open and release seeds (b, c). For the Australian Cricket Ball Hakea, *Hakea platysperma*, large woody follicles protect seeds from fire and subsequently open to release seeds following fire. Each large follicle contains a single large seed (visible in the centre of the image) that germinates post-fire (d). In panel d, the seedling, seed, and large fruit of this species are all shown. In California, many species have seeds that are stored in the soil rather than in specialized canopy structures. These seeds are cued to germinate post-fire, resulting in sometimes high densities of seedlings emerging post-fire, including for the shown species of *Ceanothus* (e). For this species, *Ceanothus megacarpus*, more than 100 seedlings germinated per metre squared at the shown site and all of the seedlings in the image are from this one species. (See Plate 5)

Source: Photos a, d, and e from Anna L. Jacobsen and photos b and c from R. Brandon Pratt.

FRIs that are shorter than the historical regime appear to be particularly threatening to sclerophyllous shrublands (i.e. **immaturity risk**), even though many species are dependent on fire for recruitment. This is especially the case for obligate seeding species that often require a fire-cue to germinate, but that must also have time to grow, mature, and produce seed during inter-fire periods to persist at a site (Figure 2.9). On other hand, some MTV species communities begin to senesce in the long absence of fire (**senescence risk**, see Chapter 7). Fire regime changes and other modern threats to MTEs are discussed in additional detail in Chapter 8.

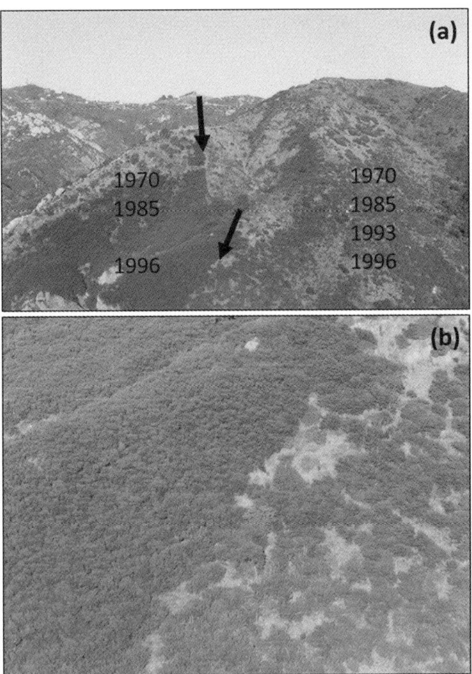

Figure 2.9 **Short-fire return intervals (FRIs) negatively impact the persistence of some species.** The clear lines in the vegetation (indicated by arrows) demarcate the location of a 1993 fire perimeter (a). This fire burned the area on the right of the images, but not on the left (the years of recent fires are included in panel a). The entire field of view burned in 1996 and the crowns of plants are thus of equal age in both areas (photograph taken in 2004). The area on the left experienced FRIs of 15 and 11 years, while the area on the right experienced FRIs of 15, 8, and 3 years. Short FRIs (8 and 3 years) precluded recruitment of a dominant obligate post-fire seeding species, *Ceanothus megacarpus*, while resprouting species are still present in both the longer and shorter FRI locations within the site (b). Canopy gaps that were created by the loss of some species were colonized by non-native annual species (b). (Jacobsen et al. 2004.)

Source: Photos from Anna L. Jacobsen.

2.2.3 Topography and geology

Comparisons of the physiography and geomorphology of the different regions were included within some mediterranean comparative works (Figure 2.10) (Thrower and Bradbury 1973; Thrower and Bradbury 1977). Focusing on landscape and terrain structure, Thrower and Bradbury (1973) suggested that some general patterns emerged across MTE landscapes. For three of the systems, most of the Mediterranean, California, and Chile, there exist coastal hills and interior high mountains. These three regions represent relatively young mountain systems, where uplift has occurred owing to folding and faulting of the Earth's crust (i.e. they are orogenic systems). This is related to the present occurrence of these regions at the boundaries of large tectonic plates and the associated occurrence of plate-associated volcanic activity (Figure 2.11).

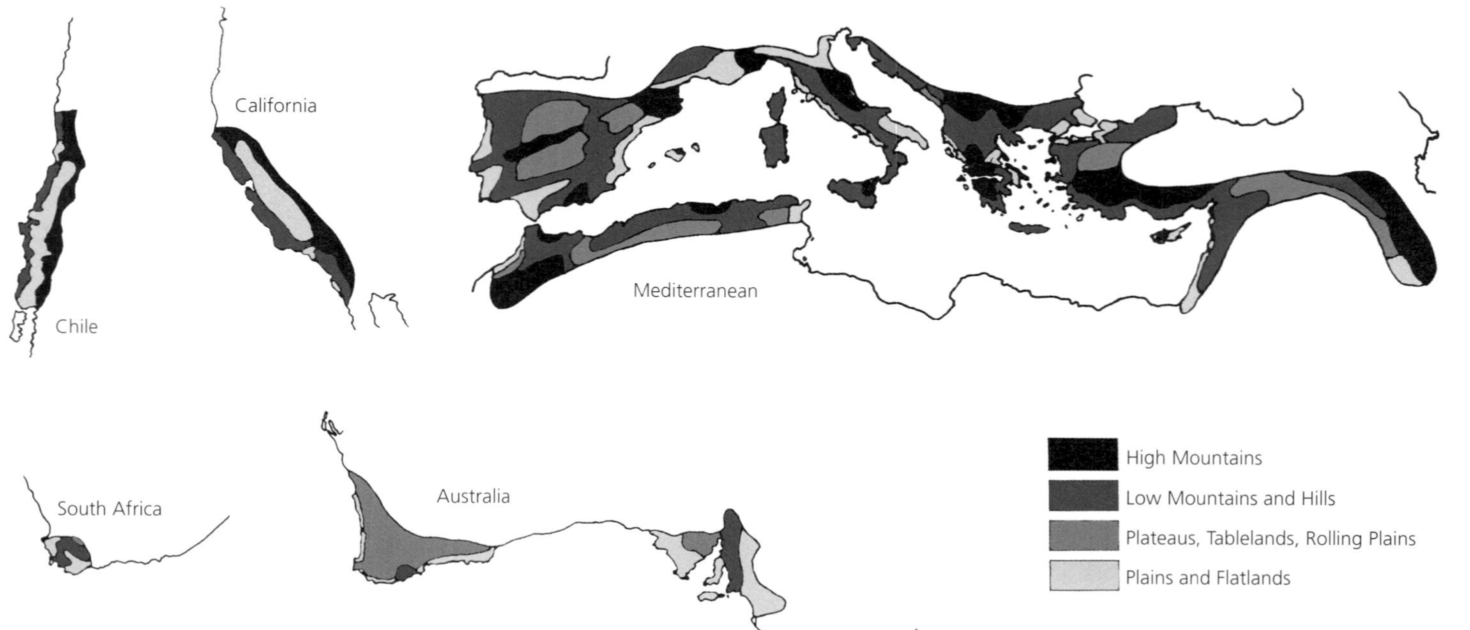

Figure 2.10 **Landscape topography of mediterranean-type ecosystems (MTEs).** The five global MTEs are shown, arranged from the most topographically diverse (Chile) to the least topographically diverse (Australia). Each region is shown at approximately the same scale. Topographical diversity is shown as divided into four landscape categories and for mediterranean-type climate (MTC) regions as most commonly defined (based on Thrower and Bradbury 1977). Note: the MTC regions as shown in these diagrams represent regions sensu stricto; excluded are the regions that have both high and low levels of mean annual precipitation. High mountains are only present in the three MTEs shown in the top row: Chile, California, and the Mediterranean.

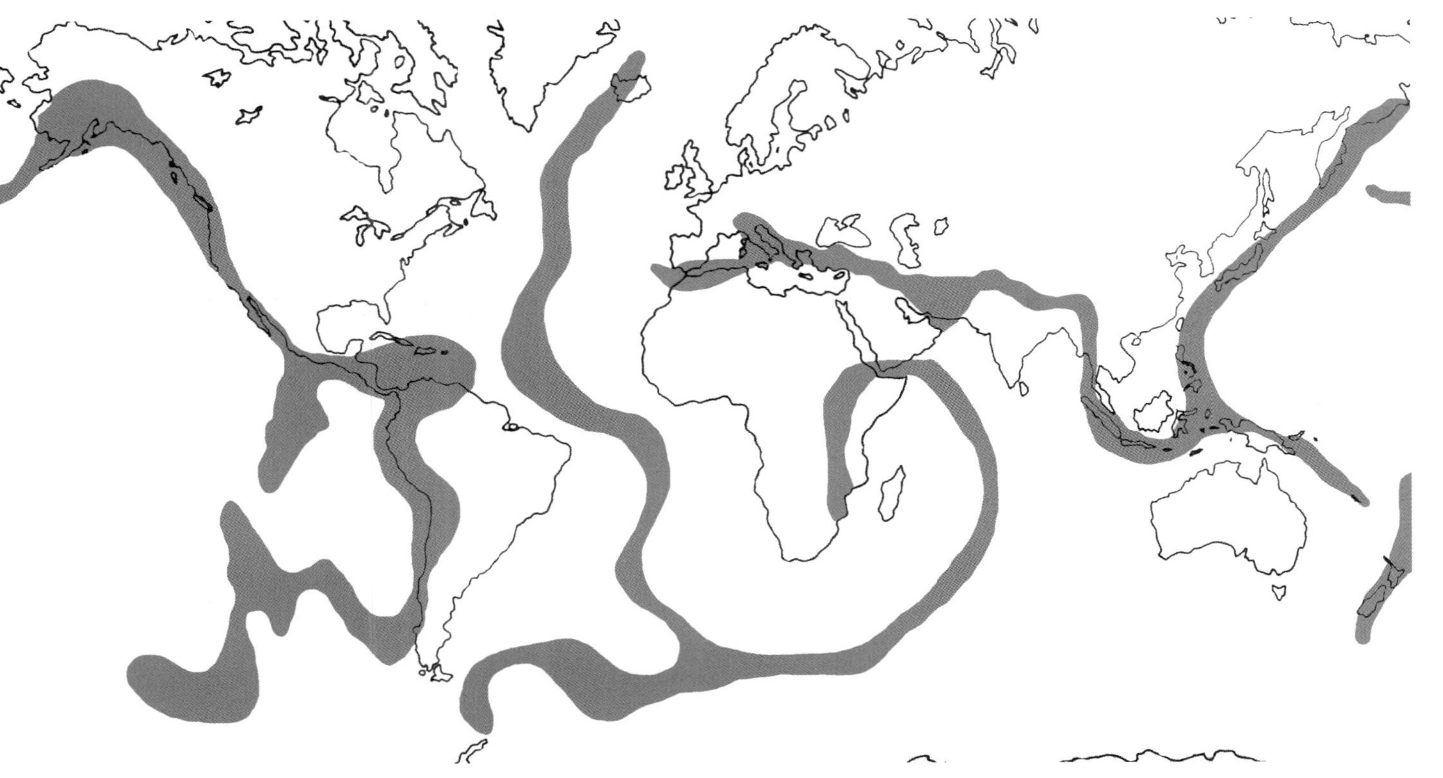

Figure 2.11 **Volcanic areas associated with tectonic plate margins.** Topographical diversity of landscapes within mediterranean-type climate (MTC) regions is linked to their association to tectonic plate boundaries. High mountains are only present in the three MTC regions that occur at the plate boundaries where there is ongoing volcanic activity: Chile, California, and the Mediterranean. Australia, which does not occur at the margin of a tectonic plate and does not have plate margin-associated volcanic activity, is the least topographically diverse of the five MTC regions. (Based on Thrower and Bradbury 1977.)

In contrast, the landscapes of South Africa and Australia are generally more mature and are less topographically diverse than the other two regions, although even these regions still contain some topographical diversity. Neither of these regions occurs along large plate boundaries and the MTC region of Australia is the least topographically diverse of the five regions.

In general, Chile and California are similar to one another in their landforms and Australia and South Africa are similar, with the Mediterranean representing an intermediate level of topographical diversity. These broad patterns are well established and have formed the basis for, or parts of, several hypotheses about diversity differences between different MTEs (see Case Study 3; Mooney 1977; di Castri et al. 1981; Cowling et al. 1996; Valente and Vargas 2013; Cowling et al. 2015). Many of these ideas are contained within the comparison of 'old, climatically buffered, infertile landscapes' (OCBILs) and 'young, often disturbed, fertile landscapes' (YODFELs) in the terminology proposed by Hopper (2009).

As an additional illustration of the variation in topography among regions, it is useful to compare variation in some of the highest points within each region:

- In Chile, the highest point near Santiago is Tupungato Mountain (6570 m) and the highest point in the country is the Nevado Ojos del Salado, located in the Andes along the Chile–Argentina border (6890 m). While neither of these peaks receives primarily winter rainfall, they are good representatives of the extreme height of the Andes, which creates much of the topographical and climatic diversity of this region.
- In California, the highest peak within the state is Mount Whitney within the California Sierra Nevada Range (4420 m), which is also the highest point in the contiguous United States of America. Mount Whitney receives primarily winter precipitation and is unequivocally part of the California MTC in this context, although the occurrence of precipitation as snow rather than rain can ameliorate the summer dry period by prolonging moisture availability during the summer during the period of snow melt.
- The Mediterranean region covers a much larger land area than the other regions. The highest points of several of the countries that form the predominant MTC areas in this region and contain significant mountain regions include the Serra de Estrela (1990 m) in Portugal that is part of the Sistema Central Mountain Ranges, the Mulhacén (3480 m) in Spain that is part of the Spanish Sierra Nevada Range, Mont Blanc (4810 m) in Italy that is part of the Alps and is the highest point in the Alps, Mount Olympus (2920 m) in Greece that is part of the Olympus Range, and Mount Demirkazik (3760 m) in the Taurus Mountains of Turkey that is the tallest peak within the western part of the country.
- In South Africa, the tallest peak within the Western Cape is Seweweek-spoortpiek (2330 m), which is part of the Cape Fold Mountains.

- In Australia, the highest peak in the Great Southern region of Western Australia is Bluff Knoll (1100 m) in the Stirling Range. The Darling Range, one of the major features of this MTC region, reaches a peak of only 580 m at Mount Cooke.

2.2.4 Soils and mineral nutrition

Soils are derived from parent materials that are subsequently altered through physical, chemical, and biological processes (Chapter 7). These processes may create soils that differ widely in their mineral content, even over relatively short distances. Additionally, soils change over time and old, weathered, and leached soils will contain fewer available nutrients than younger less weathered soils. The common substrata are generally similar across the five MTC regions (Table 2.1), although there are regional differences in the relative abundance of substrata.

MTC regions have nutrient-poor soils, but there is considerable variation both within and between regions in this pattern. Differences in the geological history and age of soils leads to differences in their mineral nutrition,

Table 2.1 Substrata and their nutrient status. Similar substrata are found between mediterranean-type ecosystems (MTEs) but they vary in their presence from occurring commonly (common), occurring in a few or more specific localities (frequent), or only in rare localities (trace) within different mediterranean-type climate regions (modified from Joffre et al. 1999 and based on data presented in Groves et al. 1983)

Substratum	Nutrients	MTE regions				
		Australia	South Africa	California	Chile	Mediterranean
Siliceous rocks (containing silica; includes quartz, chert, sandstone, and granite)	Nutrient-poor	Common	Common	Frequent	Not reported	Frequent
Argillaceous rocks (containing clays; includes shales and siltstones)	Nutrient-poor	Trace	Frequent	Common	Common	Trace
Calcareous rocks (containing calcium carbonate; includes limestone, marble, and travertine)	Nutrient-rich, high pH	Trace	Frequent	Trace	Not reported	Common
Ultramafic rocks (containing low silica and high magnesium; includes serpentine)	Nutrient-poor, Mg-rich	Not reported	Not reported	Frequent	Not reported	Trace

particularly as related to phosphorus (P) and nitrogen (N). Many authors have discussed the general pattern of Australia and South Africa containing lower nutrient soils when compared to the other three MTC regions (Chapter 7). As reviewed briefly in Lamont (1995), a broad nutrient pattern is that California, the Mediterranean, and Chile are generally N-limited and large areas in Australia and South Africa are P-limited. This idea was depicted in the classic diagram by di Castri (1991), reproduced here as Figure 2.12a.

While these general patterns of soil nutrient differences are useful to understand key differences in MTC regions, considerable variation in nutrients exists in soils over small and large spatial scales within each region. For instance, Figure 2.12b shows the nitrogen (N) and phosphorus (P) soil content from sites across the broader MTC region of southwest Australia (Specht and Moll 1983). Thus, even within this limited geographic region in the generally most nutrient-impoverished MTC region with ancient soils, there is a high level of variability in soil nutrient content. This variation in soil nutrients can be seen across the landscape through the turnover of different biological communities (Chapter 4). The role of soil nutrient

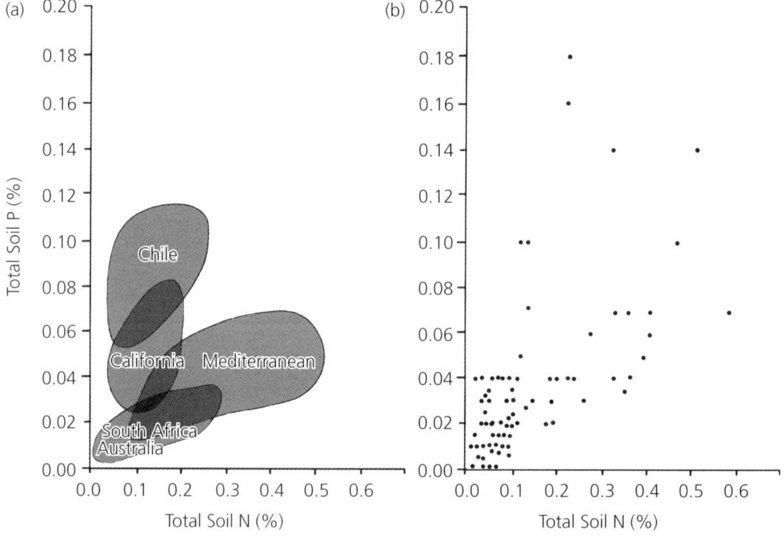

Figure 2.12 **Soil mineral nutrition of mediterranean-type ecosystems (MTEs).** Among the five global MTEs, there are differences in the level of soil nitrogen (N) and phosphorus (P) such that Australia and South Africa have the lowest soil mineral levels, California is intermediate, and Chile and the Mediterranean have higher levels (a; modified from di Castri 1991). However, there is considerable variability within a region that is not apparent from panel a. Panel b shows data gathered from sites within the MTC region of Australia and shows the high level of variability that may be present within a region. (Modified from Specht and Moll 1983.) Both panels are shown on the same axes scales.

heterogeneity and plant, microbe, and soil feedbacks is an important driver in these systems (Gallardo et al. 2000).

The nutrient status of MTE plants can be best understood in comparison to non-MTC regions. MTC regions are nutrient-poor relative to other global communities. Compared to non-MTE regions globally, MTE plants are low in N and are equivalent to other nutrient-poor systems (Vitousek 1982). Additional discussion of nutrient levels within plant tissues of MTE plants is presented in Chapters 6 and 7.

Fire has a chief role in nutrient cycling in MTEs (Chapter 7). Following fire, a large pulse of nutrients becomes available (Kruger et al. 1983); however, fire may also result in the loss of some nutrients from the system through volatilization during fire and runoff (Gray and Schlesinger 1981). This requires plants to be able to tolerate generally low nutrient conditions, but also to be able to capture nutrients when they periodically become available. Specialized adaptations for plants to absorb more N and P are discussed further in Chapters 3 and 7.

2.3 Evolutionary convergence

Convergent evolution describes the pattern of similar features independently evolving in response to similar selective pressures (Johnson 1973). As discussed in the sections above, MTC regions of the globe experience similar key ecosystem drivers that may result in the communities and organisms within them displaying similar structures or processes. Comparisons to examine evidence for evolutionary convergence typically include MTE as compared to non-MTE systems. This is because the evolutionary convergence hypothesis predicts that species and communities within MTEs should be more similar to one another than they are to species or communities from other ecosystems, i.e. as compared to non-MTE **outgroups**. The inclusion of outgroups is particularly important when examining traits that may differ between MTEs owing to small regional differences, because convergence may still be evident provided that differences between MTEs are small when compared to how different they are from other biomes. Similarly, MTEs may share some traits, but this would not be evidence of convergence if this trait were also found to be shared by non-MTE outgroups.

A complication in the study of evolutionary convergence arises from the historical complexity of ecosystems, communities, and organisms. There are numerous terms that are used to discuss the different patterns that may arise. For instance, convergence is sometimes separated from **parallel evolution**, with convergence referring to the evolution of similar features in distantly related organisms and parallel evolution referring to the evolution of similar features in close relatives (Case Study 2). Additional complications arise owing to the development of communities and ecosystems over time and through changing climates. Some traits, such as sclerophyllous leaves, pre-date MTC, but are common among modern MTV species (Chapter 6). These traits are common because they are clearly adaptive in MTC environments, and species with these traits have been able to move into newly emerging habitats and climate regions in a process termed **ecological sorting**. In this case, these traits are referred to as **exaptations** (Case Study 2). Exaptations, traits that evolved under selection for a different regime and are maintained owing to their utility in a new selective regime, are commonly contrasted to **adaptations**, traits that have been shaped directly by the new selective regime. These different processes and patterns are important in informing our understanding of which key ecosystem drivers have influenced MTE development and shaped their modern composition, structure, and function.

Convergent evolution may be apparent across different scales, ranging from ecosystem functioning, community form and function, the structure and physiology of organisms, or in the types of interactions found among species. At the broadest scale, convergence in ecosystem functioning examines the similarity of processes of energy and nutrient transfer. Because of its great scope, this is a difficult level to examine and studies have often been based on models or have been **meta-analyses**. Meta-analyses are studies that pool data from many sources to generate larger and/or broader datasets than could be generated by one group of researchers alone. Studies at this scale have found that MTC was correlated with similarities in carbon cycling and productivity among MTEs as compared to non-MTE grasslands (Paruelo et al. 1998), and have supported the importance of fire as a determinant of global biome distributions and the occurrence of shrublands in MTC regions (Bond et al. 2005).

Convergence at the community level examines similarities in species or habit densities, resource use, and the structure of communities. Cody and Mooney (1978) argued that this level and below was the scale at which convergent evolution could best be observed and this is the scale at which convergence has most often been compared. Evidence for convergence among MTEs at this level includes the finding that all five MTEs are historically dominated by closed vegetation, mostly evergreen species, with an average plant height of less than 5 m. This is usually considered to reflect similarity in plant adaptations to their environment (Cody and Mooney 1978). MTE communities have converged in their **phenology**, the timing of events, as well as their physiology for both plants and animals, especially as related to the specific features of an MTC (Cody and Mooney 1978; Jaksić and Marti 1981; Verdu et al. 2002; Jacobsen et al. 2009; Sanborn et al. 2011). Fire has also been an important factor in MTEs, especially in relation to post-fire seedling recruitment. Carrington and Keeley (1999) compared patterns of post-fire seedling establishment between mediterranean-type and non-mediterranean-type shrub communities. Their results supported a difference between MTE communities and those from non-MTE communities; the MTE sites had greater densities of post-fire seedlings and more species that were dependent on fire for successful seedling recruitment. In contrast, resprouting is prevalent in both MTE communities and non-MTE communities of similar biogeographical origin (Lloret et al. 1999; Pausas and Keeley 2014).

At the narrowest level, convergence has been compared at the organismal level, where similarities have been examined between taxa in their physiology or structure, usually within specific taxonomic groups. These types of studies, as well as many community scale studies, have often focused not on MTEs broadly, but rather on the archetypal mediterranean-type shrublands, MTV, found within MTEs as compared to non-MTE or to non-MTV communities or organisms. Studies at this level include comparisons of specific animal or plant groups, such as feeding behaviours of owls in MTE and non-MTE communities (Jaksić and Marti, 1981 and 1984) or sclerophylly in plants within the Rhamnaceae (Onstein and Linder 2016).

While the examples discussed above support the hypothesis of evolutionary convergence among MTEs, it is worth noting that this is not a unanimous perspective and several authors have argued that the MTC regions of the globe are not convergent. As an example of this perspective, Verdu et al. (2003) found that traits were similar among three MTEs and a non-MTE tropical ecosystem, and they argued that biogeographical history was the primary driver for ecosystem similarity and that the MTEs were not a separate ecosystem type. More recently, Lambers (2014) argued that soil infertility, particularly in South Africa and Australia, linked these regions more closely to other highly infertile landscapes, such as heathlands, rather than to other MTC regions. In this case, soil nutrition was suggested to be the key driver that supersedes climate in determining ecosystem function. These represent proposals for global ecosystem categories that differ from those presented in this book. These are valuable additions to the ongoing examination of the hypotheses of evolutionary convergence and the categorization of global biodiversity into functional units.

Finally, it is worth mentioning the many studies that have compared MTC regions without the inclusion of outgroups (e.g. Mooney and Dunn 1970; Cowling and Campbell 1980; Barbour and Minnich 1990; Cowling et al. 2005; Lambers et al. 2010). These studies differ from convergent evolutionary studies in that they seek primarily to elucidate mechanisms for potential differences between MTEs rather than to examine the factors that separate MTEs from other biomes. These studies are relevant in determining the generalizability of information gathered in one MTE to other MTEs, but are not as useful in examining the question of convergence.

Study of the MTC region biota and their key ecosystem drivers, particularly through comparison to non-MTC regions, is an area of active interest and discussion. The hypothesis of evolutionary convergence among MTEs remains a valuable idea to be tested and, as new data become available, will likely be an interesting area of study for current and future ecologists. To date, it appears that these regions have much in common and there is much to be gained through the examination of their shared characteristics and differences.

2.4 An overview of mediterranean-type ecosystem characteristics

Examined broadly, the five global MTC regions have some areas that share similar climates, fire regimes, and soil substrata and mineral nutrition. Of these, MTC, which is typified by a unique pattern of seasonality that includes a wet winter and a warm dry summer, is often cited as a particularly important ecosystem driver. Studies have also suggested that MTC regions are strongly shaped by their fire regimes, particularly for the dominant shrub communities of these regions. This includes experiencing periodic crown fires, usually during the dry season. Soil substrata are spatially heterogeneous within each region, but generally form nutrient-impoverished soils. Through these shared ecosystem drivers, selection has operated within each ecosystem to shape the communities and the organisms within them. This has resulted in the communities and organisms displaying similar structures and processes.

Literature cited

Abarzúa AM, Vargas C, Jarpa L, Gutiérrez NM, Hinojosa LF, Paula S. 2016. Evidence of Neogene wildfires in Central Chile: charcoal records from the Navidad formation. Palaeogeography, Palaeoclimatology, Palaeoecology 459: 76–85.

Ackerly DD. 2004. Adaptation, niche conservatism, and convergence: comparative studies of leaf evolution in the California chaparral. The American Naturalist 163: 654–71.

Agenbag L, Elser KJ, Midgley GF, Boucher C. 2008. Diversity and species turnover on an altitudinal gradient in Western Cape, South Africa: baseline data for monitoring range shifts in response to climate change. Bothalia 38: 161–91.

Alexander LV, Hope P, Collins D, Trewin B, Lynch A, Nicholls N. 2007. Trends in Australia's climate means and extremes: a global context. Australian Meteorological Magazine 56: 1–18.

Araújo MB, Nogués-Bravo D, Diniz-Filho JAF, Haywood AM, Valdes PJ, Rahbek C. 2007. Quaternary climate changes explain diversity among reptiles and amphibians. Ecography 31: 8–15.

Aravena JC, LeQuesne C, Jiménez H, Lara A, Armesto JJ. 2003. Fire history in central Chile: tree-ring evidence and modern records. Pages 343–56 in Veblen TT, Baker WL, Montenegro G, Swetham TM, eds. Fire and climatic change in temperate ecosystems of the Western Americas. Springer, New York, USA.

Armesto JJ, Arroyo MTK, Hinojosa LF. 2007. The mediterranean environment of Central Chile. Pages 184–99 in Veblen TT, Young KR, Orme AR, eds. The physical geography of South America. Oxford University Press, New York, USA.

Aschmann H. 1973. Distribution and peculiarity of mediterranean ecosystems. Pages 11–20 in di Castri F, Mooney HA, eds. Mediterranean type ecosystems: origin and structure. Springer-Verlag, Berlin.

Axelrod DI. 1958. Evolution of the Madro-Tertiary geoflora. Botanical Review 24: 433–509.

Axelrod DI. 1973. History of the mediterranean ecosystem in California. Pages 225–77 in di Castri F, Mooney HA, eds. 1973. Mediterranean type ecosystems: origin and structure. Springer-Verlag, Berlin, Germany.

Axelrod DI. 1975. Evolution and biogeography of Madrean-Tethyan sclerophyll vegetation. Annals of the Missouri Botanical Garden 62: 280–334.

Axelrod DI. 1989. Age and origin of chaparral. Pages 7–19 in Keeley SC, ed. The California chaparral: paradigms reexamined. Natural History Museum of Los Angeles County, Los Angeles, California, USA.

Ball MC, Hodges VS, Laughlin GP. 1991. Cold-induced photoinhibition limits regeneration of snow gum at tree-line. Functional Ecology 5: 663–8.

Ball MC, Egerton JJG, Leuning R, Cunningham RB, Dunne P. 1997. Microclimate above grass adversely affects spring growth of seedling snow gum (*Eucalyptus pauciflora*). Plant, Cell and Environment 20: 155–66.

Barbour MG, Minnich RA. 1990. The myth of chaparral convergence. Israel Journal of Botany 39: 453–63.

Bell DT, Heddle EM. 1989. Floristic, morphological, and vegetation diversity. Pages 53–66 in Dell B, Havel JJ, Malajczuk N, eds. The Jarrah forest: a complex Mediterranean ecosystem. Kluwer Academic Publishing, Dordrecht, The Netherlands.

Bleiweiss R. 1998. Tempo and mode of hummingbird evolution. Biological Journal of the Linnean Society 65: 63–76.

Bond WJ, Woodward FI, Midgley GF. 2005. The global distribution of ecosystems in a world without fire. New Phytologist 165: 525–38.

Boorse GC, Ewers FW, Davis SD. 1998. Response of chaparral shrubs to below-freezing temperatures: acclimation, ecotypes, seedlings vs. adults. American Journal of Botany 85: 1224–30.

Bradshaw PL, Cowling RM. 2014. Landscapes, rock types and climate of the Greater Cape Floristic Region. Pages 26–46 in Allsopp N, Colville JF, Verboom GA, eds. Fynbos: ecology, evolution and conservation of a megadiverse region. Oxford University Press, Oxford, UK.

Byrne M, Yeates DK, Joseph L, Kearney M, Bowler J, Williams MAJ, Cooper S, Donnellan SC, Keogh JS, Leys R, Melville J. 2008. Birth of a biome: insights into the assembly and maintenance of the Australian arid zone biota. Molecular Ecology 17: 4398–417.

Bytebier B, Antonelli A, Bellstedt DU, Linder HP. 2010. Estimating the age of fire in the Cape flora of South Africa from an orchid phylogeny. Proceedings of the Royal Society of London B: Biological Sciences 278: 188–95.

Carnaval AC, Hickerson MJ, Haddad CFB, Rodrigues MT, Moritz C. 2009. Stability predicts genetic diversity in the Brazilian Atlantic forest hotspot. Science 323: 785–9.

Carrington ME, Keeley JE. 1999. Comparison of post-fire seedling establishment between scrub communities in mediterranean and non-mediterranean climate ecosystems. Journal of Ecology 87: 1025–36.

Cody ML, Mooney HA. 1978. Convergence versus nonconvergence in mediterranean-climate ecosystems. Annual Review of Ecology, Evolution, and Systematics 9: 265–321.

Contreras TE, Figueroa JA, Abarca L, Castro SA. 2011. Fire regimen and spread of plants naturalized in central Chile. Revista Chilena de Historia Natural 84: 307–23.

Cowling RM, Campbell BM. 1980. Convergence in vegetation structure in the mediterranean communities of California, Chile and South Africa. Plant Ecology 43: 191–7.

Cowling RM, Rundel PW, Lamont BB, Arroyo MK, Arianoutsou M. 1996. Plant diversity in mediterranean-climate regions. Trends in Ecology and Evolution 11: 362–6.

Cowling RM, Ojeda F, Lamont BB, Rundel PW, Lechmere-Oertel R. 2005. Rainfall reliability, a neglected factor in explaining convergence and divergence of plant traits in fire-prone mediterranean-climate ecosystems. Global Ecology and Biogeography 14: 509–19.

Cowling RM, Proches S, Partridge TC. 2009. Explaining the uniqueness of the Cape flora: incorporating geomorphic evolution as a factor for explaining its diversification. Molecular Phylogenetics Evolution 51: 64–74.

Cowling RM, Potts AJ, Bradshaw P, Colville J, Arianoutsou M, Ferrier S, Forest F, Fylas N, Hopper SD, Ojeda F, Proches S, Smith RJ, Rundel PW, Vassilakis E, Zutta BR. 2015. Variation in plant diversity in mediterranean climate ecosystems: the role of climatic and topographic stability. Journal of Biogeography 42: 552–64.

Crisp M, Cook L, Steane D. 2004. Radiation of the Australian flora: what can comparisons of molecular phylogenies across multiple taxa tell us about the evolution of diversity in present-day communities? Philosophical Transactions: Biological Sciences 359: 1551–71.

Daly C, Conklin DR, Unsworth MH. 2010. Local atmospheric decoupling in complex topography alters climate change impacts. International Journal of Climatology 30: 1857–64.

Davis SD, Helms AM, Heffner MS, Shaver AR, Deroulet AC, Stasiak NL, Vaughn SM, Leake CB, Lee HD, Sayegh ET. 2007a. Chaparral zonation in the Santa Monica Mountains: the influence of freezing temperatures. Fremontia 35: 12–15.

Davis SD, Pratt RB, Ewers FW, Jacobsen AL. 2007b. Freezing tolerance impacts chaparral species distribution in the Santa Monica Mountains. Page 159–72 in Knapp DA, ed. Flora and ecology of Santa Monica Mountains: Proceedings of 32nd Annual Southern California Botanists Symposium. Southern California Botanists Special Publication No. 4, Fullerton, California, USA.

Di Castri F. 1991. An ecological overview of the five regions of the world with mediterranean climate. Pages 3–15 in Groves RH, Di Castri F, eds. Biogeography of Mediterranean invasions. Cambridge University Press, Cambridge, UK.

Di Castri F, Goodall DW, Specht RL, eds. 1981. Mediterranean-type shrublands. Elsevier, Amsterdam.

Diekmann B, Fälker M, Kuhn G. 2003. Environmental history of the south-eastern South Atlantic since the middle Miocene: evidence from the sedimentological records of ODP Sites 1088 and 1092. Sedimentology 50: 511–29.

Diffenbaugh NS, Swain DL, Touma D. 2015. Anthropogenic warming has increased drought risk in California. Proceedings of the National Academy of Sciences 112: 3931–6.

Dobrowski SZ. 2011. A climatic basis for microrefugia: the influence of terrain on climate. Global Change Biology 17: 1022–35.

Dodson JR, Ramrath A. 2001. An upper Pliocene lacustrine environmental record from south-Western Australia—preliminary results. Palaeogeography, Palaeoclimatology, Palaeoecology 167: 309–20.

Dupont LM, Linder HP, Rommerskirchen F, Schefuß E. 2011. Climate-driven rampant speciation of the Cape flora. Journal of Biogeography 38: 1059–68.

Enright NJ, Lamont BB, Miller BP. 2005. Anomalies in grasstree fire history reconstructions for south-western Australian vegetation. Austral Ecology 30: 668–73.

Fauquette S, Suc JP, Guiot J, Diniz F, Feddi N, Zheng Z, Bessais E, Drivaliari A. 1999. Climate and biomes in the West Mediterranean area during the Pliocene. Palaeogeography, Palaeoclimatology, Palaeoecology 152: 15–36.

Fuentes ER, Espinoza G. 1986. Resilience of shrublands in central Chile: a vulcanism-related hypothesis. Interciencia 11: 164–5.

Gallardo A, Rodríguez-Saucedo JJ, Covelo F, and Fernández-Alés R. 2000. Soil nitrogen heterogeneity in a Dehesa ecosystem. Plant and Soil 222: 71–82.

Ganteaume A, Jappiot M. 2015. Characterization of the large fire regime in SE France. Pages 286–9 in Keane RE, Jolly M, Parsons R, Riley K, eds. Proceedings of the large wildland fires conference; May 19–23, 2014; Missoula, MT. Proc. RMRS-P-73. Fort Collins, CO: U.S. Department of Agriculture, Forest Service, Rocky Mountain Research Station.

Gibson RK, Bradstock RA, Penman T, Keith DA, Driscoll DA. 2015. Climatic, vegetation and edaphic influences on the probability of fire across mediterranean woodlands of south-eastern Australia. Journal of Biogeography 42: 1750–60.

Givnish TJ, Barfuss MHJ, Van Ee B, et al. 2011. Phylogeny, adaptive radiation, and historical biogeography in Bromeliaceae: insights from an eight-locus plastic phylogeny. American Journal of Botany 98: 872–95.

Gould SJ, Vrba ES. 1982. Exaptation—a missing term in the science of form. Paleobiology 8: 4–15.

Graham A. 1999. Late Cretaceous and Cenozoic history of North American vegetation. Oxford University Press, Oxford.

Graham CH, Moritz C, Williams SE. 2006. Habitat history improves prediction of biodiversity in rainforest fauna. Proceedings of the National Academy of Sciences USA 103: 632–6.

Gray JT, Schlesinger WH. 1981. Nutrient cycling in mediterranean type ecosystems. Pages 259–85 in Miller PC, ed. Resource use by chaparral and matorral: a comparison of vegetation function in two mediterranean type ecosystems. Springer-Verlag, Berlin.

Gregory-Wodzicki KM. 2000. Uplift history of the Central and Northern Andes: a review. Geological Society of America Bulletin 112: 1091–105.

Groves RH, Beard JS, Deacon HJ. 1983. Introduction: the origins and characteristics of mediterranean ecosystems. Pages 1–17 in Day JA, ed. Mineral nutrients in mediterranean ecosystems. South African National Scientific Programmes Report 71. Pretoria, South Africa.

Hall M. 1984. Man's historical and traditional use of fire in southern Africa. Pages 39–52 in Booysen PdV, Tainton NM, eds. Ecological effects of fire in South African ecosystems. Springer-Verlag, Berlin, Germany.

Hallam SJ. 2014. Fire and hearth: a study of Aboriginal usage and European usurpation in south-western Australia. University of Western Australia Publishing, Crawley, Western Australia.

Hannah L, Flint L, Syphard AD, Moritz MA, Buckley LB, McCullough IM. 2014. Fine-grain modeling of species' response to climate change: holdouts, stepping-stones, and microrefugia. Trends in Ecology & Evolution 29: 390–7.

Hartley A. 2003. Andean uplift and climate change. Journal of the Geological Society 160: 7–10.

Hill RS. 1998. Fossil evidence for the onset of xeromorphy and scleromorphy in Australian Proteaceae. Australian Systematic Botany 11: 391–400.

Hoetzel S, Dupont LM, Wefer G. 2015. Miocene-Pliocene vegetation change in south-western Africa (ODP Site 1081, offshore Namibia). Palaeogeography, Palaeo-climatology, Palaeoecology 423: 102–8.

Hopper SD. 2009. OCBIL theory: towards an integrated understanding of the evolution, ecology and conservation of biodiversity on old, climatically buffered, infertile landscapes. Plant and Soil 322: 49–86.

Hopper SD, Gioia P. 2004. The southwest Australian floristic region: evolution and conservation of a global hot spot of biodiversity. Annual Review of Ecology, Evolution, and Systematics 35: 623–50.

Horrell MA. 1991. Phytogeography and paleoclimatic interpretation of the Maestrichtian. Palaeogeography, Palaeoclimatology, Palaeoecology 86: 87–138.

IPCC. 2007. Climate Change 2007. Fourth Assessment Report of the Intergovernmental Panel on Climate Change. IPCC, Geneva, Switzerland.

Jacobsen AL, Fabritius SL, Davis SD. 2004. Fire frequency impacts non-sprouting chaparral shrubs in the Santa Monica Mountains of southern California. In Arianoutsou M, Papanastasis VP, eds. Ecology, conservation and management of mediterranean climate ecosystems. Millpress, Rotterdam, Netherlands.

Jacobsen AL, Esler KJ, Pratt RB, Ewers FW. 2009. Water stress tolerance of shrubs in mediterranean-type climate regions: convergence of fynbos and succulent karoo communities with California shrub communities. American Journal of Botany 96: 1445–53.

Jaksić FM, Marti CD. 1981. Trophic ecology of Athene owls in mediterranean-type ecosystems: a comparative analysis. Canadian Journal of Zoology 59: 2331–40.

Jaksić FM, Marti CD. 1984. Comparative food habits of Bubo owls in mediterranean-type ecosystems. The Condor 86: 288–96.

Janis CM. 1993. Tertiary mammal evolution in the context of changing climates, vegetation, and tectonic events. Annual Review of Ecology, Evolution and Systematics 24: 467–500.

Joffre R, Rambal S. 2001. Mediterranean ecosystems. eLS. DOI: 10.1038/npg.els.0003196.

Joffre R, Rambal S, Damesin C. 1999. Functional attributes in mediterranean-type ecosystems. Pages 347–80 in Pugnaire FI, Valladares F, eds. Handbook of func-tional plant ecology. Marcel Dekker, Inc., New York, USA.

Johnson AW. 1973. Historical view of the concept of ecosystem convergence. Pages 3–7 in di Castri F, Mooney HA, eds. Mediterranean type ecosystems. Springer-Verlag, Berlin, Germany.

Kadereit JW, Baldwin BG. 2012. Western Eurasian–western North American dis-junct plant taxa: the dry-adapted ends of formerly widespread north temperate mesic lineages—and examples of long-distance dispersal. Taxon 61: 3–17

Keeley JE, Bond WJ. 1997. Convergent seed germination in South African fynbos and Californian chaparral. Plant Ecology 133: 153–67.

Keeley JE, Fotheringham CJ, Morais M. 1999. Reexamining fire suppression impacts on brushland fire regimes. Science 284: 1829–32.

Keeley JE, Bond WJ, Bradstock RA, Pausas JG, Rundel PW. 2012. Fire in mediterranean ecosystems: ecology, evolution and management. Cambridge University Press, New York.

Kelsey H. 1998. Juan Rodríguez Cabrillo. Huntington Library Press, San Marino, CA, USA.

Kennett JP. 1977. Cenozoic evolution of Antarctic glaciation, the circum-Antarctic Ocean, and their impact on global paleoceanography. Journal of Geophysical Research 82: 3843–60.

Klausmeyer KR, Shaw MR. 2009. Climate change, habitat loss, protected areas and the climate adaptation potential of species in mediterranean ecosystems worldwide. PLoS ONE 4: e6392.

Köppen W, Geiger R. 1936. Handbuch der Klimatologie. Gebründer Bortraeger, Berlin, Germany.

Kovar-Eder J, Kvaček Z, Martinetto E, Roiron P. 2006. Late Miocene to early Pliocene vegetation of southern Europe (7–4 Ma) as reflected in the megafossil plant record. Palaeogeography, Palaeoclimatology, Palaeoecology 238: 321–39.

Krebs P, Pezzatti GB, Massoleni S, Talbot LM, Conedera M. 2010. Fire regime: history and definition of a key concept in disturbance ecology. Theory in Biosciences 129: 53–69.

Kreft H, Jetz W. 2007. Global patterns and determinants of vascular plant diversity. Proceedings of the National Academy of Sciences USA 104: 5925–30.

Kruger FJ, Bigalke RC. 1984. Fire in fynbos. Pages 67–114 in Booysen PdV, Tainton NM, eds. Ecological effects of fire in South African ecosystems. Springer-Verlag, Berlin, Germany.

Kruger FJ, Mitchell DT, Jarvis JUM. 1983. Mediterranean-type ecosystems: the role of nutrients. Springer-Verlag, Berlin.

Lambers H. 2014. Plant life on the sandplains in Southwest Australia: a global biodiversity hotspot. The University of Western Australia Publishing, Crawley, Australia. pp. 332.

Lambers H, Brundrett MC, Raven JA, Hopper SD. 2010. Plant mineral nutrition in ancient landscapes: high plant species diversity on infertile soils is linked to functional diversity for nutritional strategies. Plant and Soil 334: 11–31.

Lamont BB. 1995. Mineral nutrient relations in mediterranean regions of California, Chile, and Australia. Pages 211–35 in Arroyo MTK, Zedler PH, Fox MD, eds. Ecology and biogeography of mediterranean ecosystems in Chile, California, and Australia. Springer-Verlag, New York, USA.

Lancaster LT, Kay KM. 2013. Origin and diversification of the California flora: re-examining classic hypotheses with molecular phylogenies. Evolution 67: 1041–54

Lara AR, Fraver S, Aravena JC, Wolodarsky-Franke A. 1999. Fire and the dynamics of *Fitzroya cupressoides* (alerce) forests of Chile's Cordillera Pelada. Ecoscience 6: 100–9.

Larcher W. 1981. Low temperature effects on Mediterranean sclerophylls: an unconventional viewpoint. Pages 259–66 in Margaris NS, Mooney HA, eds. Components of productivity in mediterranean-climate regions—basic and applied aspects. Dr W. Junk Publishers, The Hague, Netherlands.

Linder HP. 2008. Plant species radiations: where, when, why? Philosophical Transactions of the Royal Society B: Biological Sciences 363: 3097–105.

Lloret F, Verdu M, Flores-Hernandez N, Valiente-Banuet A. 1999. Fire and resprouting in mediterranean ecosystems: insights from an external biogeographical region, the mexical shrubland. American Journal of Botany 86: 1655–61.

Loarie SR, Duffy PB, Hamilton H, Asner GP, Field CB, Ackerly DD. 2009. The velocity of climate change. Nature 462: 1052–5.

Lo Presti RM, Oberprieler C. 2009. Evolutionary history, biogeography and eco-climatological differentiation of the genus *Anthemis* L. (Compositae, Anthemideae) in the circum-Mediterranean area. Journal of Biogeography 36: 1313–32.

Martin HA. 2006. Cenozoic climate change and the development of the arid vegetation in Australia. Journal of Arid Environments 66: 533–63.

Matusick G, Ruthrof KX, Brouwers NC, Hardy GStJ. 2014. Topography influences the distribution of autumn frost damage on trees in a mediterranean-type *Eucalyptus* forest. Trees 28: 1449–62.

McChesney CJ, Koch JM, Bell DT. 1995. Jarrah forest restoration in Western Australia: canopy and topographic effects. Restoration Ecology 3: 105–10.

McLoughlin S. 2001. The breakup history of Gondwana and its impact on pre-Cenozoic floristic provincialism. Australian Journal of Botany 49: 271–300.

Mensing SA, Michaelsen J, Byrne R. 1999. A 560-year record of Santa Ana fires reconstructed from charcoal deposited in the Santa Barbara Basin, California. Quaternary Research 51: 295–305.

Miller PC. 1981. Resource use by chaparral and matorral: a comparison of vegetation function in two mediterranean type ecosystems. Springer-Verlag, Berlin.

Miller PC, Hajek E. 1981. Resource availability and environmental characteristics of mediterranean type ecosystems. Pages 17–41 in Miller PC, ed. Resource use by chaparral and matorral. Springer, New York.

Mittermeier RA, Gil RP, Hoffman M, Pilgrim J, Brooks T, Mittermeier CG, Lamoreux J, Fonseca GAB. 2005. Hotspots revisited: Earth's biologically richest and most endangered terrestrial ecoregions. Cemex, Mexico City.

Moll EJ, McKenzie B, McLachlan D. 1980. A possible explanation for the lack of trees in the fynbos, Cape Province, South Africa. Biological Conservation 17: 221–8.

Montenegro G, Gómez M, Díaz F, Ginocchio R. 2003. Regeneration potential of Chilean matorral after fire: an updated view. Pages 381–409 in Veblen TT, Baker WL, Montenegro G, Swetham TM, eds. Fire and climatic change in temperate ecosystems of the Western Americas. Springer, New York, USA.

Mooney HA. 1977. Convergent evolution in Chile and California: mediterranean climate ecosystems. Dowden, Hutchinson & Ross, Stroudsburg, Pennsylvania, USA.

Mooney HA, Dunn EL. 1970. Convergent evolution of mediterranean-climate evergreen sclerophyll shrubs. Evolution 24: 292–303.

Nobel PS. 1999. Physiocochemical and environmental plant physiology. Academic Press, New York, USA.

Onstein RE, Linder HP. 2016. Beyond climate: convergence in fast evolving sclerophylls in Cape and Australian Rhamnaceae predates the mediterranean climate. Journal of Ecology 104: 665–77.

Onstein RE, Carter RJ, Xing Y, Linder HP. 2014. Diversification rate shifts in the Cape Floristic Region: the right traits in the right place at the right time. Perspectives in Plant Ecology, Evolution and Systematics 16: 331–40.

Onstein RE, Carter RJ, Xing Y, Richardson JE, Linder HP. 2015. Do mediterranean-type ecosystems have a common history? Insights from the Buckthorn family (Rhamnaceae). Evolution 69: 756–71.

Onstein RE, Jordan GJ, Sauquet H, Weston PH, Bouchenak-Khelladi Y, Carpenter RJ, Linder HP. 2016. Evolutionary radiations of Proteaceae are triggered by the interaction between traits and climates in open habitats. Global Ecology and Biogeography 25: 1239–51.

Palamarev E. 1989. Paleobotanical evidences of the Tertiary history and origin of the Mediterranean sclerophyll dendroflora. Plant Systematics and Evolution 162: 93–107.

Paruelo JM, Jobbágy EG, Sala OE, Lauenroth WK, Burke IC. 1998. Functional and structural convergence of temperate grassland and shrubland ecosystems. Ecological Applications 8: 194–206.

Pausas JG, Keeley JE. 2014. Evolutionary ecology of resprouting and seeding in fire-prone ecosystems. New Phytologist 204: 55–65.

Pratt RB, Ewers FW, Lawson MC, Jacobsen AL, Brediger M, Davis SD. 2005. Mechanisms for tolerating freeze-thaw stress of two evergreen chaparral species: *Rhus ovata* and *Malosma laurina* (Anacardiaceae). American Journal of Botany 92: 1102–13.

Raven PH. 1973. The evolution of mediterranean floras. Pages 213–24 in di Castri F, Mooney HA, eds. 1973. Mediterranean type ecosystems: origin and structure. Springer-Verlag, Berlin.

Raven PH, Axelrod DI. 1978. Origin and relationships of the California flora. University of California Publications in Botany, Vol. 72. University of California Press, USA.

Ricklefs RE. 2004. A comprehensive framework for global patterns in biodiversity. Ecology Letters 7: 1–15.

Rix MG, Edwards DL, Byrne M, Harvey MS, Joseph L, Roberts JD. 2015. Biogeography and speciation of terrestrial fauna in the south-western Australian biodiversity hotspot. Biological Reviews 90: 762–93.

Rundel PW. 1999. Disturbance in mediterranean-climate shrublands and woodlands. Pages 271–86 in Walker LR, ed. Ecosystems of disturbed ground (Vol. 16). Elsevier, Netherlands.

Salvo G, Ho SY, Rosenbaum G, Ree R, Conti E. 2010. Tracing the temporal and spatial origins of island endemics in the Mediterranean region: a case study from the citrus family (*Ruta* L., Rutaceae). Systematic Biology 59: 705–22.

Sanborn AF, Phillips PK, Heath JE, Heath MS. 2011. Comparative thermal adaptation in cicadas (Hemiptera: Cicadidae) inhabiting Mediterranean ecosystems. Journal of Thermal Biology 36: 150–5.

Sander J, Wardell-Johnson G. 2012. Defining and characterizing high-rainfall mediterranean climates. Plant Biosystems 146: 451–60.

Sauquet H, Weston PH, Anderson CL, Barker NP, Cantrill DJ, Mast AR, Savolainen V. 2009. Contrasted patterns of hyperdiversification in Mediterranean hotspots. Proceedings of the National Academy of Sciences USA 106: 221–5.

Segura AM, Holmgren M, Anabalón JJ, Fuentes ER. 1998. The significance of fire intensity in creating local patchiness in the Chilean matorral. Plant Ecology 139: 259–64.

Singh G. 1988. History of aridland vegetation and climate: a global perspective. Biological Reviews 63: 159–95.

Sniderman JMK, Jordan GJ, Cowling RM. 2013. Fossil evidence for a hyperdiverse sclerophyll flora under a non-Mediterranean-type climate. Proceedings of the National Academy of Sciences USA 110: 3423–8.

Specht RL, Moll EJ. 1983. Mediterranean-type heathlands and sclerophyllous shrublands of the world: an overview. Pages 41–65 in Kruger FJ, Mitchell DT, Jarvis JUM, eds. Mediterranean-type ecosystems. Springer, Berlin, Heidelberg.

Suc J-P. 1984. Origin and evolution of the Mediterranean vegetation and climate in Europe. Nature 307: 429–32.

Thompson JD. 2005. Plant evolution in the Mediterranean. Oxford University Press, New York.

Thrower NJ, Bradbury DE. 1973. The physiography of the Mediterranean lands with special emphasis on California and Chile. Pages 37–52 in Di Castri F, Mooney HA, eds. Mediterranean type ecosystems, origin and structure. Springer, Berlin.

Thrower NJW, Bradbury DE, eds. 1977. Chile-California mediterranean scrub atlas. Dowden, Hutchinson & Ross, Inc., Stroudsburg, Pennsylvania.

Valente LM, Vargas P. 2013. Contrasting evolutionary hypotheses between two mediterranean-climate floristic hotpots: the Cape of southern Africa and the Mediterranean Basin. Journal of Biogeography 40: 2032–46.

Valente LM, Savolainen V, Manning JC, Goldblatt P, Vargas P. 2011. Explaining disparities in species richness between Mediterranean floristic regions: a case study in *Gladiolus* (Iridaceae). Global Ecology and Biogeography 20: 881–92.

van Dam JA. 2006. Geographic and temporal patterns in the late Neogene (12–3 Ma) aridification of Europe: the use of small mammals as paleoprecipitation proxies. Palaeogeography, Palaeoclimatology, Palaeoecology 238: 190–218.

Vanniere B, Colombaroli D, Chapron E, Leroux A, Tinner W, Magny M. 2008. Climate versus human-driven fire regimes in mediterranean landscapes: the Holocene record of Lago dell'Accesa (Tuscany, Italy). Quaternary Science Reviews 27: 1181–96.

van Wilgen BW, Bond WJ, Richardson DM. 1992. Ecosystem management. Pages 345–71 in Cowling RM, ed. The ecology of fynbos: nutrients, fire and diversity. Oxford University Press, Capetown.

van Wilgen BW, Forsyth GG, de Klerk H, Das S, Khuluse S, Schmitz P. 2010. Fire management in mediterranean-climate shrublands: a case study from the Cape fynbos, South Africa. Journal of Applied Ecology 47: 631–8.

Vázquez A, Moreno JM. 1998. Patterns of lightning-, and people-caused fires in peninsular Spain. International Journal of Wildland Fire 8: 103–15.

Verboom AG, Linder HP, Forest F, Hoffmann V, Bergh NG, Cowling RM. 2014. Cenozoic assembly of the Greater Cape flora. Pages 93–118 in Allsopp N, Colville JF and Verboom AG, eds. Fynbos: ecology, evolution and conservation of a megadiverse region. Oxford University Press, Oxford.

Verdú M, Barrón-Sevilla JA, Valiente-Banuet A, Flores-Hernández N, García-Fayos P. 2002. Mexical plant phenology: is it similar to Mediterranean communities? Botanical Journal of the Linnean Society 138: 297–303.

Verdú M, Dávila P, García-Fayos P, Flores-Hernández N, Valiente-Banuet A. 2003. Convergent traits of mediterranean woody plants belong to pre-mediterranean lineages. Biological Journal of the Linnean Society 78: 415–27.

Vitousek P. 1982. Nutrient cycling and nutrient use efficiency. American Naturalist 119: 553–72.

Walter H. 1973. Vegetation of the earth in relation to climate and the eco-physiological conditions. Springer-Verlag, New York.

Wen J, Ickert-Bond SM. 2009. Evolution of the Madrean-Tethyan disjunctions and the North and South American amphitropical disjunctions in plants. Journal of Systematics and Evolution 47: 331–48.

Werneck FP, Nogueira C, Colli GR, Sites JW, Costa GC. 2012. Climatic stability in the Brazilian Cerrado: implications for biogeographical connections of South American savannas, species richness and conservation in a biodiversity hotspot. Journal of Biogeography 39: 1695–706.

White ME. 1986. The greening of Gondwana. Reed Books Pty Ltd, Australia.

Willis KJ, McElwain JC. 2014. The evolution of plants. Oxford University Press, Oxford, UK.

Wilson AM, Latimer AM, Silander JA, Gelfand AE, De Klerk H. 2010. A hierarchical Bayesian model of wildfire in a mediterranean biodiversity hotspot: implications of weather variability and global circulation. Ecological Modelling 221: 106–12.

Witter M, Taylor RS, Davis SD. 2007. Fire history and vegetation response to wildfire in the Santa Monica Mountains, California. Pages 173–94 in Knapp DA, ed. Flora and ecology of Santa Monica Mountains: Proceedings of 32nd Annual Southern California Botanists Symposium. Southern California Botanists Special Publication No. 4, Fullerton, California.

Zachos J, Pagani M, Sloan L, Thomas E, Billups K. 2001. Trends, rhythms, and aberrations in global climate 65 Ma to present. Science 292: 686–93.

Zheng B, Chenu K, Fernanda Dreccer M, Chapman SC. 2012. Breeding for the future: what are the potential impacts of future frost and heat events on sowing and flowering time requirements for Australian bread wheat (*Triticum aestivum*) varieties? Global Change Biology 18: 2899–914.

Painting the Picture: The Living Template

3 Organisms and their Interactions

Abstract

Both animal and plant species exhibit adaptive traits related to features of mediterranean-type ecosystems (MTEs). For plants, the seasonality of the MTC has been an important factor in the evolution of plant phenological traits. Root adaptive traits that improve nutrient extraction from impoverished soils are present within MTC regions, including cluster roots, root nodules, and mycorrhizal symbioses. Fire has been an important driver of plant traits, such as smoke, charate, or heat-induced seed germination or seed release (i.e. serotiny), and post-fire flowering. Adaptive traits in animals include both physiological and behavioural traits. MTC regions have been used in the study of many ecological and evolutionary patterns, particularly as related to organismal adaptations to unique soil and substrates (edaphic communities) and interactions between plants and animals, such as plant–herbivore interactions, plant–pollinator interactions, and plant–seed disperser interactions. These interactions shape many plant and animal characters within MTC regions.

3.1 Organismal adaptations

Organisms interact with their environment in numerous ways and these interactions are key factors that shape the evolution of species through time. When certain traits are maintained or selected for through the process of natural selection they may be considered an **adaptation**. An **adaptive trait**, as used here, refers to a trait that has a current beneficial function for an organism in its environment. These traits may include characters that have evolved in response to abiotic factors, such as climate, soil, or disturbance, or in response to biotic factors, such as predator–prey, plant–pollinator, or

The Biology of Mediterranean-Type Ecosystems. Karen J. Esler, Anna L. Jacobsen, and R. Brandon Pratt,
Oxford University Press (2018). © Karen J. Esler, Anna L. Jacobsen, and R. Brandon Pratt 2018.
DOI 10.1093/oso/9780198739135.001.0001

symbiotic interactions. Adaptive traits may also include characters that evolved prior to the development of mediterranean-type climate (MTC) regions, but which are currently beneficial in the regions. These types of traits are sometimes referred to as **exaptations**, or pre-adaptations. In this chapter we will discuss examples of adaptations and adaptive traits, especially as they relate to MTC regions and the organisms within them. We will then use these examples to build up to MTC region communities, groups of interacting populations of organisms (Chapter 4), before discussing how these patterns have contributed to the evolution and diversity of mediterranean-type ecosystems (MTEs) broadly (Chapter 5).

Both animal and plant species exhibit adaptive traits related to features of MTEs, chief of which are the MTC, fire, and soils. MTC is important in understanding adaptations of the organisms that inhabit these regions (Chapter 6). MTC is generally semi-arid and predictably vacillates from cool moist winters to hot and dry summers (Chapter 2). This means that organisms must be active when water is available, when temperatures are cool and limiting, and when days are short (see Figure 6.5). With the arrival of the longer and warmer days of spring and summer, water becomes increasingly limited. Other important drivers are nutrient-impoverished soils and recurrent fires. Further discussion of adaptive traits related to fire and nutrients are found in Chapters 6 and 7.

Historically, plant adaptations have been the focus of many studies in MTEs, because plants are immobile as adults and therefore must be able to tolerate the range of conditions that occur within their environment over the course of their lifetime. Adaptations to MTC environments in animals are most likely to be evident in organisms with limited ranges and organisms that do not migrate, i.e. organisms that are limited to an MTC region. Additionally, MTE adaptations are most likely to be apparent in organisms that are generalists and interact with many species rather than ones that are tied to specific resources or species. This is because adaptations in organisms with strong interactions with a limited number of other species or resources are likely to be shaped by those specific interactions rather than the more general drivers of an MTE. While these species-level interactions are undoubtedly important drivers of evolution, they are more specific than the types of adaptations we are interested in examining here.

3.1.1 Plants

3.1.1.1 Phenology and life form

For plants, the seasonality of MTC has been an important factor in the evolution of plant **phenology**, the timing of important events such as emergence, growth, and reproduction. In general, MTE plant phenology appears to be associated with unique aspects of the MTC, especially the winter wet/

summer dry climate regime. Climate may be particularly important in impacting the establishment and growth of seedlings in the reestablishment period post-fire (Treurnicht et al. 2016). For adult plants, seasonal moisture availability appears to be especially important for vegetative growth, whereas there is greater variability in reproductive phenology (Chapter 6).

Drier sites in MTC regions are often dominated by shrub species that lose their leaves during the summer dry season. These species are classified as **drought deciduous** or summer deciduous and this pattern of leaf loss differs from the winter deciduous pattern present in many temperate deciduous species. Species may be leafless or nearly leafless during the hot and dry summer, and they flush new leaves during the cool and moist winter and early spring months. Some species may also produce two sets of leaves, termed **seasonally dimorphic leaves**, with a small and tough dry-season set of leaves and a larger and thinner set of leaves that is present during the wet season (Westman 1981). This pattern of dry season deciduous or reduced leaf behaviour has been described for MTE species in the Mediterranean (Correia et al. 1992; De Lillis and Fontanella 1992; Kyparissis et al. 1997; Werner et al. 1999), California (Figure 3.1; Field and Mooney 1983; Gill and Mahall 1986; Nilsen and Muller 1981), Chile (Armesto et al. 1993; Montenegro et al. 1981), and South Africa (Cowling et al. 1999; Veste et al. 2001), but is rare in Australia.

Other plant species may limit water loss during the summer dry season through the loss of above-ground portions of the plant and perennation through underground structures. This includes species that may persist for multiple years and multiple dry seasons by relying on underground bulbs, corms, rhizomes, or other storage structures. Collectively, these types of plants are referred to as **geophytes**. Geophytes are present in all five MTEs, ranging from 5 to 30 per cent of the flora depending on the region and plant community (Mooney and Parsons 1973; Naveh and Whittaker 1980; Rundel 1996; Esler et al. 1999; Parsons and Hopper 2003; Montenegro et al. 2004; Alhamad 2006; Alhamad et al. 2010), although this range increases if communities that occur in MTEs sensu lato are also included (Table 3.1). Most geophytes are inactive during the summer dry period, although some geophytes flower during this period. For MTC region geophytes, both water limitation during the summer and temperature limitation during the cooler winter months appear to drive their phenology (DeBussche et al. 2004). The proportion of geophytes within MTC region floras far exceeds those reported for some other ecosystems, including seasonal dry tropical forest (<2 per cent; Charest et al. 2000) and rainforest (<3 per cent; Moro et al. 2016).

Geophytes are able to persist following fire, because they are protected by the soil. Some species of geophytes are adapted to the periodic wildfires that occur within MTEs and require or respond to fire-associated cues. This

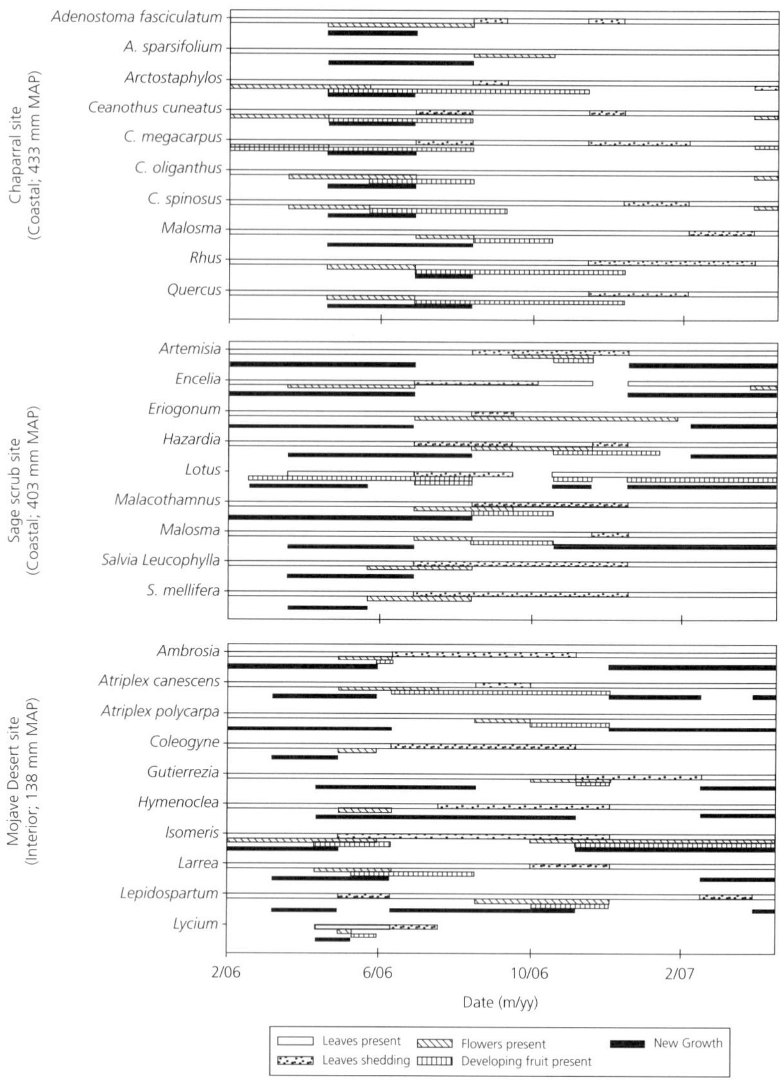

Figure 3.1 **Phenology of some California mediterranean-type ecosystem species from three different communities.** The three communities included differ in their mean annual precipitation (MAP), but all occur within winter rainfall sites in the mediterranean-type climate region of California. The sage scrub community has many species that are drought deciduous and there are also a few examples of this leaf behaviour on the Mojave Desert community. Timing for most activities, including flowering and the formation of new growth, occurs primarily during the winter and spring months (December–June). Note especially the very similar growth phenology of the evergreen shrub species in the top panel. (The three sites from which these data are derived were described in Jacobsen et al. 2007 and Jacobsen et al. 2008.)

Table 3.1 The proportion (%) of different Raunkiaer life forms across different mediterranean-type climate (MTC) regions. Included are the five predominant life-form categories for mediterranean-type ecosystems (MTEs; geophytes, therophytes [annuals], hemicryptophytes, chamaephytes, and phanerophytes). Definitions for some of the different life-form categories are provided, as are terms that are comparable and more generally used (annuals, subshrubs, trees, and shrubs). Where prior studies have omitted to report on the prevalence of a category, this is indicated as 'not reported' (n.r.) and this could indicate either that this life form is not present or was not included in the study

| | Annuals | | | Subshrubs | Trees and shrubs | |
| | Buds below ground (bulbs, tubers, rhizomes, etc.) | | Perennial with buds at soil surface | Perennial with low aerial buds (<0.25 m) | Perennial with high aerial buds (>0.25 m) | |
Region	Geophytes	Therophytes	Hemicryptophytes	Chamaephytes	Phanerophytes	Climbing plants
Australia	1.4	40.4	22.8	9.3	25.6	0.5
Australia	12.8	0.0	12.8	12.8	59.0	2.6
Australia	19.0	4.1	17.4	14.9	44.6	0.0
Australia	7.1	19.0	33.3	n.r.	40.5	n.r.
Australia average	**10.1**	**15.9**	**21.6**	**12.3**	**42.4**	**1.0**
California	3.5	18.1	30.4	13.1	32.6	2.4
California	0.7	32.0	50.0	6.0	11.3	n.r.
California	3.6	68.4	16.5	3.0	7.4	1.2
California	7.0	40.0	33.0	5.0	15.0	n.r.
California	12.5	6.0	12.5	9.0	60.0	0.0
California average	**5.4**	**32.9**	**28.5**	**7.2**	**25.3**	**1.2**
Chile	5.8	53.3	15.0	11.7	14.2	n.r.
Chile	15.2	23.0	13.7	13.7	34.3	n.r.

Continued

Table 3.1 Continued

Region	Annuals		Subshrubs	Trees and shrubs		
	Buds below ground (bulbs, tubers, rhizomes, etc.)		Perennial with buds at soil surface	Perennial with low aerial buds (<0.25 m)	Perennial with high aerial buds (>0.25 m)	
	Geophytes	Therophytes	Hemicryptophytes	Chamaephytes	Phanerophytes	Climbing plants
Chile average	**10.5**	**38.2**	**14.3**	**12.7**	**24.3**	
Mediterranean	7.5	73.6	1.9	n.r.	7.5	9.4
Mediterranean	6.9	59.3	12.1	n.r.	16.7	4.9
Mediterranean	7.5	54.4	20.2	6.9	9.6	1.5
Mediterranean	1.5	5.5	33.5	20.0	27.5	12.0
Mediterranean	7.5	64.3	12.8	5.1	6.5	3.8
Mediterranean	11.3	33.0	38.1	7.2	10.3	n.r.
Mediterranean average	**7.1**	**48.3**	**19.8**	**9.8**	**13.0**	**6.3**
South Africa	10.3	34.9	23.3	9.4	21.1	1.1
South Africa	8.2	53.7	n.r.	38.1	n.r.	n.r.
South Africa	16.5	28.7	17.0	31.8	5.9	n.r.
South Africa average	**11.7**	**39.1**	**20.2**	**26.5**	**13.5**	**1.1**
MTE average	**9.0**	**34.9**	**20.9**	**13.7**	**23.7**	**2.4**
Non-mediterranean-type ecosystems	Cryptophytes (including geophytes)					

Brazil, seasonally dry tropical	1.9	40.3	12.6	18.6	26.6	n.r.
Canada, boreal forest	19.0	0.0	32.0	12.0	37.0	n.r.
Brazil, rainforest	2.3	0.0	5.7	4.0	89.5	n.r.
Great Smoky Mtns, USA, temperate forest	15.1	11.5	52.1	1.8	19.5	n.r.
India, alpine	29.2	5.8	48.0	10.5	6.4	n.r.
New York, USA, prairie	24.9	29.9	33.1	0.9	11.1	n.r.
Sweden	23.1	4.9	53.3	10.6	8.2	n.r.
Venezuela, savanna	n.r.	23.1	23.5	n.r.	33.2	20.2

n.r., Not reported.

Sources: Naveh and Whittaker 1980; Specht 1969; Pettit et al. 1995; Shmida 1981; Esler and Rundel 1999; Mooney and Parsons 1973; Montenegro et al. 2003; Hadar et al. 1999; Buide et al. 1998; Struck 1994a.

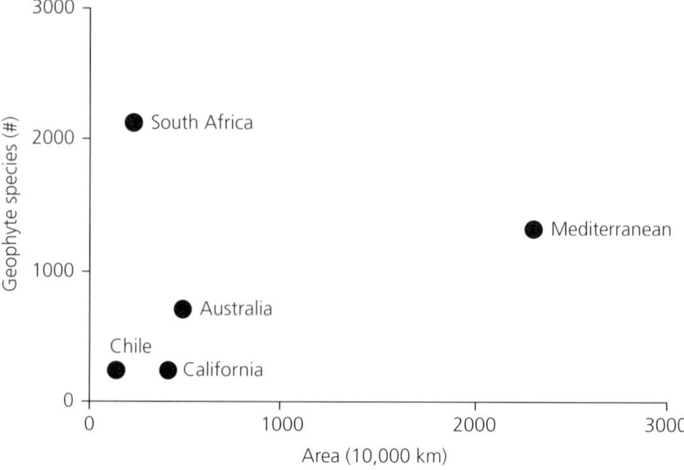

Figure 3.2　**Geophyte diversity in mediterranean-type climate (MTC) regions.** There are many geophyte species within each of the MTC regions, with a particularly high level of geophyte diversity in South Africa. (Figure modified from Procheş et al. 2006.)

includes many examples of fire-stimulated flowering in geophyte species across MTEs (LeMaitre and Brown 1992; Keeley 1993; Arianoutsou 1998; Montenegro et al. 2003; Tyler and Borchert 2003; Lamont and Downes 2011). Diverse geophyte species are found in all of the MTC regions (Table 3.1), and are particularly diverse in South Africa (Figure 3.2; Procheş et al. 2005; Procheş et al. 2006).

Annual species represent another growth form that is adaptive in the context of MTC. Like geophytes, annuals can avoid the summer dry period by timing growth and reproduction to the winter and spring months. They then persist at a site through the dry season as dormant seeds. This is also a life form that is relatively common in MTC regions and which is not as common in some other ecosystems, such as temperature forest (<12 per cent; Cain et al. 1956), boreal forest (<5 per cent; Cain et al. 1956; Gustafsson 1994), alpine (<6 per cent; Nautiyal et al. 2001), and rainforest (<1 per cent; Moro et al. 2016).

It is interesting to note that each of the plant groups described above, drought deciduous perennial plants, geophytes, and annuals, are all adaptive plant forms to various drivers present in MTC environments, yet they have often been overlooked in MTE comparisons that focus on the archetypal evergreen sclerophyllous shrub species and communities of MTEs (however, see, for example, Mooney et al. 1970; Gulmon 1977; Parsons and Hopper 2003; Jacobsen et al. 2009). The diverse communities that occur within MTEs and differences in their dominant plant forms and function are examined in Chapter 4. These communities vary across MTC landscapes, depending on mean annual precipitation of a site, the presence or absence of freezing, differences in soil nutrition, fire regime, and aspect.

The evergreen sclerophyllous shrubs found in MTEs have received much attention. The adaptive nature of this growth form has been considered in the context of carbon acquisition, use, and turnover and competition with deciduous species (Mooney and Dunn 1970; Chapter 6). Mild and moist winters allow evergreen leaves to photosynthesize all year round except when water is limited during the peak of the summer rainless season. This leaf form is also opportunistic and is able to capitalize rapidly on pulses of rainfall because new leaves do not have to be grown, as would be required of deciduous species. Because of this, the evergreen leaf can acquire more carbon than a deciduous one over the course of the lifetime of a leaf. Evergreen shrub physiology is discussed in detail in Chapter 6.

Across all plant life forms, MTE plants flower predominantly in the spring and early summer, although some flowering occurs year round in most MTE communities (Mooney et al. 1974; Figure 6.11). Data from two MTC regions, California and the Mediterranean, are shown in Figure 3.3, but these same patterns are also found in species from Chile (Arroyo et al. 1981), Australia (Specht and Rayson 1957), and South Africa (Struck 1994a; Johnson 1993). This pattern is consistent with flowering occurring when both soil moisture and cool temperatures are not limiting. This seasonal pattern of flowering differs from other ecosystems. For instance, in a wet tropical system flowering occurred throughout the year (Figure 3.3c), and in a temperate prairie, where winter temperatures are very cold and rainfall occurs during the summer, all species flowered during a relatively short period during the summer (Figure 3.3d).

3.1.1.2 Roots, soils, and symbioses

Soils that are nutrient-poor have also likely contributed to the growth forms represented in the MTC regions. Within MTC areas, soils tend to be relatively nutrient-poor and plants may exhibit many different responses to low nutrient availability. The impacts of soils on organism traits are termed **edaphic effects**, which is a character that is influenced by soil or substrate characteristics. At the leaf level, there is strong evidence that poor soil nutrition selects for sclerophyllous leaves (Ordoñez et al. 2009; Salleo and Nardini 2000; Lambers et al. 2010; Chapter 6). Evergreen succulent leaves may also occur in arid low nutrient sites (Cowling and Campbell 1980).

Root adaptive traits that improve nutrient extraction from impoverished soils are present within MTEs (Lamont 1983b). Both cluster roots and root nodules are found in many species (Figure 3.4). Cluster roots are particularly prevalent in Australia and South Africa in members of the Proteaceae and occur in plants growing in phosphorus- and nitrogen-deficient soils (Case Study 5; Chapter 7).

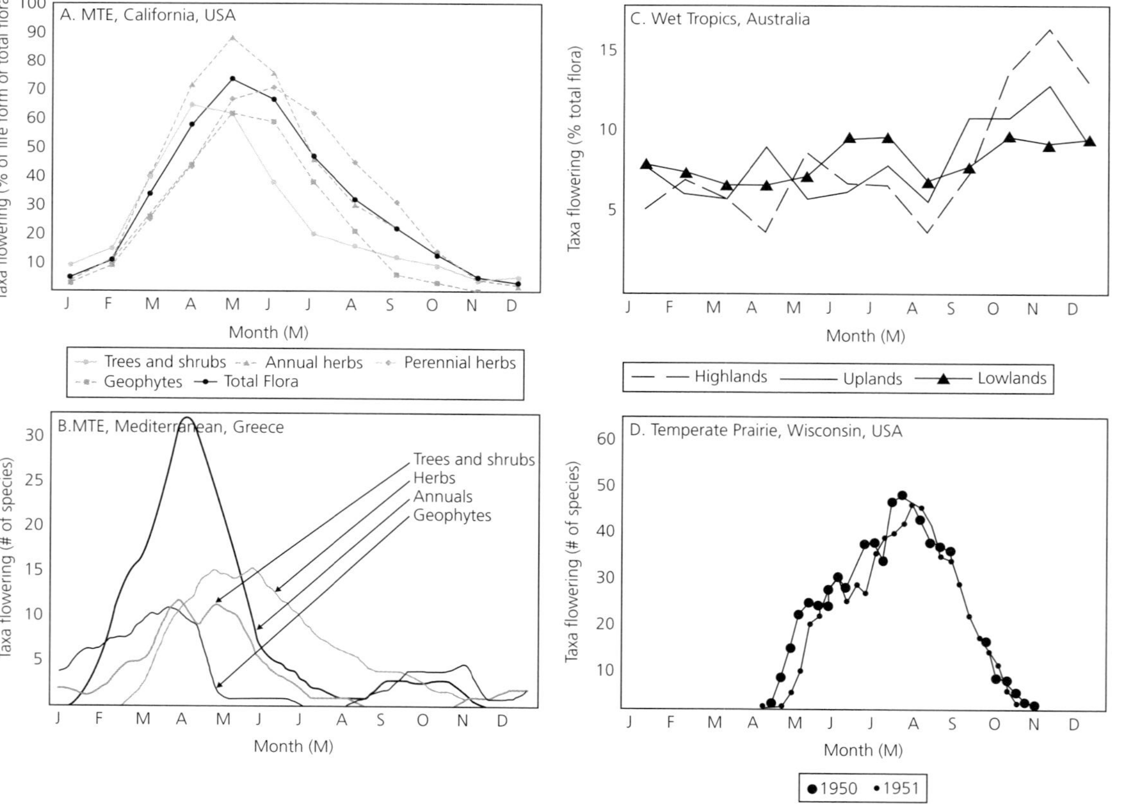

Figure 3.3　**The proportion of species flowering by month.** Within mediterranean-type ecosystems (MTEs), there is a peak in flowering in the spring and early summer (panels a and b) and this peak is apparent across all plant life forms. In MTEs, there is some flowering all year round. This contrasts with other ecosystems, which have different peaks or more dispersed flowering. As examples, the proportion of taxa that flower by month in the tropics (panel c) is spread throughout the year and flowering is limited to the summer and fall months for all species within a temperate prairie (panel d). (Data are compiled or figures modified from Mooney and Parsons 1973; Petanidou et al. 1995; Boulter et al. 2006; and Anderson and Adams 1981.)

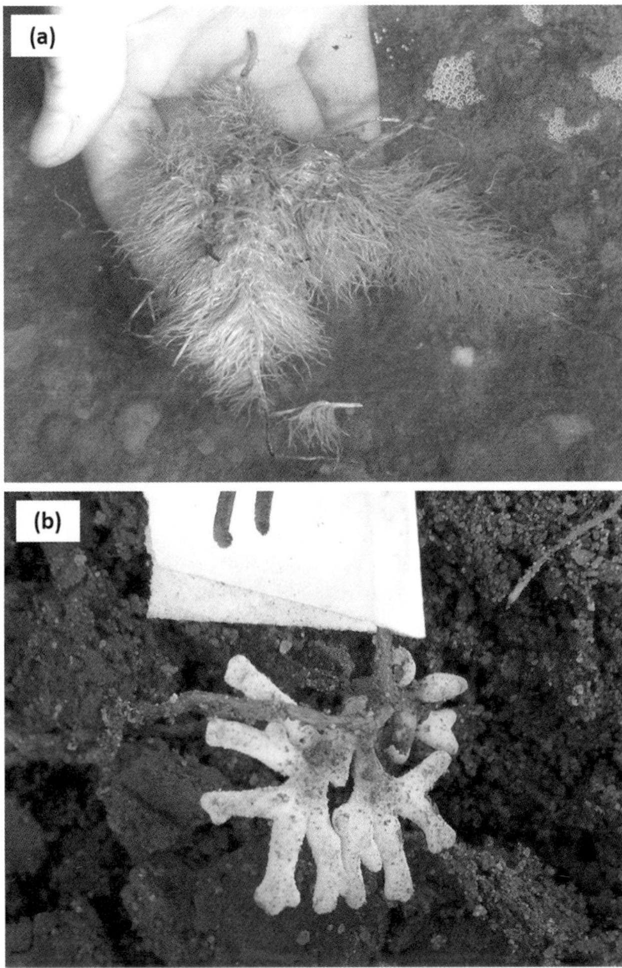

Figure 3.4   **Root adaptations to nutrient-poor soils.** Within mediterranean-type ecosystems, some species exhibit proteoid root clusters (a) or root nodules (b). The cluster roots in panel a are shown after being rinsed clear of soil in a stream. The stellate root nodule in panel b was tagged (white tape) for study. Both cluster roots and nodules enable plants to increase the nutrients available to them, especially phosphorus for cluster roots and nitrogen for root nodules.

Source: Photos from Anna L. Jacobsen.

Symbioses, close associations between two different organisms, occur between plant roots and soil bacteria or fungi for many MTC region species. These associations are particularly important in the nutrient-impoverished soils that predominate across MTC regions and can significantly increase the access that plants have to nutrients and their ability to efficiently extract them from their environment.

Case Study 5 Pioneering research on proteoid root clusters in three mediterranean regions

Byron B. Lamont, Department of Environment and Agriculture, Curtin University, Perth, Australia

With assistance from María Pérez-Fernández, Ecology Área, University Pablo de Olavide, Sevilla, Spain

I needed a break from study, having gone straight from school (I was Dux of Perth's Wesley College in 1962) to undertake a 4-year degree in agricultural science at the University of Western Australia (I was an average student). So I spent 2 years working in my father's landscape gardening business. Realizing by 1968 that the business world and hard physical work were not for me, I wandered around Kings Park Botanic Garden looking for inspiration, as I knew by then that my destiny lay in research on native plants. *Hakea* seemed an ideal subject: such striking flowers and ornate woody fruits and a myriad of sculptured leaves. I asked the only academic person in the Botany Department I knew, Alison Baird, if there was anything worth studying about *Hakea*? She pointed to a paper by Helen Purnell, published in 1960, on the unique root clusters she had discovered in the Proteaceae, including *Hakea*, that she had termed proteoid roots (see some of my examples of proteoid roots and early proteoid root development in Case Study 5 Figure). This seemed right up my alley: a background in soil science and microbiology and adept at growing plants. I started in 1969 (self-funded) and produced a 212-page thesis by the end of the year (I only slept 6 hours a night and never got ill that year!). I received the equivalent of first class honours and was awarded an Australian postgraduate award. By 1972 I had produced five papers as sole author (I resented my supervisor making even the smallest changes to my text as it then had to be retyped by a professional typist!) and in 1974 was the first person in Western Australia (the world?) to receive a PhD based only on published papers. The PhD committee was reluctant to accept my thesis without an overview chapter summarizing my findings; now, pre-published/in press chapters are the goal of every PhD student. Nevertheless, it is the only thesis that I know of that required no changes before being passed!

Growers who were aware of these strange roots thought they might be a form of mycor-rhiza, but I confirmed Purnell's observations that they lacked endogenous bacteria or fungi, unlike the other two types of specialized roots: nodules and mycorrhizas (Lamont 1982). Nevertheless, Purnell suggested a microbial stimulus might be required for their formation, so in 1972 I set up a glasshouse trial to test that. I grew *Hakea* seedlings in 1-L jars filled at the bottom with sand under sterile (axenic) conditions by autoclaving

Case Study 5 Figure (Opposite). Cluster roots. Upper: young (simple) root cluster of *Leucadendron salicifolium* (from South Africa), 10 mm long; mid-left: mature compound root cluster (made up of many simple clusters), 50 mm long, of *Banksia (Dryandra) armata* (southwest Australia) with root-lets densely covered in long root hairs; mid-right: transection of root cluster of *Viminaria juncea* (southwest Australia) before rootlets have emerged from parent root, 2 mm diameter; lower: transection of mature root cluster of *Hakea prostrata* showing rootlets, 0.15 mm diameter, covered in root hairs and spreading out through the soil. Note two rootlets often arise from the same position opposite the xylem arches to give up to 1000 rootlets per 10 mm, the densest of all clusters.

Source: Photos from Byron B. Lamont.

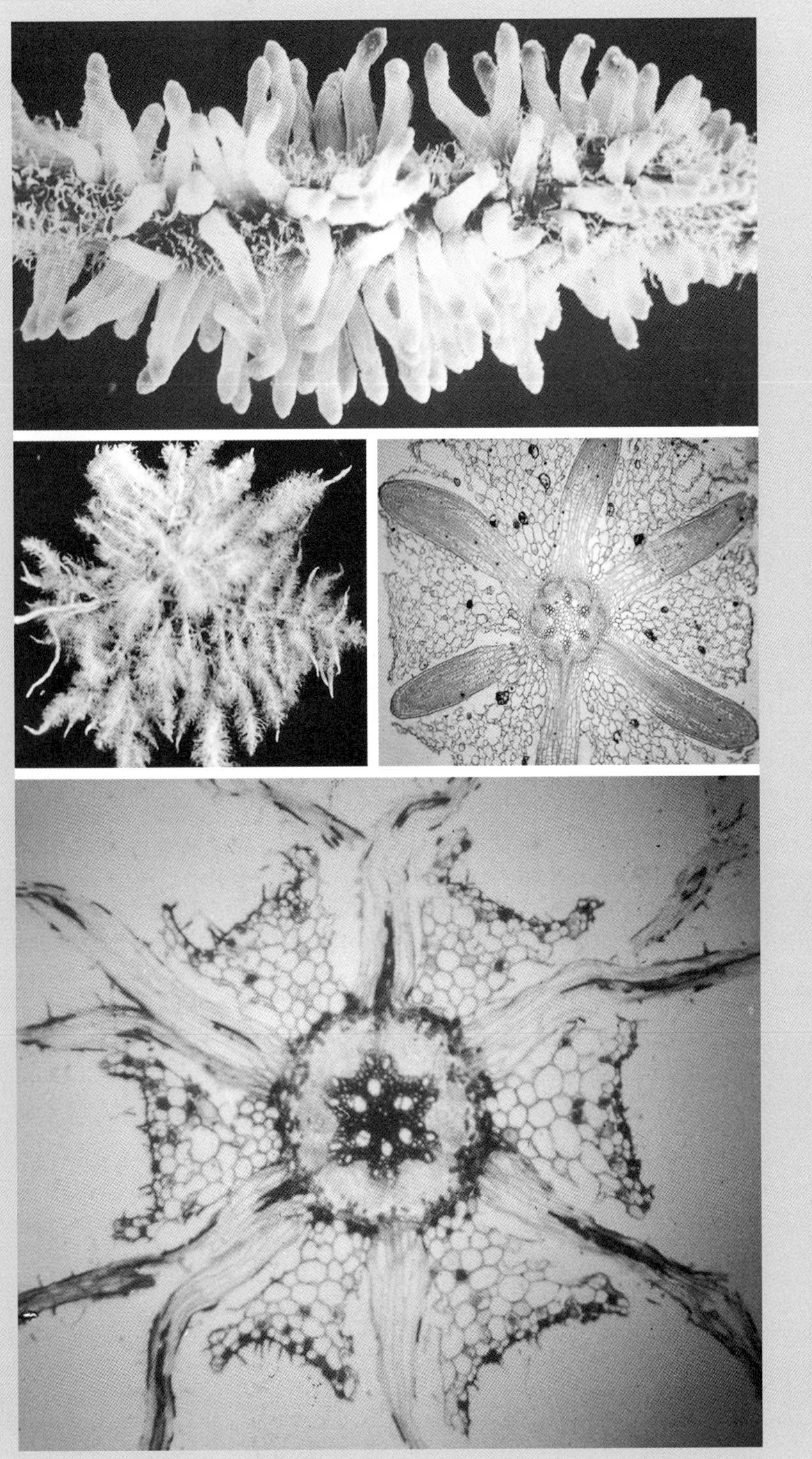

Continued

Case Study 5 (*Continued*)

everything first and sterilizing the seeds with sodium hypochlorite solution. I then added an autoclaved soil wash to half the jars and untreated soil wash to the other half and kept the set-up going in the glasshouse for 18 months. I shared the lab with another PhD student, Nick Malajzuk, who was intrigued by what I was doing and advised me on sterile techniques. He went to the CSIRO Soils Division in Adelaide in 1973, ostensibly to undertake further work on his PhD topic, the root pathogen, *Phytophthora cinnamomi*. A few months before my paper, entitled 'Microorganisms and the formation of proteoid roots', was published in 1974 (it had been accepted some 6 months before), a paper by Nick and his colleague in Adelaide appeared in the fast-track journal *Nature*, entitled 'Proteoid roots are microbially induced'. One knows science is highly competitive where the imperative is to be the first to discover something, but cooperative work is equally prized and in this case I felt cheated by my friend who had completely outsmarted me (he used the much quicker water culture technique), especially as he was not upfront with his colleague in Adelaide about his motivation for the trial, as I found out later.

In 1974, I was appointed as a lecturer in biology at WAIT, later Curtin University, even though I had not quite finished compiling my PhD (the seven published papers helped). In 1980, I went on study leave to South Africa with my young family and worked on proteoid root clusters in the Cape and showed that probably all 350 species of Proteaceae there produce them (Lamont 1983a). More importantly, I spent 6 months preparing a review entitled 'Mechanisms for enhancing nutrient uptake in plants, with particular reference to mediterranean South Africa and Western Australia' that has been widely cited (>270 cites, August 2015). Over the past 40 years there has been conflicting support for the Lamont/Malajczuk findings, but by 2000 the consensus among those interested in root clusters was that a microbial stimulus was not required. I reviewed everything known about them in 2003 and concluded that the case against a microbial stimulus had not been satisfactorily proved, even though the editor wanted me to withdraw the section on microorganisms because of the prevailing view. I even suggested that the Malajczuk results were an indirect response to microbes as the plants grew much better in their presence (a plant-size effect), unlike my results where the growth of plants with and without clusters was identical. But at that stage no-one had actually added a specific microbe to axenic culture and induced the production of root clusters, so even I was having doubts.

I retired from teaching/administration/supervision duties in 2010, giving me more time to catch up with some of my old research interests. In late 2013, I discussed the microbial-stimulus problem with a close colleague in Spain, María Pérez-Fernández. Her specialty is microbial ecology and she had spent time with me at Curtin in 1996–2003 working on the nodulation of Australian and Spanish legumes. María said she had the ideal water culture set-up at the University of Pablo de Olavide, including a culture collection of known plant-growth-promoting soil bacteria (rhizobacteria). She obtained seeds of *Hakea laurina* (Proteaceae) and *Viminaria juncea* (Fabaceae) from southwest Australia— I was the first person to show (in 1972) that root clusters could be produced by taxa other than the Proteaceae (nine families are now known). Working with her research assistant, Jesus Rodríguez-Sánchez, they set up and maintained an elaborate sterile, water culture trial with two rhizobacterial species, *Bradyrhizobium elkanii* and *Bacillus megaterium*, plus controls, at two levels of nitrogen. Within 3 months, Jesus had harvested the plants and counted root clusters per plant. I then analysed the data, taking parent root length into account as a covariable. By late 2014, the manuscript entitled 'Soil bacteria hold the key to root cluster formation' was published online in the prestigious journal, *New Phytologist*, the fastest turnaround from planning the study to publication in my 45 years of research. Ironically, the paper was reviewed immediately by *Nature Plants.*

Inspired by our success, María set up another trial to include South African (*Leucadendron*) and Mediterranean Basin (*Lupinus*) species, known to produce clusters, as well as *Viminaria* from southwest Australia used previously (Case Study 5 Table). The results confirmed the complex interaction between host plant species, microbial species/strain and soil nutrient status (Lamont and Pérez-Fernández 2016). In fact, these two sets of data cover the full range of results, from no microbial effect to no clusters in the absence of microbes, as obtained over the previous 40 years (Lamont et al. 2015). The explanation is that the hormone, indole acetic acid, is the transducer that controls root cluster production, and it can have positive, nil or negative effects, depending on its concentration at the induction site, which is the net supply from the plant and microbes. I have selected just three rhizobacteria of the seven examined and the low nitrogen treatment only to illustrate the great range of results that can occur. Note that no clusters form in the absence of microbes in *Leucadendron salicifolium* and *Lupinus albus* but they are abundant in *Viminaria juncea*. Further, the three bacteria have vastly different effects on both cluster production and general plant growth, including their complete suppression by *Bradyrhizobium* in the case of *Viminaria*, whereas cluster production in *Leucadendron* in the presence of *Pseudomonas* is more than 10 times that in the presence of *Bradyrhizobium* for a doubling in total plant mass. So after many years of uncertainty we now know that root cluster production is the outcome of the biochemical interaction between the plant and the bacteria in its rhizosphere, and both are affected by the soil nutrient status, especially nitrogen, even though the main function of root clusters is to enhance phosphorus uptake from nutrient-impoverished soils.

Case Study 5 Table. Production and growth of cluster roots. Production of proteoid-type root clusters and total growth (in g) by 4-month-old plants of three species native to three mediterranean regions in the presence of three soil bacteria plus control. Plants were grown in water culture in a balanced nutrient solution but with nitrogen and phosphorus omitted. Values in italics are the lowest cluster values for each species and those in bold are the highest. Note that two species do not produce root clusters in the absence of soil bacteria. Whether soil bacteria stimulate or inhibit cluster production and plant growth depends on the plant species to whose roots they are attached. From Lamont, BB, and Pérez-Fernández, M, unpublished (2015)

Mediterranean-type region:	South Africa		Australia		Mediterranean	
Species:	*Leucadendron salicifolium*		*Viminaria juncea*		*Lupinus albus*	
Family:	Proteaceae		Fabaceae		Fabaceae	
Growth response:	Root clusters	Plant weight	Root clusters	Plant weight	Root clusters	Plant weight
Control	*0.00*	1.50	10.78	14.42	*0.00*	3.89
Bradyrhizobium elkanii	1.40	1.49	*0.00*	10.72	3.57	2.02
Bacillus megaterium	1.20	0.78	**17.11**	14.28	2.60	10.12
Pseudomonas putida	**15.67**	2.81	5.00	6.94	**7.60**	7.30

Root nodules represent symbioses between plants and several different types of nitrogen-fixing bacteria. Root nodules occur in a phylogenetically restricted group of plants that includes members of the Rosid I clade (Gualtieri and Bisseling 2000; see Table 7.3 for a list of common MTC region genera that form root nodules). The formation of nodules is at least partially dependent on low levels of soil nitrogen (Streeter and Wong 1988; Brockwell et al. 1989), and nodule formation and fixation rates are both diminished in the presence of increased soil nitrogen levels. Nodule formation is also influenced by the carbon balance of the plant (Pratt et al. 2012) and soil moisture (Pratt et al. 1997). While root nodules are not unique to MTEs, they are formed by species in some of the iconic genera of MTEs, including *Ceanothus* in California and numerous *Acacia* species in Australia, and are an indication of the generally low nitrogen levels found across MTE soils.

Many MTC region plant species form fungus–plant associations through mycorrhizal symbioses, thereby enhancing the nutrient extraction capability of plants from soils (Cruz et al. 2008; Sánchez-Castro et al. 2012). Mycorrhizae are particularly useful in increasing the access of their associated plant species to mineral nutrients and water (Chapter 7). For these species, plant adaptations to seasonal water stress may be linked to the responses of their associated symbiotic species (Goicoechea et al. 2004).

3.1.1.3 Fire

Fire has been an important driver in MTEs and there has been much research on fire-associated adaptive traits in plants within these systems (reviewed in Keeley et al. 2012; see Chapter 6). While resprouting is prevalent, it is likely that this trait existed prior to MTEs and in some cases may have evolved in response to non-fire-related disturbances; species that utilize this post-fire recovery strategy were likely able to persist in regions following the emergence of MTC (Bellingham and Sparrow 2000; Bond and Midgley 2003; Pausas and Verdú 2005). In contrast, fire-associated traits, such as smoke, charate, or heat-induced seed germination or seed release (i.e. serotiny), and post-fire flowering, may represent traits that have been selected for by fire regimes and are of considerable interest in research on the evolution of fire-associated traits in MTEs (Paula et al. 2009; Keeley et al. 2011). Studies that have examined these traits have found evidence of their increased prevalence in MTEs and that they appear to have emerged coincident to the onset of fire. For instance, a study of fire-stimulated flowering in species from Australasia and South Africa found that 71 per cent of species that exhibited fire-stimulated flowering occurred within regions experiencing MTC (Lamont and Downes 2011) and, in Australia, the evolution of serotiny in *Banksia* occurred coincident with a seasonally dry climate and fire (He et al. 2011).

The role of fire in shaping plant trait adaptations in MTC environments has been controversial (Bradshaw et al. 2011, with response by Keeley et al. 2011). Many of the same issues apply to other traits described above as well.

Multiple approaches have been used to investigate potential adaptations in response to different evolutionary pressures in an attempt to separate adaptive traits from MTE-specific adaptations. One approach is to reconstruct trait changes over time and then to examine evidence for these changes occurring coincident with changing environments, for instance, with the onset of fire. To date, the most common approach is the comparison of rates of occurrence of certain traits across different systems, with the implication that evolutionary processes have driven these differences either through adaptation or through community assembly processes that have filtered certain adaptive traits. These, as well as additional, issues that may influence how we understand and study evolution of MTEs are discussed in Chapters 5 and 6. Within MTC regions, there continues to be great interest in the study of species adaptations to the drivers of modern MTEs. There is also much interest in the adaptive traits associated with lineages that pre-date these conditions and which have persisted within these regions from long before the emergence of modern MTE drivers (Case Study 6).

Case Study 6   Central Chile compared to California: the enigma of the Chilean mediterranean-type climate flora

Mary T.K. Arroyo, Departamento de Ciencias Ecológicas, Instituto de Ecología y Biodiversidad (IEB), Universidad de Chile, Santiago, Chile

The central Chilean mediterranean-type flora is less species-rich than other mediterranean-type climate (MTC) floras, including the California Floristic Province, to which it is closely climatically matched (Arroyo et al. 1995; Cowling et al. 1996). Compared with California, central Chile has more woody species and genera on an equal area basis. Why should this be so? The answer could lie in the fact that Chile's tree flora is richer in old woody plant lineages, legacies of the ancient Gondwana flora, which extended from southern South America across to Australia and New Zealand via Antarctica, and the Eocene and early Miocene forests, which extended across southern South America from the Pacific to the Atlantic before significant uplift of the Andes had taken place.

There are numerous examples of species that reflect Gondwanan floristic origins. Take, for example, three of the most abundant and widespread sclerophyllous trees within the Chilean MTC region: *Peumus boldus*, *Quillaja saponaria* and *Lithraea caustica* (Case Study 6 Figure). The monotypic *Peumus* clade has Gondwanan roots. It split from the Australian *Palmeria* clade at 57 million years ago (Ma; Renner et al. 2010). There are other Gondwanan links, as seen in *Caldcluvia*, *Gevuina*, *Lomatia*. A particularly fascinating Gondwana legacy is the endemic MTC tree *Nothofagus alessandri*, the sole South American representative of the Fuscospora clade, which split from the Australasian part of clade minimally at around 30 Ma (Sauquet et al. 2012).

With regard to neotropical legacies, *Quillaja* is monotypic and is the sole member of the family Quillajeae, with only one other species in eastern South America. *Lithraea* is likewise disjunct across South America with two additional extra-Andean species.

Continued

Case Study 6 (*Continued*)

Case Study 6 Figure. Gondwanan floristic elements of the Chilean flora. *Peumus boldus* (a and b) and *Lithraea caustica* (c and d) are both examples of ancient lineages that are represented within the flora of the mediterranean-type climate region of Chile.

Source: Photos from Anna L. Jacobsen.

Another originally transcontinental lineage that became disjunct is *Myrceugenia*, which split into a Chilean and a Brazilian clade around 16–13 Ma contingent on the Andean uplift (Murillo et al. 2012) to undergo significant radiations in both areas.

Such old legacies have likely survived in central Chile on account of the moderate oceanic climate buffering them against extinction and thereby increasing the woody component in the flora (Arroyo et al. 2005). However, there could be another explanation. The herbaceous flora of central Chile mostly diversified as of the mid-Miocene times. Over this period, central Chile became increasingly cut off from potential sources of lineages to the east by the emerging Andean mountains and to the north by the Atacama Desert. Thus the higher proportion of woody species and genera in central Chile's flora could be an artefact of a relatively poorer herbaceous flora. That fewer annual species have evolved in central Chile than in California could also be relevant. Disentangling these hypotheses reflecting complex historical processes constitutes a major challenge in mediterranean ecosystem research and underscores the need for more biome-level comparisons using state-of-the-art phylogenetic approaches. Ultimately, of course, all these factors may turn out to be relevant.

3.1.2 Animals

Adaptive traits in animals may be either physiological or behavioural. Behavioural traits are related to the ability of mobile animals to alter their environmental conditions through behavioural responses, such as moving to a new location or altering the timing of their activity. For instance, animals may alter the temperatures that they experience by moving into or out of the sun. These behavioural traits mitigate their need to evolve physiologically based stress resistance traits. In other cases, physiological adaptations, for instance, to tolerate dehydration, may be critical.

Animals often interact indirectly with the ecosystem drivers that were discussed above for plants. For instance, herbivores will respond to the traits of the plants that they consume; herbivores do not directly experience selection to low soil nutrition, but may experience selection to cope with low leaf nutrition levels that are themselves the result of low soil nutrition. Similarities in soil animal traits among regions appears to be related to similarities in soil humus, particularly the characteristics of the sclerophyllous leaf litter (Di Castri 1973). Insect abundance and timing of activity has also been shown to be strongly linked to plant phenology in a Mediterranean shrub community (Petanidou et al. 1995).

Animal adaptations may vary among taxa and involve diverse types of responses. Adaptations will be most apparent in vertebrates rather than invertebrates. This is because vertebrates will generally interact with a broader array of species and resources, whereas insects are commonly more limited in their interactions to a smaller set of species, such as being specialized to eating the foliage of a single plant species (Cody and Mooney 1978).

The seasonality of MTC has shaped animal traits, including through selection to the particular temperature and moisture regimes of MTEs, especially low water availability during the hottest months of the year. Many organisms time their activity to periods of favourable temperatures and water availability. For instance, soil macroinvertebrates are largely inactive in upper soil layers during the summer period when these layers become very dry (Sgardelis et al. 1995). Mediterranean land snails have been reported to have higher temperature tolerances than a non-Mediterranean species from a more temperate environment (Scheil et al. 2011). Mediterranean land snail dormancy was linked to changes in humidity (Lazaridou-Deimitriadou and Saunders 1986) and the seasonality of precipitation (Iglesias et al. 1996), such that they were able to avoid seasonally dry periods. In the common dormouse, Mediterranean populations do not exhibit the winter dormancy of more temperate populations and instead undergo a period of summer dormancy during the summer dry period (Panchetti et al. 2004).

In response to fire, animals may be grouped based on their type of response (reviewed in van Mantgem et al. 2015). Animals may persist in an area through a fire by surviving underground, particularly for invertebrates, which may have a dormant phase. Vertebrates, such as rodents and tortoises, may also successfully survive through a burn, either through underground shelter or in unburned patches, such as rocky areas, within larger burned areas. Still other animals, often those that are larger or mobile, such as large mammals and birds, may be able to leave a burn area and return following the fire. Some species may be killed by fires and rely on recolonization from adjacent unburned areas. All of these animal responses have been described within MTC regions (e.g. Quinn 1979; Friend 1993; Brotons et al. 2005). There are also secondary impacts on animal species through alterations in the habitat types that are available to animals during the post-fire recovery period (Potts et al. 2003; Arnan et al. 2006; Elia et al. 2012). Important habitat types in this context are aquatic ones that may be highly modified by sedimentation following fire. These changes can alter population demography, size, and trophic interactions following a fire (Gamradt and Kats 1997; Kerby and Kats 1998). Additionally, temperature and dehydration tolerance may be particularly important for MTE animals in the post-fire period (Santos and Cheylan 2013). The response of diverse animal groups to fire regimes within MTEs remains an area of interest, but has rarely been a topic of experimental study and has been cited as an area where knowledge is particularly limited (Parr and Chown 2003; Driscoll et al. 2010; van Mantgem et al. 2015).

3.1.3 Microbes

Microorganisms are critically important for the function of ecosystems, particularly in their role in nutrient cycling. These organisms also exhibit adaptive traits that enable them to persist within MTEs. Microbes must have physiological traits that allow them to tolerate seasonally dry conditions and to time their activity to times when temperatures and water are not limiting (Schaefer 1973).

Studies on MTC region microbial species have focused on how microbial communities change in response to specific stresses and how they recover. These studies often rely on the identification of species from extracted DNA and are therefore not linked to the specific physiological responses of organisms, but rather how different types of microbes change in presence and abundance over time. Following fire in MTEs, soil microbial communities are altered, but recover over time (Torres and Honrubia 1997; Cilliers et al. 2005; Goberna et al. 2012; Pizarro-Tobías et al. 2015). Microbial communities also vary seasonally, particularly in response to soil drying (Papatheodorou 2008). An exciting area of future research is comparison of microbial traits that may be unique adaptations to MTE characteristics.

3.2 Diversity and endemism

In examination of adaptive traits, there are two different indices that may be useful in identifying groups that exhibit traits that are particularly beneficial in MTC regions. The first is by looking at groups of organisms that have high levels of diversity within MTEs; that is, taxa that contain large numbers of species. High diversity may indicate the presence of a **key adaptation**, also termed an adaptive breakthrough, within a lineage. In these lineages, it may be that a novel trait or suite of traits has evolved that equips these organisms to be particularly successful within their environment and/or to utilize a resource to which other organisms do not have access.

Numerous traits have been suggested to have been key adaptations for different lineages that have been successful and diversified within MTEs (Linder 2005). It has been suggested that the evolution of fire-dependent recruitment (i.e. post-fire seeding) and the loss of resprouting are both key innovations in plants occurring in fire-prone systems, such as MTC environments (He et al. 2011; Pausas and Keeley 2014). Others have suggested that the evolution of drought resistance coupled with post-fire seedling recruitment was a key adaptation that allowed plant taxa to diversify into arid sites within MTEs (Pratt et al. 2010; Keeley et al. 2012). Increased longevity of tissues, for instance, long leaf life span, is suggested to be a key adaptation to low nutrient soils (Craine et al. 2012). In Mediterranean lizards, the key traits of large egg and hatchling sizes and early rapid growth were reported to be key adaptations to low productivity related to low nutrition and aridity (Iraeta et al. 2013). These examples of key adaptations within MTEs highlight the common selective stresses that have already been discussed and the need for organisms within MTEs to be able to tolerate seasonal water deficit, recurrent fire, and low soil mineral nutrition.

Endemic taxa are also informative in the identification of traits that are important within a particular habitat (Rundel et al. 2016). Endemic taxa

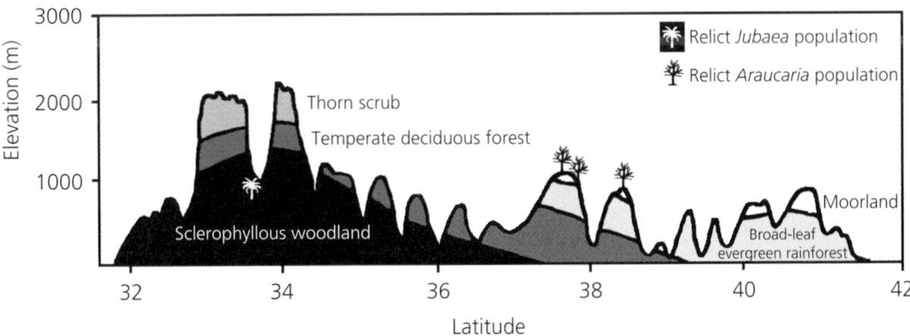

Figure 3.5   **Relict populations.** A transect profile is shown through the coast range of central toward southern Chile. Populations of relictual species are indicated for *Jubaea* and *Araucaria*. These species have narrow ranges and occur predominantly in protected and isolated regions (valleys and mountain tops). (Redrawn from Villagrán 1995.)

Table 3.2 Plant and vertebrate diversity and endemism in mediterranean-type ecosystems (MTEs). Each of the different MTE regions, along with their area, number of vascular plant and vertebrate species, number of endemic plant and vertebrate species, and the plant and vertebrate species diversity are reported as the number of species per land area. Values include an across-study mean, with the range of values reported across studies in parentheses below the mean. There are large ranges for some of these values depending on how strict or lax authors were in their definition of each MTE and variations in taxonomy of species. Plant diversity numbers represent broad-scale data from entire MTE regions and do not account for non-linearity of species-area relationships as discussed in Rosenzweig (1995). For a comparison of regional diversity see Keeley and Fotheringham (2003). Sources include: Mooney 1988; Goldblatt 1997; Cowling and Hilton-Taylor 1997; Simonetti 1999; Médail and Quézel 1999; Blondel and Aronson 1999; Myers et al. 2000; Beard et al. 2000; Cox and Underwood 2011; Mittermeier et al. 2011; and Burge et al. 2016

MTE region	Area (km^2)	Vascular plants (#)	Endemic vascular plants (#)	Plant diversity (# spp. per km^2)	% Endemic plants	Vertebrates (#)	Endemic Vertebrates (#)	Animal Diversity (# spp. per km^2)	% Endemic vertebrates
Australia	417,786	7347	4764	0.019	66	605	100	0.0012	22
	(309,840–802,523)	(5469–8451)	(3000–6000)	(0.010–0.026)	(53–79)	(456–754)	–	(0.0009–0.0015)	–
California	311,085	4973	1977	0.016	39	536	71	0.0023	12
	(176,425–324,000)	(4300–6927)	(1505–2612)	(0.012–0.024)	(30–48)	(487–584)	–	(0.0018–0.0028)	–
Chile	169,877	2657	935	0.015	32	273	56	0.0016	18
	(140,000–300,000)	(2400–3429)	(552–1605)	(0.011–0.017)	(23–47)	(198–335)	(51–61)	(0.0011–0.0024)	(18–18)
Mediterranean	2,303,382	26,000	12,625	0.011	48	927	248	0.0004	27
	(2,000,000–2,777,778)	(25,000–30,000)	(12,500–13,000)	(0.011–0.013)	(42–52)	(770–1091)	(235–261)	(0.0003–0.0005)	(24–31)
South Africa	128,595	10,594	6673	0.087	64	779	65	0.0062	8
	(74,000–201,212)	(8200–15,356)	(5623–8649)	(0.070–0.111)	(56–75)	(562–1034)	(45–98)	(0.0056–0.0076)	(5–9)

occur only within a specific region and therefore have traits that allow them to persist within a unique area. Endemism may be the result of habitat loss or a failure to adapt to changing conditions. Many endemic species represent formerly widespread species that now have a much reduced distribution; this type of population distribution is called relictual. There are several well-known MTE species with relictual distributions, including the Giant Sequoia (*Sequoiadendron giganteum*) of California and Sugar Palm (*Jubaea chilensis*) of Chile (Figure 3.5). Endemism may also result from local adaptations to specific habitats with narrow distributions often in species that have only diverged recently, i.e. 'neoendemics'. In these types of endemic taxa, ranges were never large and species are presumably limited to a specific habitat because the traits that allow them to successfully compete within their endemic range preclude them from successfully competing outside of it. Species that are endemic to MTEs or to habitats within MTEs, particularly species that are non-relictual endemics, may be informative in identifying traits that are adaptations to MTEs. Endemism is particularly prevalent within MTEs (Table 3.2).

There are some common features among endemic taxa that may be useful in identification of adaptive traits to MTCs. Adaptations to special soils as well as slightly different habitats created by topographical and climatic heterogeneity are linked to endemism across MTEs (Cowling et al. 1996; Poot and Lambers 2003; Thompson et al. 2005; Burge et al. 2011; Molina-Venegas et al. 2017). Many endemic plant species tend to be woody, evergreen sclerophyllous species (Beard et al. 2000) and to be obligate post-fire seeding species (Wells 1969). Consistent with this, studies that have developed profiles for the most common traits of endemic plant species in South Africa found that endemic taxa are most likely to be non-resprouting shrubs with soil-stored seeds that have limited seed dispersal distances (Cowling and Holmes 1992; McDonald and Cowling 1995).

3.3 Ecological and evolutionary context

Many different potentially adaptive traits have been examined within MTEs. This includes examination of general adaptations to the key drivers of MTEs that are described above, as well as more specific studies of organisms and their ecological and evolutionary interactions. Below we briefly discuss some important ecological and evolutionary topics of study within MTEs as related to organismal adaptations. The following are not meant to be an exhaustive representation of the ecological and evolutionary interactions that occur within MTE communities, but rather a brief introduction to some of the interactions that occur, especially those that have been of particular interest.

3.3.1 Edaphic communities

Edaphic communities are those that are primarily determined by features of the soil, such as particularly poor mineral nutrition, unique soil textures or depths, or soils containing unique mineral compositions, often including metals or other soil components that are toxic to many plants. As mentioned above, many MTE endemic species are associated with specialized soils and several different types of edaphic communities are found within MTEs (Table 3.3). Some have suggested that one of the advantages of growing on unique soils is that few species can tolerate them, which leads to lower plant densities and increased water availability for the plants that occur on them (Poot and Lambers 2003). Increased access to and decreased competition for soil moisture may be particularly advantageous in seasonally dry MTCs.

3.3.2 Animal and plant interactions

Animals and plants interact in many ways and these interactions have shaped the features of organisms. Of particular research importance within MTEs has been studies of plant–herbivore interactions, plant–pollinator interactions, and plant–seed disperser interactions (Arianoutsou and Groves 1994). These interactions shape many plant and animal characters.

Table 3.3 Edaphic communities in mediterranean-type climate regions. Examples of some different edaphic soil types and communities from mediterranean-type ecosystems are provided below along with the unique features of the soil of each community and sources describing them within each region. These specialized communities contain many endemic species. Regions may contain additional types of edaphic communities in addition to those listed here

Region	Edaphic soil type or community	Unique soil feature(s)	Source(s)
Australia	Ironstone winter wetlands	Seasonally wet and shallow	Gibson et al. 2000; Poot and Lambers 2003
	Granite outcrops and inselbergs	Shallow and rocky	Hopper et al. 1997; Sander and Wardell-Johnson 2011
California	Serpentine	High in magnesium and low in calcium	Harrison 1999; Safford et al. 2005
	Gabbro	High in both magnesium and iron	Schoenherr 1992; Medeiros et al. 2015
Mediterranean	Gypsum (Spain)	High in calcium and sulfate	Mota et al. 2009; Martínez-Hernández, et al. 2015
	Serpentine (Turkey)	High in magnesium and low in calcium	Stevanović et al. 2003; Adıgüzel and Reeves 2012
South Africa	Quartz fields	Shallow and sometimes saline	Schmiedel and Jürgens 1999
	Limestone	Alkaline	Willis et al. 1996

3.3.2.1 Plants and herbivores

Plants have evolved many types of defence traits in response to herbivores (Hanley et al. 2007). In MTC regions, defence of plant tissues may be particularly important because of low soil nutrition, which makes loss of tissues especially costly. Plant defences may include physical defences, such as hairs, spines, or sclerophyllous leaves, chemical defences, such as toxins, or colouration, particularly of fruit and new leaves, that may make them less apparent to herbivores. Examples of all of these traits are found in all MTC regions and some examples of these follow.

An interesting example of plant defence traits and the **co-evolution** of plants and their herbivores is found in Australia. *Gastrolobium* and *Oxylobium* species (Fabaceae) produce a toxin, fluoroacetate, as a herbivore deterrent. For these species, the toxin is their primary defence and they lack common types of physical defences that characterize many species (Twigg and Socha 1996). Native mammals have evolved the ability to tolerate the toxin and this is an example of the co-evolution between these interacting species (Twigg and King 1991). This pattern is referred to as co-evolutionary, because direct interactions between the plants and herbivores have shaped one another's traits over time, such that increased toxicity of the plant selects for increased animal tolerance, which in turn favours more toxic plants in what is called an 'evolutionary arms race'. In contrast to the native animals, introduced mammals are not able to tolerate fluoroacetate (Twigg and King 1991). This has provided the opportunity for the use of this local plant toxin in controlling populations of introduced mammals through baiting with reduced potential for harm of native species (Kinnear et al. 2010).

Many MTC region plants produce spines, thorns, or other similar physical defence structures (Figure 3.6; see Plate 5; Hanley et al. 2007). These are particularly common in species that produce associations with nitrogen-fixing root bacteria, presumably because their increased nitrogen makes them a favoured target, but are also present in other species as well. The most common example of armed, nitrogen-fixing species within MTEs are the acacias as well as other Fabaceae, but many members of the Rosaceae and Rhamnaceae also fix nitrogen and display defensive structures.

Defensive traits may also extend beyond protection of foliage and to the protection of seeds. Animals that feed on seeds are termed granivores, and there is evidence that selection to avoid seed loss to granivores has been important in some MTC regions. For example, in Australia, the presence of granivorous birds that can consume seeds that are not strongly protected has impacted seed and fruit traits in *Hakea*. For seeds stored in the canopy (serotinous) to survive long inter-fire periods, they need to be strongly protected from granivores and this has selected for increased seed defence traits, such as extremely thick woody fruits, in some species (Lamont et al. 2016). In the Mediterranean, selection for the production of deceptive fruits

Figure 3.6 **Plant physical defence structures related to herbivory.** Within mediterranean-type ecosystems, many species exhibit herbivory defensive traits, including thorns, spines, hairs, and sclerophylly. Some examples of these include the glandular hairs found on *Acacia denticulosa*, Australia (a), thorns on *Acacia caven*, Chile (b), thorns on *Euphorbia acanthothamnos*, Mediterranean (c), thorns on *Rhamnus oleoides*, Mediterranean (d), thorns on *Ceanothus spinosus*, California (e), and spiny leaves on *Cliffortia*, South Africa (f). (See Plate 6)

Source: Photos from Anna L. Jacobsen.

in *Pistacia* appears to be driven by selection to avoid seed loss to granivorous birds (Verdú and García-Fayos 2001).

3.3.2.2 Plants and pollinators

Interactions between plants and their pollinators have long been of interest and was the topic of one of the earliest work on co-evolution (Darwin 1862). The vast majority of flowering plants are pollinated by animals (>87 per cent; Ollerton et al. 2011). MTE animal pollinators most commonly include insects and birds, as well as some mammals. However, animal pollinators within MTEs are diverse and have also been reported to include other animals such as arthropods (Collins and Rebelo 1987) and lizards (Pérez-Mellado and Casas 1997).

Animal-pollinated plants may form flowers that are highly specialized for a specific type of pollinator or flowers that appeal to a wide array of generalist pollinators. Specialized flowers may have specific displays, colours, scents, shapes, or rewards, such as pollen or nectar, that make them particularly attractive or available to specific pollinators. Similarly, animal pollinators may be generalists, and visit a wide range of flower types, or may be specialized to a specific type of flower.

In MTC regions, most plants and pollinators are generalists and the most common pollinators are insects and birds (Figure 3.7; see Plate 6; Herrera 1988; Petanidou and Ellis 1993; Struck 1994b; Bell 1994; Petanidou and Potts 2006). This has been suggested to be the result of seasonally variable climates within MTC regions that make the ability to forage broadly advantageous for pollinators, so that by switching between species flowering at different times they are able to increase their activity across more of the year. This also increases the pool of pollinators available to plants, especially in years where pollinators may be limited due to fire or climatic variability.

While specialists are generally rare within MTC regions, there are still many examples of tightly interacting and co-evolved specialist plant–pollinator interactions. As an example, in the California winter-rainfall Mojave Desert, the iconic Joshua tree (*Yucca brevifolia*) has co-evolved with obligate pollinating yucca moths (*Tegeticula* spp.; Godsoe et al. 2008). Across MTEs, many orchid species have evolved with specialized pollinators. Specificity in the pollination of orchids has been examined in the Mediterranean (Paulus 2006; Cozzolino and Scopece 2008), Australia (Phillips et al. 2011), South Africa (Johnson and Steiner 2003; Pauw 2006), and Chile (Lehnebach and Riveros 2003).

The interactions between plants and their pollinators have not been studied in great detail for many species within MTEs. Increased knowledge of the specificity of plant–pollinator interactions may be particularly important in identifying species at risk to changes in climate. While species with generalist interactions may be relatively robust to changes in climate, highly specific plant–pollinator interactions may be highly susceptible to perturbations that impact the nature or timing of their interactions (Bond 1995; Hegland et al. 2009).

3.3.2.3 Plants and seed dispersers

Animals disperse the seeds from many different types of plants, and the abundance of plant species within communities is influenced by dispersal

Figure 3.7 **Flowers may be suited for generalist or specialist pollinators.** Examples are shown of generalist insect flowers and pollinators from Chile (a), California (b), and the Mediterranean (d), bird pollinators from South Africa (c) and Australia (e), and an example of a flower from South Africa that is rodent-pollinated (f). Rodent pollination in *Protea amplexicaulis* (shown in panel f) was described by Wiens (1978). (See Plate 7)

Source: Photos a, b, d, e, and f from Anna L. Jacobsen and photo c from R. Brandon Pratt.

traits (Azcárate et al. 2002). In MTEs, studies have examined the roles of many different types of animals, including mammals, birds, and insects, as seed dispersers. Dispersal by vertebrates may be particularly important for long-distance dispersal (Herrera 1989; Calviño-Cancela et al. 2006). Both plants and their dispersers influence one another; a study focusing on bird dispersal in the Mediterranean found that birds were particularly important agents of seed dispersal and this interaction has shaped plant fruit traits and phenology as well as bird traits, such as their foraging behaviour and tolerance of plant toxins (Herrera 1995).

Of particular interest in MTEs has been the prevalence of **myrmecochory**, the dispersal of seeds by ants. Plant species that are ant-dispersed often produce seeds that contain specialized structures, such as elaiosomes, that are attached to the seed and provide a nutrient reward for the ants. Dispersal distances are relatively short in myrmecochorus species (Gómez and Espadaler 1998), which may limit gene flow between populations and be a source of some of the high levels of diversity found in MTEs. Myrmecochory appears to be particularly prevalent in Australia and South Africa (Milewski and Bond 1982; Hoffmann and Armesto 1995) and occurs mostly in shrub species with small dry fruits; however, myrmecochory also occurs in other MTEs, is prevalent in both the northern and southern hemispheres (Gómez and Espadaler 1998), and can also be important for plant species with fleshy fruits (Aronne and Wilcock 1994). Seed dispersal by ants has been reported in all MTEs: Australia (Berg 1975), California (Holway and Suarez 2006), the Mediterranean (Espadaler and Gómez 1996; Quilichini and Debussche 2000), and South Africa (Holmes and Newton 2004), and it is present, but rarer, in Chile (Stacey 2011).

Literature cited

Adıgüzel N, Reeves RD. 2012. Important serpentine areas of Turkey and distribution patterns of serpentine endemics and nickel accumulators. Bocconea 24: 7–17.

Alhamad MN. 2006. Ecological and species diversity of arid Mediterranean grazing land vegetation. Journal of Arid Environments 66: 698–715.

Alhamad MN, Oswald BP, Bataineh MM, Alrababah MA, Al-Gharaibeh MM. 2010. Relationships between herbaceous diversity and biomass in two habitats in arid Mediterranean rangeland. Journal of Arid Environments 74: 277–83.

Anderson RC, Adams DE. 1981. Flowering patterns and production on a central Oklahoma grassland. Pages 232–5 in Ohio Biological Survey. Proceedings of the sixth North American prairie conference. Ohio State University, Columbus, Ohio, USA.

Arianoutsou M. 1998. Aspects of demography in post-fire Mediterranean plant communities of Greece. Pages 273–95 in Rundel PW, Montenegro G, Jaksic F, eds. Landscape disturbance and biodiversity in mediterranean-type ecosystems. Springer-Verlag, Berlin.

Arianoutsou M, Groves RH, eds. 1994. Plant-animal interactions in mediterranean-type ecosystems. Springer Science+Business Media, Dordrecht.

Armesto JJ, Vidiella PE, Gutiérrez JR. 1993. Plant communities of the fog-free coastal desert of Chile: plant strategies in a fluctuating environment. Revista Chilena de Historia Natural 66: 271–82.

Arnan X, Rodrigo A, Retana J. 2006. Post-fire recovery of Mediterranean ground ant communities follows vegetation and dryness gradients. Journal of Biogeography 33: 1246–58.

Aronne G, Wilcock CC. 1994. First evidence of myrmecochory in fleshy-fruited shrubs of the Mediterranean region. New Phytologist 127: 781–8.

Arroyo MTK, Armesto JJ, Villagran C. 1981. Plant phenological patterns in the high Andean Cordillera of central Chile. Journal of Ecology 69: 205–23.

Arroyo MTK, Cavieres L, Marticorena C, Muñoz-Schick M. 1995. Convergence in the Mediterranean floras in central Chile and California: insights from comparative biogeography. Pages 43–88 in Arroyo MTK, Zedler PH, Fox MD, eds. Ecology and biogeography of Mediterranean ecosystems in Chile, California, and Australia. Springer, New York, USA.

Arroyo MTK, Matthei O, Muñoz-Schick M, et al. 2005. Flora of four biological reserves in the coastal range of the VII region (35°–36°S) and their relevance for the conservation of regional biodiversity. Pages 237–260 in Smith-Ramírez C, Armesto JJ, Valdovinos C (Eds) Historia, biodiversidad y ecología de los bosques costeros de Chile. Editorial Universitaria, Santiago, Chile.

Azcárate FM, Sánchez AM, Arqueros L, Peco B. 2002. Abundance and habitat segregation in Mediterranean grassland species: the importance of seed weight. Journal of Vegetation Science 13: 159–66.

Beard JS, Chapman AR, Gioia P. 2000. Species richness and endemism in the Western Australian flora. Journal of Biogeography 27: 1257–68.

Bell DT. 1994. Plant community structure in southwestern Australia and aspects of herbivory, seed dispersal and pollination. Pages 63–70 in Arianoutsou M, Groves RH, eds. 1994. Plant-animal interactions in mediterranean-type ecosystems. Springer Science+Business Media, Dordrecht.

Bellingham PJ, Sparrow AD. 2000. Resprouting as a life history strategy in woody plant communities. Oikos 89: 409–16.

Berg RY. 1975. Myrmecochorous plants in Australia and their dispersal by ants. Australian Journal of Botany 23: 475–508.

Blondel J, Aronson J. 1999. Biology and wildlife of the Mediterranean region. Oxford University Press, USA.

Bond WJ. 1995. Effects of global change on plant–animal synchrony: implications for pollination and seed dispersal in mediterranean habitats. Pages 181–202 in Moreno J, Oechel WC, eds. Global change and mediterranean-type ecosystems, Springer, New York.

Bond WJ, Midgley JJ. 2003. The evolutionary ecology of sprouting in woody plants. International Journal of Plant Sciences 164: S103–14.

Boulter SL, Kitching RL, Howlett BG. 2006. Family, visitors and the weather: patterns of flowering in tropical rain forests of northern Australia. Journal of Ecology 94: 369–82.

Bradshaw SD, Dixon KW, Hopper SD, Lambers H, Turner SR. 2011. Little evidence for fire-adapted plant traits in mediterranean climate regions. Trends in Plant Science 16: 69–76.

Brockwell J, Gault RR, Morthorpe LJ, Peoples MB, Turner GL, Bergersen FJ. 1989. Effects of soil nitrogen status and rate of inoculation on the establishment of populations of *Bradyrhizobium japonicum* and on the nodulation of soybeans. Crop and Pasture Science 40: 753–62.

Brotons L, Pons P, Herrando S. 2005. Colonization of dynamic Mediterranean landscapes: where do birds come from after fire? Journal of Biogeography 32: 789–98.

Buide ML, Sánchez JM, Guitián J. 1998. Ecological characteristics of the flora of the Northwest Iberian Peninsula. Plant Ecology 135: 1–8.

Burge DO, Erwin DM, Islam MB, et al. 2011. Diversification of Ceanothus (Rhamnaceae) in the California Floristic Province. International Journal of Plant Sciences 172: 1137–64.

Burge DO, Thorne JH, Harrison SP. 2016. Plant diversity and endemism in the California Floristic Province. Madroño 63: 3–206.

Cain SA, de Oliveira Castro GM, Pires JM, da Silva NT. 1956. Application of some phytosociological techniques to Brazilian rain forest. American Journal of Botany 43: 911–41.

Calviño-Cancela M, Dunn RR, Van Etten EJB, Lamont BB. 2006. Emus as non-standard seed dispersers and their potential for long-distance dispersal. Ecography 29: 632–40.

Charest R, Brouillet L, Bouchard A, Hay S. 2000. The vascular flora of Terra Nova National Park, Newfoundland, Canada: a biodiversity analysis from a biogeographical and life form perspective. Canadian Journal of Botany 78: 629–45.

Cilliers CD, Botha A, Esler KJ, Boucher C. 2005. Effects of alien plant management, fire and soil chemistry on selected soil microbial populations in the Cape Peninsula National Park, South Africa. South African Journal of Botany 71: 211–20.

Cody ML, Mooney HA. 1978. Convergence versus nonconvergence in mediterranean-climate ecosystems. Annual Review of Ecology, Evolution, and Systematics 9: 265–321.

Collins BG, Rebelo T. 1987. Pollination biology of the Proteaceae in Australia and southern Africa. Austral Ecology 12: 387–421.

Correia OA, Martins AC, Catarino FM. 1992. Comparative phenology and seasonal foliar nitrogen variation in Mediterranean species of Portugal. Ecologia Mediterranea 18: 7–18.

Cowling RM, Campbell BM. 1980. Convergence in vegetation structure in the mediterranean communities of California, Chile and South Africa. Plant Ecology 43: 191–7.

Cowling RM, Hilton-Taylor C. 1997. Phytogeography, flora and endemism. Pages 43–61 in Cowling RM, Richardson DM, Pierce SM, eds. Vegetation of Southern Africa. Cambridge University Press, UK.

Cowling RM, Holmes PM. 1992. Endemism and speciation in a lowland flora from the Cape Floristic Region. Biological Journal of the Linnean Society 47: 367–83.

Cowling RM, Esler KJ, Rundel PW. 1999. Namaqualand, South Africa–an overview of a unique winter-rainfall desert ecosystem. Plant Ecology 142: 3–21.

Cowling RM, Rundel P, Lamont BB, Arroyo MTK, Arianoutsou M. 1996. Plant diversity in mediterranean-climate regions. Trends in Ecology and Evolution 11: 362–6.

Cox RL, Underwood EC. 2011. The importance of conserving biodiversity outside of protected areas in Mediterranean ecosystems. PLoS One 6(1): e14508.

Cozzolino S, Scopece G. 2008. Specificity in pollination and consequences for post-mating reproductive isolation in deceptive Mediterranean orchids. Philosophical Transactions of the Royal Society of London B: Biological Sciences 363: 3037–46.

Craine JM, Engelbrecht BM, Lusk CH, McDowell NG, Poorter H. 2012. Resource limitation, tolerance, and the future of ecological plant classification. Frontiers in Plant Science 3: 1–10.

Cruz C, Correia P, Ramos A, Carvalho L, Bago A, Loução MAM. 2008. Arbuscular mycorrhiza in physiological and morphological adaptations of Mediterranean plants. Pages 733–52 in Varma A, ed. Mycorrhiza, Springer, Berlin Heidelberg.

Darwin C. 1862. On the various contrivances by which British and foreign orchids are fertilised by insects: and on the good effects of intercrossing. London: J. Murray.

Debussche M, Garnier E, Thompson JD. 2004. Exploring the causes of variation in phenology and morphology in Mediterranean geophytes: a genus-wide study of Cyclamen. Botanical Journal of the Linnean Society 145: 469–84.

De Lillis M, Fontanella A. 1992. Comparative phenology and growth in different species of the Mediterranean maquis of central Italy. Pages 83–96 in Romaine F, Terradas J, eds. *Quercus ilex* L. ecosystems: function, dynamics and management, Springer, Netherlands.

Di Castri F. 1973. Soil animals in latitudinal and topographical gradients of mediterranean ecosystems. Pages 171–90 in di Castri F, Mooney HA, eds. Mediterranean type ecosystems: origin and structure. Springer-Verlag, Berlin.

Driscoll DA, Lindenmayer DB, Bennett AF, Bode M, Bradstock RA, Cary GJ, Clarke MF, Dexter N, Fensham R, Friend G, Gill M. 2010. Fire management for biodiversity conservation: key research questions and our capacity to answer them. Biological Conservation 143: 1928–39.

Elia M, Lafortezza R, Tarasco E, Colangelo G, Sanesi G. 2012. The spatial and temporal effects of fire on insect abundance in Mediterranean forest ecosystems. Forest Ecology and Management 263: 262–7.

Esler KJ, Rundel PW. 1999. Comparative patterns of phenology and growth form diversity in two winter rainfall deserts: the Succulent Karoo and Mojave Desert ecosystems. Plant Ecology 142: 97–104.

Esler KJ, Rundel PW, Vorster P. 1999. Biogeography of prostrate-leaved geophytes in semi-arid South Africa: hypotheses on functionality. Plant Ecology 142: 105–20.

Espadaler XT, Gómez C. 1996. Seed production, predation and dispersal in the Mediterranean myrmecochore *Euphorbia characias* (Euphorbiaceae). Ecography 19: 7–15.

Field C, Mooney HA. 1983. Leaf age and seasonal effects on light, water, and nitrogen use efficiency in a California shrub. Oecologia 56: 348–55.

Friend GR. 1993. Impact of fire on small vertebrates in mallee woodlands and heathlands of temperate Australia: a review. Biological conservation 65: 99–114.

Gamradt SC, Kats LB. 1997. Impact of chaparral wildfire-induced sedimentation on oviposition of stream-breeding California newts (*Taricha torosa*). Oecologia 110: 546–9.

Gibson N, Keighery G, Keighery B. 2000. Threatened plant communities of Western Australia. 1. The ironstone. Journal of the Royal Society of Western Australia 83: 1–11.

Gill DS, Mahall BE. 1986. Quantitative phenology and water relations of an evergreen and a deciduous chaparral shrub. Ecological Monographs 56: 127–43.

Goberna M, García C, Insam H, Hernández MT, Verdú M. 2012. Burning fire-prone Mediterranean shrublands: immediate changes in soil microbial community structure and ecosystem functions. Microbial Ecology 64: 242–55.

Godsoe W, Yoder JB, Smith CI, Pellmyr O. 2008. Coevolution and divergence in the Joshua tree/yucca moth mutualism. The American Naturalist 171: 816–23.

Goicoechea N, Merino S, Sánchez-Díaz M. 2004. Contribution of arbuscular mycorrhizal fungi (AMF) to the adaptations exhibited by the deciduous shrub *Anthyllis cytisoides* L. under water deficit. Physiologia Plantarum 122: 453–64.

Goldblatt P. 1997. Floristic diversity in the Cape flora of South Africa. Biodiversity and Conservation 6: 359–77.

Gómez C, Espadaler X. 1998. Myrmecochorous dispersal distances: a world survey. Journal of Biogeography 25: 573–80.

Gualtieri G, Bisseling T. 2000. The evolution of nodulation. Plant Molecular Biology 42: 181–94.

Gulmon SL. 1977. A comparative study of the grassland of California and Chile. Flora 166: 261–78.

Gustafsson L. 1994. A comparison of biological characteristics and distribution between Swedish threatened and non-threatened forest vascular plants. Ecography 17: 39–49.

Hadar L, Noy-Meir I, Perevolotsky A. 1999. The effect of shrub clearing and grazing on the composition of a Mediterranean plant community: functional groups versus species. Journal of Vegetation Science 10: 673–82.

Hanley ME, Lamont BB, Fairbanks MM, Rafferty CM. 2007. Plant structural traits and their role in anti-herbivore defence. Perspectives in Plant Ecology, Evolution and Systematics 8: 157–78.

Harrison S. 1999. Local and regional diversity in a patchy landscape: native, alien, and endemic herbs on serpentine. Ecology 80: 70–80.

He T, Lamont BB, Downes KS. 2011. Banksia born to burn. New Phytologist 191: 184–96.

Hegland SJ, Nielsen A, Lázaro A, Bjerknes AL, Totland Ø. 2009. How does climate warming affect plant-pollinator interactions? Ecology Letters 12: 184–95.

Herrera CM. 1989. Frugivory and seed dispersal by carnivorous mammals, and associated fruit characteristics, in undisturbed Mediterranean habitats. Oikos 55: 250–62.

Herrera CM. 1995. Dispersal systems in the Mediterranean: ecological, evolutionary, and historical determinants. Annual Review of Ecology and Systematics 26: 705–27.

Herrera J. 1988. Pollination relationships in southern Spanish Mediterranean shrublands. Journal of Ecology 76: 274–87.

Hoffmann AJ, Armesto JJ. 1995. Modes of seed dispersal in the mediterranean regions in Chile, California, and Australia. Pages 289–310 in Arroyo MTK, Zedler PH, Fox MD, eds. Ecology and biogeography of mediterranean ecosystems in Chile, California, and Australia. Springer, New York, USA.

Holmes PM, Newton RJ. 2004. Patterns of seed persistence in South African fynbos. Plant Ecology 172: 143–58.

Holway DA, Suarez AV. 2006. Homogenization of ant communities in mediterranean California: the effects of urbanization and invasion. Biological Conservation 127: 319–26.

Hopper SD, Brown AP, Marchant NG. 1997. Plants of Western Australian granite outcrops. Journal of the Royal Society of Western Australia 80: 141–58.

Iglesias J, Santos M, Castillejo J. 1996. Annual activity cycles of the land snail *Helix aspersa* Müller in natural populations in North-Western Spain. Journal of Molluscan Studies 62: 495–505.

Iraeta P, Salvador A, Díaz JA. 2013. Life-history traits of two Mediterranean lizard popu-lations: a possible example of countergradient covariation. Oecologia 172: 167–76.

Jacobsen AL, Pratt RB, Davis SD, Ewers FW. 2007. Cavitation resistance and seasonal hydraulics differ among three arid Californian plant communities. Plant, Cell and Environment 30: 1599–609.

Jacobsen AL, Pratt RB, Davis SD, Ewers FW. 2008. Comparative community physiology: non-convergence in water relations among three semi-arid shrub communities. New Phytologist 180: 100–13.

Jacobsen AL, Esler KJ, Pratt RB, Ewers FW. 2009. Water stress tolerance of shrubs in Mediterranean-type climate regions: convergence of fynbos and succulent karoo communities with California shrub communities. American Journal of Botany 96: 1445–53.

Johnson SD. 1993. Climatic and phylogenetic determinants of flowering seasonality in the Cape flora. Journal of Ecology 81: 567–72.

Johnson SD, Steiner KE. 2003. Specialized pollination systems in southern Africa: review article. South African Journal of Science 99: 345–8.

Keeley JE. 1993. Smoke-induced flowering in the fire-lily *Cyrtanthus ventricosus*. South African Journal of Botany 59: 638.

Keeley JE, Fotheringham CJ. 2003. Species–area relationships in Mediterranean-climate plant communities. Journal of Biogeography 30: 1629–57.

Keeley JE, Pausas JG, Rundel PW, Bond WJ, Bradstock RA. 2011. Fire as an evolution-ary pressure shaping plant traits. Trends in Plant Science 16: 406–11.

Keeley JE, Bond WJ, Bradstock RA, Pausas JG, Rundel PW. 2012. Fire in mediterranean ecosystems: ecology, evolution and management. Cambridge University Press, New York.

Kerby JL, Kats LB. 1998. Modified interactions between salamander life stages caused by wildfire-induced sedimentation. Ecology 79: 740–5.

Kinnear JE, Krebs CJ, Pentland C, Orell P, Holme C, Karvinen R. 2010. Predator-baiting experiments for the conservation of rock-wallabies in Western Australia: a 25-year review with recent advances. Wildlife Research 37: 57–67.

Kyparissis A, Grammatikopoulos G, Manetas Y. 1997. Leaf demography and photosynthesis as affected by the environment in the drought semi-deciduous Mediterranean shrub *Phlomis fruticosa* L. Acta Oecologica 18: 543–55.

Lambers H, Brundrett MC, Raven JA, Hopper SD. 2010. Plant mineral nutrition in ancient landscapes: high plant species diversity on infertile soils is linked to functional diversity for nutritional strategies. Plant and Soil 334: 11–31.

Lamont B. 1982. Mechanisms for enhancing nutrient uptake in plants, with particu-lar reference to mediterranean South Africa and Western Australia. The Botanical Review 48: 597–689.

Lamont BB. 1983a. Proteoid roots in the South African Proteaceae. Journal of South African Botany 49: 103–23.

Lamont BB. 1983b. Strategies for maximizing nutrient uptake in two Mediterranean eco-systems of low nutrient status. Pages 246–73 in Kruger FJ, Mitchell DT, Jarvis JUM, eds. Mediterranean-type ecosystems: the role of nutrients. Springer-Verlag, Berlin.

Lamont BB, Downes KS. 2011. Fire-stimulated flowering among resprouters and geophytes in Australia and South Africa. Plant Ecology 212: 2111–25.

Lamont BB, Pérez-Fernández M, Rodríguez-Sánchez J. 2015. Soil bacteria hold the key to root cluster formation. New Phytologist 206: 1156–62.

Lamont BB, Pérez-Fernández M. 2016. Total growth and root-cluster production by legumes and proteas depends on rhizobacterial strain, host species and nitrogen level. Annals of Botany 118: 725–732.

Lamont BB, Hanley ME, Groom PK, He T. 2016. Bird pollinators, seed storage and cockatoo granivores explain large woody fruits as best seed defense in *Hakea*. Perspectives in Plant Ecology, Evolution and Systematics 21: 55–77.

Lazaridou-Dimitriadou M, Saunders DS. 1986. The influence of humidity, photoperiod, and temperature on the dormancy and activity of *Helix lucorum* L. (Gastropoda, Pulmonata). Journal of Molluscan Studies 52: 180–9.

Lehnebach C, Riveros M. 2003. Pollination biology of the Chilean endemic orchid *Chloraea lamellata*. Biodiversity and Conservation 12: 1741–51.

Le Maitre DC, Brown PJ. 1992. Life cycles and fire-stimulated flowering in geophytes. Pages 145–60 in van Wilgen BW, Richardson DM, Kruger FJ, van Hensbergen HJ, eds. Fire in South African Mountain fynbos: ecosystem, community and species response at Swartsboskloof. Springer, Berlin.

Linder HP. 2005. Evolution of diversity: the Cape flora. Trends in Plant Science 10: 536–41.

Martínez-Hernández F, Mendoza-Fernández AJ, Pérez-García FJ, Martínez-Nieto MI, Garrido-Becerra JA, Salmerón-Sánchez E, Merlo ME, Gil C, Mota JF. 2015. Areas of endemism as a conservation criterion for Iberian gypsophilous flora: a multi-scale test using the NDM/VNDM program. Plant Biosystems 149: 483–93.

McDonald DJ, Cowling RM. 1995. Towards a profile of an endemic mountain fynbos flora: implications for conservation. Biological Conservation 72: 1–12.

Médail F, Quézel P. 1999. Biodiversity hotspots in the Mediterranean Basin: setting global conservation priorities. Conservation Biology 13: 1510–13.

Medeiros ID, Rajakaruna N, Alexander EB. 2015. Gabbro soil-plant relations in the California Floristic Province. Madroño 62: 75–87.

Milewski AV, Bond WJ. 1982. Convergence of myrmecochory in mediterranean Australia and South Africa. Pages 89–98 in Buckley RC, ed. Ant–plant interactions in Australia. Springer, Netherlands.

Mittermeier RA, Turner WR, Larsen FW, Brooks TM, Gascon C. 2011. Global biodiversity conservation: the critical role of hotspots. Pages 3–22 in Zachos FE, Habel JC, eds. Biodiversity hotspots: distribution and protection of conservation priority areas. Springer-Verlag, Berlin.

Molina-Venegas R, Aparicio A, Lavergne S, Arroyo J. 2017. Climate and topographical correlates of plant palaeo- and neoendemism in a mediterranean biodiversity hotspot. Annals of Botany 119: 229–38.

Montenegro G, Aljaro ME, Walkowiak A, Saenger R. 1981. Seasonality, growth and net productivity of herbs and shrubs of the Chilean matorral. Pages 135–41 in Conrad CE, Oechel WC, eds. Proceedings of the symposium on dynamics and management of mediterranean-type ecosystems; 1981 June 22–26; San Diego, CA. Gen. Tech. Rep. PSW- 58. Berkeley, CA. U.S. Department of Agriculture, Forest Service, Pacific Southwest Forest and Range Experiment Station.

Montenegro G, Gómez M, Díaz F, Ginocchio R. 2003. Regeneration potential of Chilean matorral after fire: an updated view. Pages 381–409 in Veblen TT, Baker WL, Montenegro G, Swetham TM, eds. Fire and climatic change in temperate ecosystems of the Western Americas. Springer, New York, USA.

Montenegro G, Ginocchio R, Segura A, Keeley JE, Gomez M. 2004. Fire regimes and vegetation responses in two mediterranean-climate regions. Revista Chilena de Historia Natural 77: 455–64.

Mooney HA. 1988. Lessons from mediterranean-climate regions. Pages 157–65 in Wilson EO, Peter FM, eds. Biodiversity. National Academy Press, Washington, DC.

Mooney HA, Dunn EL. 1970. Photosynthetic systems of Mediterranean-climate shrubs and trees of California and Chile. The American Naturalist 104: 447–53.

Mooney HA, Dunn EL, Shropshire F, Song L. 1970. Vegetation comparisons between the mediterranean climatic areas of California and Chile. Flora 159: 480–96.

Mooney HA, Parsons DJ. 1973. Structure and function of the California chaparral—an example from San Dimas. Pages 83–112 in di Castri F, Mooney HA, eds. 1973. Mediterranean type ecosystems: origin and structure. Springer-Verlag, Berlin.

Mooney HA, Parsons DJ, Kummerow J. 1974. Plant development in mediterranean climates. Pages 255–67 in Leith H, ed. Phenology and seasonality modeling. Springer, Berlin Heidelberg.

Moro MF, Lughadha EN, Araújo FS, Martins FR. 2016. A phytogeographical metaanalysis of the semiarid Caatinga domain in Brazil. The Botanical Review 82: 91–148.

Mota JF, Gómez PS, Calvente MEM, Rodríguez PC, Lumbreras EL, De la Cruz Rot M, Reyes FBN, Gallardo FM, Esteban CB, Labarga JMM, Ollero HS. 2009. Aproximación a la checklist de los gipsófitos ibéricos. Anales de Biología 31: 71–80.

Murillo AJ, Ruiz PE, Landrum LR, Stuessy TF, Barfuss MH. 2012. Phylogenetic relationships in *Myrceugenia* (Myrtaceae) based on plastid and nuclear DNA sequences. Molecular Phylogenetics and Evolution 62: 764–76.

Myers N, Mittermeier RA, Mittermeier CG, de Fonseca GAB, Kent J. 2000. Biodiversity hotspots for conservation priorities. Nature 403: 853–8.

Nautiyal MC, Nautiyal BP, Prakash V. 2001. Phenology and growth form distribution in an alpine pasture at Tungnath, Garhwal, Himalaya. Mountain Research and Development 21: 168–74.

Naveh Z, Whittaker RH. 1980. Structural and floristic diversity of shrublands and woodlands in northern Israel and other mediterranean areas. Plant Ecology 41: 171–90.

Nilsen ET, Muller WH. 1981. Phenology of the drought-deciduous shrub *Lotus scoparius*: climatic controls and adaptive significance. Ecological Monographs 51: 323–41.

Ollerton J, Winfree R, Tarrant S. 2011. How many flowering plants are pollinated by animals? Oikos 120: 321–6.

Ordoñez JC, Van Bodegom PM, Witte JPM, Wright IJ, Reich PB, Aerts R. 2009. A global study of relationships between leaf traits, climate and soil measures of nutrient fertility. Global Ecology and Biogeography 18: 137–49.

Panchetti F, Amori G, Carpaneto GM, Sorace A. 2004. Activity patterns of the common dormouse (*Muscardinus avellanarius*) in different Mediterranean ecosystems. Journal of Zoology 262: 289–94.

Papatheodorou EM. 2008. Responses of soil microbial communities to climatic and human impacts in Mediterranean regions. Pages 63–87 in Liu T-Z, ed. Soil ecology research developments. Nova Science Publishers, Inc, New York.

Parr CL, Chown SL. 2003. Burning issues for conservation: a critique of faunal fire research in Southern Africa. Austral Ecology 28: 384–95.

Parsons RF, Hopper SD. 2003. Monocotyledonous geophytes: comparison of south-western Australia with other areas of mediterranean climate. Australian Journal of Botany 51: 129–33.

Paula S, Arianoutsou M, Kazanis D, Tavsanoglu Ç, Lloret F, Buhk C, Ojeda F, Luna B, Moreno JM, Rodrigo A, Espelta JM. 2009. Fire-related traits for plant species of the Mediterranean Basin. Ecology 90: 1420.

Paulus HF. 2006. Deceived males: pollination biology of the Mediterranean orchid genus *Ophrys* (Orchidaceae). Journal Europäischer Orchideen 38: 303–53.

Pausas JG, Keeley JE. 2014. Evolutionary ecology of resprouting and seeding in fire-prone ecosystems. New Phytologist 204: 55–65.

Pausas JG, Verdú M. 2005. Plant persistence traits in fire-prone ecosystems of the Mediterranean basin: a phylogenetic approach. Oikos 109: 196–202.

Pauw A. 2006. Floral syndromes accurately predict pollination by a specialized oil-collecting bee (*Rediviva peringueyi*, Melittidae) in a guild of South African orchids (Coryciinae). American Journal of Botany 93: 917–26.

Pérez-Mellado V, Casas JL. 1997. Pollination by a lizard on a Mediterranean island. Copeia 3: 593–5.

Petanidou T, Ellis WN. 1993. Pollinating fauna of a phryganic ecosystem: composition and diversity. Biodiversity Letters 1: 9–22.

Petanidou T, Potts SG. 2006. Mutual use of resources in mediterranean plant–pollinator communities: how specialized are pollination webs. Pages 220–44 in Waser NM, ed. Plant-pollinator interactions: from specialization to generalization. University of Chicago Press.

Petanidou T, Ellis WN, Margaris NS, Vokou D. 1995. Constraints on flowering phenology in a phryganic (East Mediterranean shrub) community. American Journal of Botany 82: 607–20.

Pettit NE, Froend RH, Ladd PG. 1995. Grazing in remnant woodland vegetation: changes in species composition and life form groups. Journal of Vegetation Science 6: 121–30.

Phillips RD, Brown AP, Dixon KW, Hopper SD. 2011. Orchid biogeography and factors associated with rarity in a biodiversity hotspot, the Southwest Australian Floristic Region. Journal of Biogeography 38: 487–501.

Pizarro-Tobías P, Fernández M, Niqui JL, Solano J, Duque E, Ramos JL, Roca A. 2015. Restoration of a Mediterranean forest after a fire: bioremediation and rhizoremediation field-scale trial. Microbial Biotechnology 8: 77–92.

Poot P, Lambers H. 2003. Are trade-offs in allocation pattern and root morphology related to species abundance? A congeneric comparison between rare and common species in the south-western Australian flora. Journal of Ecology 91: 58–67.

Potts SG, Vulliamy B, Dafni A, Ne'eman G, O'toole C, Roberts S, Willmer P. 2003. Response of plant-pollinator communities to fire: changes in diversity, abundance and floral reward structure. Oikos 101: 103–12.

Pratt SD, Konopka AS, Murry MA, Ewers FW, Davis SD. 1997. Influence of soil moisture on the nodulation of post fire seedlings of *Ceanothus* spp. growing in the Santa Monica Mountains of Southern California. Physiologia Plantarum 99: 673–9.

Pratt RB, North GB, Jacobsen AL, Ewers FW, Davis SD. 2010. Xylem root and shoot hydraulics is linked to life history type in chaparral seedlings. Functional Ecology 24: 70–81.

Pratt RB, Jacobsen AL, Hernandez J, Ewers FW, North GB, Davis SD. 2012. Allocation tradeoffs among chaparral shrub seedlings with different life history types (Rhamnaceae). American Journal of Botany 99: 1464–76.

Procheş Ş, Cowling RM, du Preez DR. 2005. Patterns of geophyte diversity and storage organ size in the winter-rainfall region of southern Africa. Diversity and Distributions 11: 101–9.

Procheş Ş, Cowling RM, Goldblatt P, Manning JC, Snijman DA. 2006. An overview of the Cape geophytes. Biological Journal of the Linnean Society 87: 27–43.

Quilichini A, Debussche M. 2000. Seed dispersal and germination patterns in a rare Mediterranean island endemic (*Anchusa crispa* Viv., Boraginaceae). Acta Oecologica 21: 303–13.

Quinn RD. 1979. Effects of fire on small mammals in the chaparral. . Transactions of the Western Section of the Wildlife Society 15: 125–33.

Renner SS, Strijk JS, Strasberg D, Thébaud C. 2010. Biogeography of the Monimiaceae (Laurales): a role for East Gondwana and long distance dispersal, but not West Gondwana. Journal of Biogeography 37: 1227–38.

Rosenzweig ML. 1995. Species Diversity in Space and Time. Cambridge University Press, Cambridge.

Rundel PW. 1996. Monocotyledonous geophytes in the California flora. Madroño 43: 355–68.

Rundel PW, Arroyo MT, Cowling RM, Keeley JE, Lamont BB, Vargas P. 2016. Mediterranean biomes: evolution of their vegetation, floras, and climate. Annual Review of Ecology, Evolution, and Systematics 47: 383–407.

Safford HD, Viers JH, Harrison SP. 2005. Serpentine endemism in the California flora: a database of serpentine affinity. Madroño 52: 222–57.

Salleo S, Nardini A. 2000. Sclerophylly: evolutionary advantage or mere epiphenomenon? Plant Biosystems 134: 247–59.

Sánchez-Castro I, Ferrol N, Barea JM. 2012. Analyzing the community composition of arbuscular mycorrhizal fungi colonizing the roots of representative shrubland species in a mediterranean ecosystem. Journal of Arid Environments 80: 1–9.

Sander J, Wardell-Johnson G. 2011. Fine-scale patterns of species and phylogenetic turnover in a global biodiversity hotspot: implications for climate change vulnerability. Journal of Vegetation Science 22: 766–80.

Santos X, Cheylan M. 2013. Taxonomic and functional response of a mediterranean reptile assemblage to a repeated fire regime. Biological Conservation 168: 90–8.

Sauquet H, Ho SY, Gandolfo MA, Jordan GJ, Wilf P, Cantrill DJ, Bayly MJ, Bromham L, Brown GK, Carpenter RJ, Lee DM, Murphy DJ, Sniderman JM, Udovicic F. 2012. Testing the impact of calibration on molecular divergence times using a fossil-rich group: the case of *Nothofagus* (Fagales). Systematic Biology 61: 289–313.

Schaefer R. 1973. Microbial activity under seasonal conditions of drought in mediterranean climates. Pages 191–8 in di Castri F, Mooney HA, eds. Mediterranean type ecosystems: origin and structure. Springer-Verlag, Berlin.

Scheil AE, Köhler HR, Triebskorn R. 2011. Heat tolerance and recovery in mediterranean land snails after pre-exposure in the field. Journal of Molluscan Studies 77: 165–74.

Schmiedel U, Jürgens N. 1999. Community structure on unusual habitat islands: quartz-fields in the Succulent Karoo, South Africa. Plant Ecology 142: 57–69.

Schoenherr AA. 1992. A natural history of California. University of California Press, Oakland, California.

Sgardelis SP, Pantis JD, Argyropoulou MD, Stamou GP. 1995. Effects of fire on soil macroinvertebrates in a Mediterranean phryganic ecosystem. International Journal of Wildland Fire 5: 113–21.

Shmida A. 1981. Mediterranean vegetation in California and Israel: similarities and differences. Israel Journal of Botany 30: 105–23.

Simonetti JA. 1999. Diversity and conservation of terrestrial vertebrates in mediterranean Chile. Revista Chilena de Historja Natural 72: 493–50.

Specht RL. 1969. A comparison of the sclerophyllous vegetation characteristic of mediterranean type climates in France, California, and Southern Australia. I. Structure, morphology, and succession. Australian Journal of Botany 17: 277–92.

Specht RL, Rayson P. 1957. Dark Island heath (Ninety-Mile Plain, South Australia). I. Definition of the ecosystem. Australian Journal of Botany 5: 52–85.

Stacey LM. 2011. The impacts and spread of the Argentine ant (*Linepithema humile* Mayr) invasion in coastal sclerophyllous forests of Chile. Dissertation, University of Washington.

Streeter J, Wong PP. 1988. Inhibition of legume nodule formation and N2 fixation by nitrate. Critical Reviews in Plant Sciences 7: 1–23.

Stevanović V, Tan K, Iatrou G. 2003. Distribution of the endemic Balkan flora on serpentine I.–obligate serpentine endemics. Plant Systematics and Evolution 242: 149–70.

Struck M. 1994a. Flowering phenology in the arid winter rainfall region of southern Africa. Bothalia 24: 77–90.

Struck M. 1994b. Flowers and their insect visitors in the arid winter rainfall region of southern Africa: observations on permanent plots. Insect visitation behaviour. Journal of Arid Environments 28: 51–74.

Thompson JD, Lavergne S, Affre L, Gaudeul M, Debussche M. 2005. Ecological differentiation of Mediterranean endemic plants. Taxon 54: 967–76.

Torres P, Honrubia M. 1997. Changes and effects of a natural fire on ectomycorrhizal inoculum potential of soil in a *Pinus halepensis* forest. Forest Ecology and Management 96: 189–96.

Treurnicht M, Pagel J, Esler KJ, Schutte-Vlok A, Nottebrock H, Kraaij T, Rebelo AG, Schurr FM. 2016. Environmental drivers of demographic variation across the global geographical range of 26 plant species. Journal of Ecology 104: 331–42.

Twigg LE, King DR. 1991. The impact of fluoroacetate-bearing vegetation on native Australian fauna: a review. Oikos 61: 412–30.

Twigg LE, Socha LV. 1996. Physical versus chemical defence mechanisms in toxic *Gastrolobium*. Oecologia 108: 21–8.

Tyler C, Borchert M. 2003. Reproduction and growth of the chaparral geophyte, *Zigadenus fremontii* (Liliaceae), in relation to fire. Plant Ecology 165: 11–20.

Van Mantgem EF, Keeley JE, Witter M. 2015. Faunal responses to fire in chaparral and sage scrub in California, USA. Fire Ecology 11: 128–48.

Verdú M, García-Fayos P. 2001. The effect of deceptive fruits on predispersal seed predation by birds in *Pistacia lentiscus*. Plant Ecology 156: 245–8.

Veste M, Herppich WB, von Willert DJ. 2001. Variability of CAM in leaf-deciduous succulents from the Succulent Karoo (South Africa). Basic and Applied Ecology 2: 283–8.

Villagrán CM. 1995. Quaternary history of the mediterranean vegetation of Chile. Pages 3–20 in Arroy MTK, Zedler PH, Fox MD, eds. Ecology and biogeography of mediterranean ecosystems in Chile, California, and Australia. Springer-Verlag, Berlin, Germany.

Wells PV. 1969. The relation between mode of reproduction and extent of speciation in woody genera of the California chaparral. Evolution 23: 264–7.

Werner C, Correia O, Beyschlag W. 1999. Two different strategies of mediterranean macchia plants to avoid photoinhibitory damage by excessive radiation levels during summer drought. Acta Oecologica 20: 15–23.

Westman WE. 1981. Seasonal dimorphism of foliage in Californian coastal sage scrub. Oecologia 51: 385–8.

Wiens D. 1978. Rodent pollination in southern African Protea spp. Nature 276: 71–3.

Willis CK, Cowling RM, Lombard AT. 1996. Patterns of endemism in the limestone flora of South African lowland fynbos. Biodiversity & Conservation 5: 55–73.

4 Diversity and Community Structure

Abstract

Similar drivers across mediterranean-type climate (MTC) regions have selected for plants with comparable traits (Chapters 3 and 6). This has resulted in the assembly of MTC region plant communities (groups of interacting populations of organisms) that are broadly similar, both structurally and functionally. However, there are also many different types of communities within MTC regions that may be quite divergent from one another even though they occur within a single region. These differences in community types also reveal both similarities and differences among the different MTC regions. This chapter briefly introduces and describes the composition and structure of the main types of plant communities that are found within MTC regions, including the iconic evergreen sclerophyllous shrublands, known as mediterranean-type vegetation (MTV), and other types of shrublands (drought-deciduous soft-leaved shrubs, desert shrublands along arid margins), grasslands, woodlands, and forests.

4.1 Assembling plant communities

We have noted in earlier chapters that, despite floristic (and other) differences across regions, similar climatic and other drivers across mediterranean-type climate (MTC) regions (Chapter 2) have selected for plants with comparable traits (Chapter 3). This has resulted in MTC region plant communities (groups of interacting populations of organisms) that are broadly similar, both structurally and functionally. However, there are also many different types of communities within MTC regions and these communities may be quite divergent from one another even though they occur within a

The Biology of Mediterranean-Type Ecosystems. Karen J. Esler, Anna L. Jacobsen, and R. Brandon Pratt,
Oxford University Press (2018). © Karen J. Esler, Anna L. Jacobsen, and R. Brandon Pratt 2018.
DOI 10.1093/oso/9780198739135.001.0001

single region. These differences in community types also reveal both similarities and differences among the different MTC regions.

Plant communities within MTC regions result from many different **community assembly** processes. Community assembly processes describe how different species come to co-occur and the factors that influence their abundance within the community. There are multiple factors that influence community assembly, such as biogeographic history (lineages present in the total species pool and the environments from which they come), dispersal (ability of a species to arrive at a site), **environmental filters** (species ability to survive abiotic site conditions, termed **abiotic filters**), and internal dynamics (species remaining after biotic interactions, termed **biotic filters**). These factors are dependent on scale, and they may act simultaneously such that, at any particular site, a series of nested processes (i.e. 'filters') determine which species are able to coexist, and which are precluded. These filters ultimately determine community structure and shape the plant communities we recognize today as typically 'mediterranean' (i.e. mediterranean-type vegetation; MTV), but there are also other communities present in these regions that have been shaped by different factors.

In South Africa, for example, geology and soil type (**edaphic filters**) are strongly aligned to elevation gradients, establishing a template reflected in the plant communities that occur there (Rebelo et al. 2006). No such alignment of geological substrate with elevation exists in California, where plant communities are more closely aligned with moisture/temperature gradients (**climatic filters**) (Ackerly et al. 2014). The shared Gondwanan biogeographic histories of South Africa and Australia have resulted in common lineages, such as the family Restionaceae, which has no functional analogue elsewhere, and yet which forms distinct structural communities (e.g. restioid fynbos; Rebelo et al. 2006). The abundance of annuals and herbaceous species in the Mediterranean and Californian species pool means that grassland communities feature in these MTC regions, but are less common in other regions. Drought-deciduous soft-leaved shrubland communities are largely absent in South Africa (Rebelo et al. 2006) and Australia (Groom and Lamont 2015), whilst distinct post-fire ephemeral communities do not feature in Chilean matorral, likely because of the modern absence of a predictable natural source of fire in this system (Keeley et al. 2012b). The aim of this chapter is briefly to introduce and describe the composition, structure, and function of the main types of plant communities that are found within MTC regions.

Much MTC region research, including most comparative work, has focused on **evergreen sclerophyll shrublands** (see also Chapter 6). This is the iconic vegetation type associated with MTC regions, and is normally what we refer to as MTV. This was the vegetation that Schimper (1903) drew early attention to when he proposed his ideas on convergent evolution (see Chapter 1, Figure 1.4). Early on, Warming (1909), expanding on the work of Schimper, noted that these sclerophyllous shrublands were not uniform. He recognized four different vegetation communities ('formations') within the sclerophyllous shrublands. These included: 1) garrigue, an open vegetation with perennial herbs and subshrubs occurring in rocky areas; 2) tomillares, with 'suffrutescent' vegetation dominated mostly by mint species (Labiatae) in dry areas; 3) maqui, a closed canopy evergreen shrubland in moister areas; and 4) sclerophyllous forests in 'favoured localities', with the tree component varying between oaks, eucalyptus, acacia, Proteaceae, or *Sequoia sempervirens* depending on the region. Profiles illustrating spatial patterning in these communities variously emphasize the roles of elevation (tied to climate) and/or soil in understanding these patterns (Figure 4.1).

Depending on how MTC regions are defined, areas that are wetter, drier, or colder may also be included in regional delimitations (Chapter 2). A range of other communities falls within broadly defined MTC regions, including grasslands, shrubland communities dominated by drought-deciduous soft-leaved shrubs, desert shrublands along arid margins, and coniferous or deciduous woodland and forest communities mostly (but not always) in the wetter regions. Comparative work across these communities is less well developed and is an area of future research opportunity.

The level, detail, and approach to plant community classification varies considerably between regions (Table 4.1). Broadly speaking, plant community types can be grouped into four structural/functional categories—**grassland, shrubland, woodland**, and **forest**. Across MTC regions, these four broad community types occur in various proportions. In some cases, depending on how MTC regions are defined, some categories are largely absent (Figure 4.2). This chapter is structured around these broad community categories. Azonal riparian and wetland vegetation, which may also have a distinct mediterranean character, also occurs throughout MTC regions and is mentioned, since much needs to be learned about the ecology of these systems. Dallman (1998), in his book *Plant life in the world's mediterranean climates*, did not describe grasslands and included a category of coastal scrub when describing the main forms of vegetation in MTC regions. The latter category included coastal sage scrub (California), coastal matorral (Chile), strandveld (South Africa), kwongan (Australia), and garrigue (Mediterranean). Many of these communities we have instead included in our broadly inclusive 'shrubland' category. This is an illustration that, to some extent, these categories are relatively arbitrary. We have not tried to present a definitive classification of the diverse communities of MTC regions. Rather, we have sought to illustrate the breadth of communities within these regions, especially those that have often been overlooked because they are not MTV.

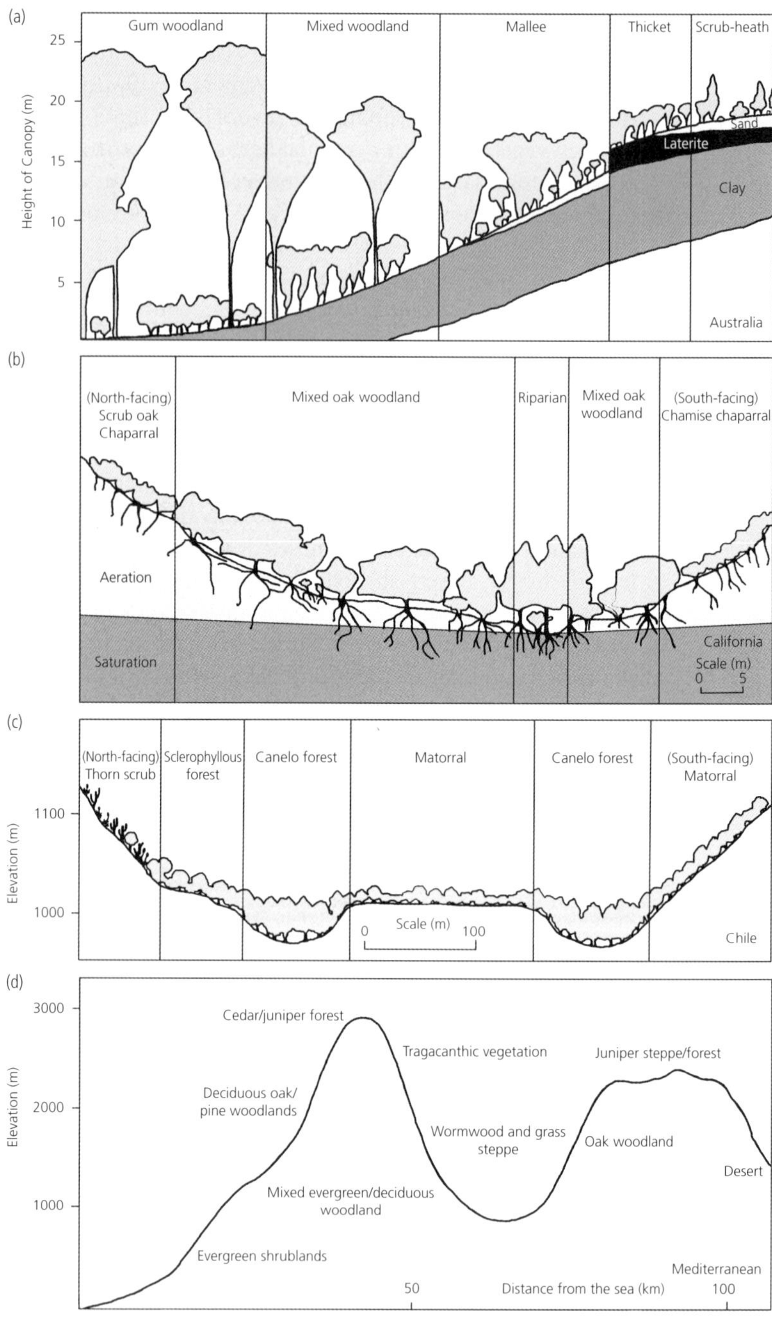

Figure 4.1 **Vegetation profiles from mediterranean-type climate regions.** Communities vary across the landscape at both smaller and larger spatial scales. This includes changes owing to soil properties (a), access to ground water (b), topography (c), and elevation and climatic gradients (d). (Modified from Beard 1983; Mooney and Parsons 1973; Mooney et al. 1977; Blondel et al. 2010.)

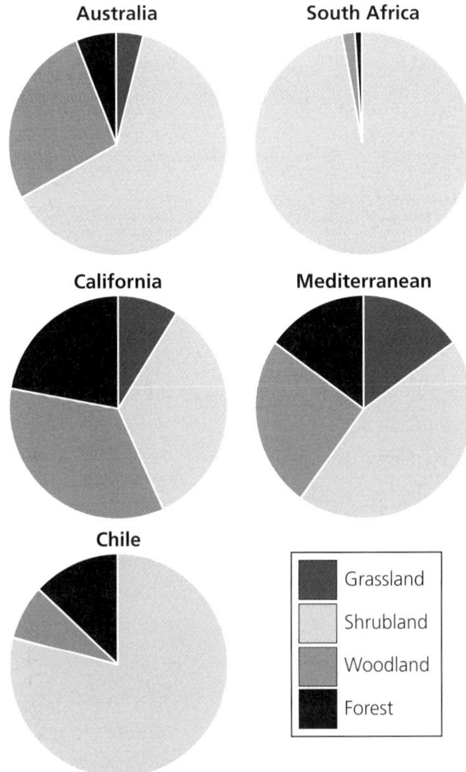

Figure 4.2   **The relative proportions (%) of plant community types in mediterranean-type climate (MTC) regions of the world.** Data sourced from Underwood et al. (2009) and Keeley et al. (2012). These sources defined MTC in the narrowest sense (see Chapter 2) and they differ from some analyses of MTC regions when more broadly considered; nevertheless, these data are a good example of how the abundance of different community types may vary across regions. See Table 4.1 for a list of communities that is based on a more inclusive analysis and which is based on a definition of MTC regions in the broadest sense.

Table 4.1 Examples of vegetation community types found in each mediterranean-type climate (MTC) region. There are many community types within each of the MTC regions and they vary in their extent and specifications according to broad local classifications. A sampling of the most common communities is included in this table. Characteristics and key families or species that are indicative of these example communities are included

Region/type	Name	Key families or species present/characteristic
Australia (Beard 1979; Mitchell and Wilcox 1994; Morton et al. 2011; Groom and Lamont 2015; Lamont and Keith 2017; B. Lamont, pers. comm.)		
Forest	Karri forest	*Eucalyptus diversicolor*; tall forest on fertile soils
	Northern and southern jarrah forest	*E. marginata* forest with some *E. wandoo* woodlands, sparse sclerophyllous understorey

Continued

Table 4.1 Continued

Region/type	Name	Key families or species present/characteristic
Woodland	Eucalypt woodlands—tuart, marri, wandoo	*E. gomphocephala*, *Corymba calophylla*, *E. wandoo* on less leached soils
	Eucalypt woodlands—York gum and salmon gum	*E. loxophleba*, *E. salmonophloia* woodland
	Mixed dry woodlands	Mixed eucalypt species, e.g. *E. transcontinentalis*
	Banksia low woodland	*Banksia* low woodland on leached sandy soils
	Acacia–casuarina thickets and scrub	*Acacia–Allocasuarina* thickets, especially *Acacia acuminata, A. aneura, Allocasuarina campestris*
	Mallee woodland/shrubland	Mallees are short eucalypts with many stems arising from a lignotuber (swollen rootstock)

Shrubland	Heathland (kwongan)
	  Geraldton-Swan sandplains
	  Esperance sandplains
	  Avon wheatbelt
		Myrtaceae (e.g. *Melaleuca*), Proteaceae (e.g. *Hakea*), Fabaceae (e.g. *Daviesia*), Ericaceae (e.g. *Leucopogon*) on sandy soils, sometimes laterite, limestone, quartzite

	Kwongan–Scrub heath	Myrtaceae, Proteaceae, Fabaceae, Ericaceae. On sandy soils

	Arid shrublands
	  Acacia shrubland
	  Chenopod shrubland
		Acacia spp., *Atriplex* spp. Acacia shrubland: mulga shrubland or low woodland in New South Wales, South Australia. Chenopod shrubland: Nullarbor Plain, chenopod plains, stony deserts on calcarosols, sodosols

Grassland	May occur as ground cover beneath tuart or York gum (or banksia woodland after fire), also clumped spinifex among acacia open shrubland	*Amphipogon*, *Neurachne* (perennial grasses), *Austrostipa* (post-fire annuals), *Triodia* (spinifex)

South Africa (Rebelo et al. 2006; Bergh et al. 2014)

Region/type	Name	Key families or species present/characteristic
Forest	Southern Afrotemperate forest	Podocarpaceae, *Afrocarpus*; multilayered, well developed low tree, shrub and herb layer
	Southern coastal forest	*Sideroxylon inerme*; dense, low-medium height, unlayered canopy
Woodland (thicket)	Subtropical thicket	*Portulacaria afra, Crassula ovata;* dense canopy of mostly evergreen, broad-leaved, spiny or succulent shrubs and low trees. Sparse understorey
	Western strandveld	*Searsia glauca, Euclea racemosa, Pterocelastrus tricuspidatus*; species-poor

Shrubland	Fynbos
	  Sandstone fynbos
	  Quartzite fynbos
		Ericaceae, Restionaceae, Proteaceae; low treeless evergreen 'heathland' on a range of nutrient-poor soils

	Sand fynbos
	  Shale fynbos
	  Fynbos shale band vegetation
	  Silcrete, ferricrete and conglomerate fynbos
		Classifications that take on a more management-related approach to community classification (e.g. Esler et al. 2014) refer to mountain fynbos, foothill fynbos, lowland fynbos, grassy fynbos

Region/type	Name	Key families or species present/characteristic
	Alluvium fynbos	
	Granite fynbos	
	Limestone fynbos	
	Renosterveld	*Elytropappus rhinocerotis, Eriocephalus africanus*; fine-leaved evergreen shrubland on relatively nutrient-rich soils, may have grasses
	Shale renosterveld	
	Granite and dolerite renosterveld	
	Alluvium renosterveld	
	Silcrete and limestone renosterveld	
	Succulent karoo	Mesembryanthemaceae; arid shrubland bordering fynbos, renosterveld and subtropical thicket
Grassland	Although grasses are present in South Africa, true grassland is restricted, but can occur in the eastern parts of the MTC region	*Cynodon dactylon, Cymbopogon popischilii, Themeda triandra*. Tussock grasslands and grazing lawns occur as alternative structural states of renosterveld shrubland. Their historical extent is debated. In the Eastern Cape, grassy fynbos on fine-textured higher nutrient soils with less summer drought is replaced by grassland on dry, inland north-facing slopes

California (Whittaker 1960; Barbour et al. 2007; North et al. 2016; Davis et al. 2016)

Region/type	Name	Key families or species present/characteristic
Forest	Coniferous forest	*Pinus ponderosa, P. jeffreyii, Pinus lambertiana, Abies concolor, Calocedrus decurrens, Pseudotsuga menziesii*, Pinaceae, Cupressaceae. Most common montane forest ecosystem. Forest types influenced by temperature and precipitation gradients; mountains above 2000 m
	Mixed conifer	
	Klamath mixed conifer	
	White fir	
	Montane hardwood-conifer (mixed evergreen)	
	Giant sequoia	
	Ponderosa, Jeffrey, and 'Eastside' pine	
	Broad-leaved evergreen forest	*Lithocarpus densiflora, Quercus chrysolepis, Arbutus menziesii, Umbellularia californica*. Transitional between sclerophyll shrublands and coniferous forest. Upper tree stratum of needle-leaved evergreen or coniferous species; lower tree stratum of broad-leaved evergreen/sclerophyllous species, with strata varying considerably in density
Woodland	Oak woodland/savanna	*Quercus* spp., *Pinus sabiniana, Aesculus californica*. Open canopy of scattered deciduous, evergreen, semi-evergreen trees. Herbaceous understory of exotic annual grasses, native and exotic forbs, and often scattered shrubs
	Southern coastal oak woodlands (*Q. agrifolia; Q. engelmannii*)	
	Interior foothill woodlands (*Q. wislizenii var wislizenii; Q. lobata*)	

Continued

Table 4.1 Continued

Region/type	Name	Key families or species present/characteristic
	Oregon white oak woodlands (*Q. garryana*)	
	Montane oak woodland (*Q. chrysolepis; Q. kelloggii*)	
	Riparian woodland	*Salix* spp., *Platanus racemosa*, *Alnus* spp., *Acer* spp., *Fraxinus* spp., *Populus* spp.
	Joshua tree woodland	*Yucca brevifolia*. Pinyon *Pinus edulus* and juniper *Juniperus californica* as codominants

Shrubland	Chaparral
	  Maritime
	  Serpentine
	  Montane
	  Mesic
	  Xeric
	  Chamise (*Adenostoma fasciculatum*)
	  Ceanothus (*Ceanothus* spp.)
	  Manzanita (*Arctostaphylos* spp.)
	  Scrub-oak (*Quercus dumosa*)
	  Red-shank (*Adenostoma sparsifolium*)
		California's most extensive plant community type. Evergreen sclerophyllous shrubland ranging from San Francisco (USA) south to Baja California. More xeric (dry) sites like south-facing slopes will be dominated by one of few species such as chamise and *Ceanothus* spp. while more mesic chaparral is often called 'mixed chaparral' and contains more species

	California sage scrub	*Artemisia californica, Salvia* spp. Seasonally dimorphic soft-leaved species (Chapter 3)
	Mojave desert scrub	*Larrea tridentata, Atriplex* spp., and *Yucca brevifolia*. Winter rainfall, high elevation desert
Grassland	Grassland	Poaceae; often disturbed communities dominated by forbs and non-native annual Mediterranean grasses; wild oats *Avena barbata* and *A. fatua*, brome, *Bromus* spp; rye grass *Lolium*. Native grasslands now disjunct fragments; Purple needle grass *Stipa pulchra*

Chile (Rundel 1981; Armesto et al. 2007; Specht et al. 2015; Porqueddu et al. 2016)

Forest	Canelo forest	*Drimys winteri, Cryptocarya alba*; in moist areas
	Olivio forest	*Aextoxicon punctatum*; relict coastal forest with climbers and epiphytes
	Nothofagus montane forest	*Nothofagus obliqua, N. glauca*; higher elevation sclerophyllous forest
	Maulino forest/Valdivian temperate rainforest	*Nothofagus glauca, N. alessandrii, N. dombeyi, Podocarpus saligna*; dominant community in south, has great growth form diversity
	Palm forest	*Jubaea chilensis*; mixed evergreen sclerophyllous forest with abundant endemic palm
	Sclerophyllous forest	*Peumus boldus*; relatively dry forest

Region/type	Name	Key families or species present/characteristic
Woodland	*Acacia caven* savanna	*Acacia caven*; xerophytic open woodland with dense herbaceous layer of introduced European annual herbs and grasses
	Andean montane woodland	*Kageneckia angustifolia*; above 1500 m, upland sclerophyllous woodland
	Sclerophyllous matorral	*Cryptocarya alba, Quillaja saponaria, Lithrea caustica*; sclerophyll woodland can reach 10–15 m in height. On mesic sites termed hygrophilous forest
Shrubland	Steppe matorral/summer-deciduous matorral with succulents	*Retanilla trinerva, Flourensia thurifera, Trichocereus, Eulychnia*; xeric vegetation with combination of deciduous shrubs present, in valleys and Andean slopes away from marine influence
	Coastal matorral	*Lithraea caustica, Bahia ambrosioides, Schinus latifolius, Escallonia pulverulenta*; *Puya chilensis*; evergreen sclerophyllous shrubland on slopes of coastal range. Seasonal spring flowering herbaceous strata
	Xerophytic thorn scrub	A combination of deciduous shrubs, such as *Trevoa trinervis* (Rhamnaceae), *Flouresia thurifera* (Asteraceae), and *Colliguaja odorifera* (Euphorbiaceae), and often includes succulent species such as *Puya* spp. and columnar cacti, *Echinopsis* and *Eulychnia* spp.
Grassland	Diverse grassland of annual (native and invasive) and perennial (mostly native) species, often associated with espinal	Flatland species: *Trisetum spicatum, Hordeum berteroanum, Plantago firma, Leontodon leysseri, Deschampsia berteroana, Parentucellia latifolia*; slope species *Bromus mollis, Stipa laevissima, Erodium botrys, Dichondra repens, Plantago hispidula, Trisetobromus hirtus*

Mediterranean (Tomaselli 1981; Moreno and Pulido 2008; Porqueddu et al. 2016; F. Ojeda, pers. comm.)

Region/type	Name	Key families or species present/characteristic
Forest	Broadleaved evergreen forest	*Quercus* spp.
	Coniferous forest	*Abies pinsapo*
	Semi-deciduous oak forest	*Quercus pubescens*
Woodland	Oak woodlands	*Quercus ilex* subsp. *ballota*, *Q. suber*, *Q. faginea*, *Q. canariensis*
	Wild olive tree woodlands 'acebuchales'	*Olea europea*
	Agrosilvopastural savanna	*Quercus ilex* subsp. *ballota*, *Q. suber*, annual herbs; of anthropogenic origin; 'Dehesas' (Spain) or 'Montados' (Portugal); tree density reduced, matorral cover eliminated, grass layer favoured by grazing and cropping
	Acacia–argania dry woodlands	*Pinus, Tetraclinis, Juniperus, Acacia, Pistacia, Argania, Haloxylon, Calligonum, Tamarix*; Morocco and Algerian North Sahara

Continued

Table 4.1 Continued

Region/type	Name	Key families or species present/characteristic
Shrubland	Matorral High matorral (= maquis, espinal, macchia alta); middle matorral (= garrigue, jaral) Low matorral (= phrygana, batha, Mediterranean heathlands, tomillares)	High (>2 m): *Olea oleaster, Ceratonia siliqua, Pistacia lentiscus,* sometimes with pines or conifers in upper strata. Middle (<2– 0.6 m): *Quercus coccifera, Cistus* spp., *Erica* spp., *P. lentiscus,* sometimes with pines or confers in upper strata. Low (<0.6 m): *Chamaerops humilis, Olea oleaster, P. lentiscus, Cistus* spp. Evergreen sclerophyllous shrublands of various combinations of height, cover, and structure
	Semi-desert	In Morocco and the Algerian North Sahara and inland from the coast, succulent thickets (stands of *Euphorbia*) occur
Grassland	Steppe/grassland	*Avena* spp., *Bromus* spp., *Triticum* spp., *Hordeum* spp. Long-term impact of cultivation and domestic livestock has converted original grassland to degraded forms dominated by annual spp. Grasslands occur throughout the more arid areas of southern and eastern Mediterranean basin, including Spain, South France, Algeria, Egypt, Israel, Libya, Morocco, Syria, Tunisia

4.2 Shrublands

4.2.1 Evergreen sclerophyll shrublands

It is the closed-canopy evergreen sclerophyll shrubland, with analogues in all five mediterranean-type ecosystems (MTEs) that hosts the greatest diversity of plant species (see Chapter 1, Figure 1.1). Known variously as kwongan or mallee (Australia), fynbos or renosterveld (South Africa), maquis, matorral, macchia or garrigue (Mediterranean), matorral (Chile), and chaparral (California), this community type is dominated by shrubs, generally plants around 2 m in stature with woody stems (multi- and single-stemmed). These shrublands have drawn considerable attention partly because at all spatial scales they host higher levels of biodiversity and endemism than any other terrestrial ecosystem outside the wet tropics (Cowling et al. 2015; Rundel et al. 2016). MTC region boundaries are not consistently agreed upon by all authors (Chapters 1 and 2), and large areas of structurally and sometimes floristically related MTV (often in a less species-rich form) fall into aseasonal or summer rainfall zones in South Africa (Killick 1979), Australia (Lamont and Keith 2017), and North America (Keeley et al. 2012a). The processes maintaining such vegetation outside of MTC regions may well be different to those in MTC regions (Case Study 7), a point also well discussed in Keeley et al. (2012a).

Case Study 7 Mediterranean-type vegetation outside of mediterranean-type climate regions

Philip W. Rundel, Department of Ecology and Evolutionary Biology, University of California, Los Angeles (UCLA), Los Angeles, California, USA

Mediterranean-type vegetation (MTV) is defined as the sclerophyllous-leaved, closed-canopy shrublands and woodlands that characterize the five mediterranean-type climate (MTC) regions of the world. However, structurally and floristically similar vegetation can be found well outside of the occurrence of classic MTC regions where edaphic factors outweigh climate regime in structuring communities. Although similar in appearance to MTV, the absence of an MTC regime suggests caution in assuming that ecosystem processes in these communities operate in the same way.

The South African Cape Floristic Region (CFR) presents a good example of MTV outside of the typical winter-rainfall and summer-drought climate regime characteristic of MTC. As traditionally defined, the CFR extends across the southern coast of South Africa as far east as Port Elizabeth, with the western half of the region exhibiting a classic MTC regime with winter-rainfall dominance. In its eastern half, however, there is an increasing component of summer rainfall moving eastward, producing an aseasonal rainfall regime (Bradshaw and Cowling 2014). The continued dominance of fynbos shrublands across this rainfall gradient is driven by the edaphic influence of the Table Mountain sandstone, which crosses the gradient. Fynbos structure and floristic composition at the generic level in the Eastern Cape is similar to that of the Western Cape and this MTV and its floristic and community structure drive the inclusion of this area as part of the CFR outside of the classic MTC region. In the wetter areas of the Eastern Cape fire frequency is an important mediator of fynbos distribution at the expense of afro-temperate forest and subtropical thicket communities, which can maintain dominance in fire-free areas (Bergh et al. 2014).

The MTC region of southwest and South Australia include a mix of communities in the form of sclerophyllous-leaved shrublands, woodlands, and forests. The classic MTV communities are the kwongan heathlands and mallee shrublands occurring on nutrient-poor substrates. As in South Africa, these forms of MTV are driven not only by climate but also by edaphic factors in their distribution. Beyond the MTC region in southern and eastern Australia there are widespread heathlands closely associated with oligotrophic landscapes on granitic uplands, sandstone, laterite, or quartzite plateaus, and leached coastal sandplains (Lamont and Keith 2017). The distribution of these heathlands is often decoupled from climatic conditions, with rainfall regimes varying from about 300 mm in southwest Australia to 3000 mm in cold temperate southwest Tasmania. Rainfall seasonality varies from a typical MTC to that of aseasonal rainfall in the southeast to tropical monsoonal climates with a strong peak of rainfall in summer. Although less rich in species than the heathlands of southwest Australia, these examples of MTV are often centres of endemism, as in heathlands on sandstone around Sydney, in the Blue Mountains, and Tasmania. Despite this range of climatic regimes, plant regeneration in these heathlands is driven by fire, although fire regimes differ from those of the kwongan heathlands.

A third example of MTV present outside an MTC region can be seen in components of California chaparral that extend eastward into biseasonal rainfall areas of Arizona and

Continued

Case Study 7 (*Continued*)

areas of northeastern Mexico with a summer-rainfall regime (Case Study 7 Figure). Despite the change in climate regime, Arizona chaparral shares a number of important shrub species with California, with several of these ranging into northeastern Mexico (Keeley et al. 2012a). Despite the shared shrub species, the overall structure of the summer rainfall chaparral ecosystems is quite distinct from chaparral, with these differences an outcome of the reversed seasonality of moisture availability. The warm summer-rainfall regime and open stand structure in Arizona results in a rich herbaceous community and a post-fire flora, including fewer species of annuals and herbaceous perennials with a large representation of C_4 grasses (Keeley et al. 2012b).

In addition to these cases of MTV outside of MTC regions, the opposite pattern also exists, with areas with an MTC supporting vegetation very different from MTV. This occurs at the arid end of MTC regions, where winter-rainfall deserts have been traditionally excluded owing to strong community and floristic changes (Le Houerou 2004), and in south-central Chile and the northern margin of the California Floristic Province where very wet temperate forests with a pronounced winter rainfall seasonality are excluded. Distinctly non-MTV is also present in southwestern Argentina where a classic MTC extends across the lower Andes and supports woodlands of *Nothofagus* and *Austrocedrus*.

Case Study 7 Figure. Mediterranean-type vegetation in mediterranean-type and non-mediterranean-type climate regions. Components of Californian chaparral (black) occur extensively in California where there is a mediterranean-type climate, but also extend eastward into biseasonal rainfall areas of Arizona and areas of northeastern Mexico with a summer-rainfall regime.

This water-limited community-type is comprised of species from diverse and distinct clades of various biogeographic origins (Chapter 5). Despite such diverse origins, the shared traits that organisms possess in this community type (also see Chapter 6), suggest that environmental filtering is a dominant factor shaping these communities. Indeed, evidence of such filtering has been established for coastal California communities (Cornwell and Ackerly 2009) and fynbos Proteaceae communities (Thuiller et al. 2004; Yates et al. 2010). Specifically, soil fertility has been suggested as a strong environmental filter leading to the dominance of lineages adapted to low soil fertility in South Africa and Australia (Stock and Verboom 2012), although Araya et al. (2011) demonstrated that fynbos plants can also segregate along fine-scale hydrological gradients.

Since observed absence in a community does not necessarily reflect environmental limitation (i.e. a species might be absent from a community for other reasons such as dispersal limitation, land-use history, or past environmental limitations), transplant experiments are used to test for how strongly environmental conditions limit species distributions. Transplant experiments across gradients have, however, yielded mixed results. In *Protea* transplant experiments, Carlson et al. (2011) found heritable trait differences associated with gradients of rainfall seasonality, cold stress and drought stress, but not strong effects of soil fertility, while Latimer et al.'s (2009) experiments suggested that abiotic limitation did not fully explain *Protea* distribution limits under current environmental conditions. Local adaptation is therefore not always evident (also see Richards et al. 1997; Silvertown et al. 2012), which has led to suggestions that edaphic (soil-related) niche occupancy may be mediated through competitive interactions (Silvertown et al. 2012) and/or fire (Esler et al. 2015).

At a local scale, variable evidence for **biotic filtering** is noted, with some studies finding it to be important in structuring communities (e.g. Cornwell and Ackerly 2009) and others not (e.g. Potts et al. 2011). The rapid expansion of DNA sequence data and the development of novel techniques to infer species histories has facilitated a deeper understanding of the relative roles of different filters in structuring the communities we encounter today (see, for example, Verboom et al. 2014). Our current understanding is that multiple factors have influenced community assembly in the different MTC regions.

Notwithstanding the overt structural and functional similarities between evergreen sclerophyll shrublands, there are still some key differences between MTC regions, and these differences become more apparent at finer scales. Structurally, mature chaparral tends to consist of a dense, impenetrable closed-canopy layer of tall (1.5 to >2 m) shrubs, with very little understory. Mature Mediterranean shrublands are also dense, but more variable in height (1.5–4 m), with some low trees (e.g. *Quercus coccifera*) also present.

In Chile, shrubs and trees are less dense with herbaceous species interspersed, while in South Africa (>2 m) and Australia (>2 m), the shrublands are relatively open and both an understorey and an overstorey are present. These structural distinctions reflect differences in soil fertility and variable fire-return intervals (FRIs) across MTC regions. For example, the relatively infertile landscapes of South Africa and Australia do not tend to support closed-canopy communities, so shrubs of different stature can co-occur in these diverse communities. Cowling and Gxaba (1990) argue that overstorey spatial and temporal (post-fire) variability is key to maintaining understorey species richness in proteoid fynbos in South Africa. In California and the Mediterranean longer FRIs and moderately fertile soils result in the development of closed-canopy shrublands and woodlands (Cowling et al. 1996). Life history characteristics of shrubs in response to fire and drought are also important factors (Marais et al. 2014).

Because of varying interactions between climate, geology, and soils, evergreen sclerophyll shrublands also vary considerably within individual MTC regions, and various levels of community classification exist. Within South Africa, for example, renosterveld (on relatively nutrient-rich soils) and fynbos (on nutrient-poor soils) occur on different soil types and differ floristically and ecologically (Bergh et al. 2014). Within fynbos, nine recognized community subtypes are defined by edaphic conditions, while within renosterveld four edaphic community subtypes are recognized (Rebelo et al. 2006). These community subtypes are even further divided into 81 (fynbos) and 29 (renosterveld) vegetation units, reflecting not only the lack of constant dominance across communities (Rebelo et al. 2006) and the exceptional beta and gamma diversity of the flora (Cowling 1990) but also the degree to which different areas have been sampled (Rebelo et al. 2006). Californian chaparral also demonstrates steep turnover in species composition, and relatively few shrub species (with the exception of *Adenostoma fasciculatum*) are widespread and dominant. Chaparral communities are not easily explained by any one factor (Keeley 2000), although soil moisture gradients associated with aspect, elevation, and distance from the ocean are strong determinants of community structure (Keeley and Davis 2007). Here, various authors have classified communities according to combined environmental factors (e.g. xeric south-slope chaparral, mesic north-slope chaparral, montane chaparral, etc.), dominance characteristics (e.g. *Ceanothus* chaparral, manzanita chaparral) and/or floristic associations (summarized by Keeley and Davis 2007). In Chile, subtypes are also identified based on dominance characteristics (e.g. matorral de trevo, matorral de chagual y quisco, matorral de chagualillo, and matorral de mira con maicillo) (Francioli and Muñoz 2009). The majority of heathlands in Australia occur in three Australian winter-rainfall provinces (accounting for 75 per cent of total heathland area) and are differentiated by geography and/or regional environmental gradients. Mostly associated with oligotrophic landscapes, they occur on a range of substrates (e.g. sandstones, granites, laterites) (Lamont and Keith 2017). Tomaselli (1981) describes fifteen types of matorral vegetation in the Mediterranean. These types are differentiated by geography and/or regional environmental gradients, but are also shaped by people's long association with this region.

4.2.2 Heathlands—a component of evergreen sclerophyll shrublands

Small-stature fine-leaved evergreen sclerophyllous species are an important component of MTV, particularly in South Africa, Australia, and the western Mediterranean. Historically, South African fynbos ('fine bush') was classified as a **heathland** (e.g. Specht 1979; Specht and Moll 1983), since fine-leaved 'ericoid' species (from the families Ericaceae, Asteraceae, Rhamnaceae, Thymelaeaceae, and Rutaceae) can be dominant, and these authors wished to distinguish fynbos from neighbouring renosterveld. Other authors preferred to reserve the term for specific ericaceous (order Ericales) communities (e.g. Cowling and Holmes 1992). While many MTC region communities may conform to the definition of a heathland, these are not all dominated by Ericales.

While there are heathland elements within the shrubland communities of MTC regions, heathlands also occur outside of these regions. Structurally and floristically related ericaceous heathlands occur along mountains beyond the South African MTC region northwards and as far as the Ethiopian highlands (Rebelo et al. 2006). Heathlands also occur beyond the Australian kwongan, associated with nutrient-poor soils (see Case Study 7, Lamont and Keith 2017). Finally, heathlands with relatively high levels of species richness and endemism are widespread in the western Mediterranean, and are a distinctive vegetation type (known locally as herriza) occurring on acid, nutrient-poor Oligo-Miocene sandstone-derived soils (Ojeda et al. 1995; 1996; 2000). Although some authors regard herriza as MTV, others regard it as an incursion into MTV of the more widespread temperate European Atlantic heathland (e.g. Loidi et al. 2010).

Despite their prominence in MTC regions, heathlands are widespread beyond these regions (Figure 4.3) and occur over an extensive geographic range, from the wet tropics to temperate zones to the subarctic and from coastal areas to mountain tops. This suggests that their broader occurrence and community assembly is not climatically constrained, but rather reflects adaptation to highly infertile soils, hydrological gradients over the landscape, and recurring intense fires (Lamont and Keith 2017). The particular association with nutrient-impoverished soils has led to suggestions that soil may be more important than climate in determining MTEs (Lambers 2014), although in all likelihood a unimodal explanation is unlikely. Regardless of their broader affiliations, it does appear that heathlands occurring within MTC regions

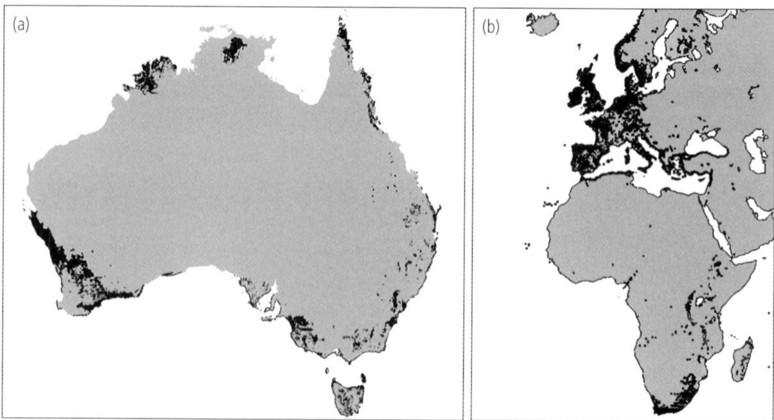

Figure 4.3 **Heathland occurrence in mediterranean-type climate (MTC) and non-MTC regions of Australia (a) and Europe–Africa (b).** Areas where heathland occurs are indicated in black. Despite their prominence in MTC regions, heathlands and related shrublands are widespread beyond MTC regions and occur over an extensive geographic range. Heathlands are evident as being prominent within southwestern Australia, where the MTC region occurs (a) as well as in the Mediterranean and South Africa (b) where MTC regions are located. (Panel a modified from Lamont and Keith 2017, panel b modified from Pirie et al. 2016.)

tend to possess higher levels of species richness, compared to relatively low levels of diversity in heathlands outside of MTC regions. This community type within MTC regions warrants far greater research attention, in particular in relation to patterns of heathland convergence (see Chapter 5, Case Study 9).

4.2.3 Drought-deciduous soft-leaved shrublands

In California and Chile, evergreen shrublands are replaced by soft-leaved **drought-deciduous** shrublands along the more arid (but non-desert) interior and coastal margins. Many of these species have **seasonally dimorphic** leaves in that they produce less sclerophyllous leaves in the wet winter/spring, with more sclerophyllous leaves produced at the end of the wet season that remain on the plant through the dry season. Called coastal sage scrub or California sage scrub in California, and coastal matorral in Chile, winter leaves are softer but drought-deciduous (e.g. sage brush *Artemisia californica* and black sage *Salvia mellifera* in California; tevo *Trevoa trinervis* and palhuén *Adesmia microphylla* in Chile) and plants tend to be of lower stature than in neighbouring evergreen shrublands. Many of these species maintain some more sclerophyllous leaves year-round, although a large portion of the canopy sheds in classic seasonal dimorphism. Species in this category include the California sage shrub species of *Salvia mellifera*, *S. apiana*, and *S. leucophylla*. Relative to sclerophyllous shrublands, a few evergreen shrubs occur sporadically, but native (and non-native) herbs are relatively common in these systems (Rundel 2007; Jacobsen et al. 2009). These drought-deciduous communities are also more shallow-rooted on the whole than neighbouring evergreen sclerophyll communities, and under extreme summer drought, this is a viable life-history strategy (Chapter 6).

Structurally related communities that include many drought-deciduous or seasonally dimorphic species but also evergreen shrubs are present in the Mediterranean, referred to as phrygana (Greece), tomillares (Spain), and batha (Israel). Notably, the important genus *Cistus* contains many obligate seeding species that have seasonally dimorphic leaves. There are no analogous plant communities in South Africa and Australia, as deciduous shrub communities are largely lacking in these areas. This may be associated with the greater nutrient use efficiencies of long-lived evergreen leaves (Chapter 6).

4.2.4 Desert shrublands along arid margins

At the arid margins of MTC regions, where rainfall is below 250 mm, **winter-rainfall desert shrublands** occur. Here, steppe or steppe-like vegetation grades into mediterranean desert at the hyperarid end of the winter-rainfall gradient. These desert shrublands are found in the winter-rainfall areas of the northern Sahara and Arabian deserts of the Mediterranean, the Mojave Desert and western portion of the Sonoran Desert in California, the Atacama Desert in Chile, and the succulent karoo in South Africa (Figure 4.4; see Plate 7). In Western Australia, forests and woodlands of the mesic south coast merge into arid sandplains of the interior (although generally not referred to as a winter-rainfall desert, these shrublands do receive less than 250 mm of winter rainfall). Morton et al. (2011) describe six vegetation types in Australian deserts that receive winter rainfall, including two shrubland types—an Acacia shrubland in New South Wales, South Australia and a chenopod shrubland on calcarosols and sodosols located on the Nullarbor Plain, chenopod plains, and stony deserts. Winter-rainfall desert shrublands are generally not fire-prone (they lack a continuous fuel structure to carry fire), although Morton et al. (2011) cite fire return times of 30–100 years in arid Australian winter-rainfall vegetation types. Winter-rainfall deserts have not generally been included in discussions of MTEs, although they experience an MTC (see Le Houerou 2004).

There are clear phytogeographic associations with traditionally recognized MTV and the taxa that occur within the winter-rainfall arid communities (Bergh et al. 2014; Verboom et al. 2014). In South Africa, for example, the boundary of the Cape Floristic Region has been extended to include adjacent semi-arid succulent karoo in what is known as the Greater Cape Floristic Region (Born et al. 2007). This is regarded as a more meaningful

floristic unit to explain patterns of diversity and endemism (reviews by Verboom et al. 2014; Ellis et al. 2014). Occurring at the arid interface with fynbos and renosterveld (Table 4.1), succulent karoo shares similarities with these MTV types in that they are all diverse and dominated by lineages that diversified in situ, albeit at different times. Renosterveld (a sclerophyllous, fire-prone shrubland) and succulent karoo are younger systems that have diversified much more recently, possibly owing to the emergence of novel soil surfaces during a time of marked aridification and cooling in the Middle Miocene (Verboom et al. 2014). Thus, succulent karoo and renosterveld are structurally and floristically more closely related to each other than to fynbos, despite the latter two vegetation types sharing fire as an ecological driver (Bergh et al. 2014).

Some comparisons of these MTC deserts have been attempted. Esler and Rundel (1999) and Esler et al. (1999) compared patterns of plant growth and structural diversity in the South Africa succulent karoo and the California Mojave Desert. They suggested that divergent patterns of resource use associated with variation in rainfall seasonality may have driven striking differences in structural diversity between these two systems. This observation was supported by Alvarado-Cárdenas et al. (2013), who compared niche similarities of 'globular' succulents across regions. Both regions have MTCs, but inter- and intra-annual variation in rainfall is far greater in the Mojave Desert (Esler et al. 1999). Succulent karoo is a structurally homogeneous shrubland dominated by shallow-rooted leaf-succulent shrubs of the family Mesembryanthemaceae (Aizoaceae) and features an exceptionally diverse suite of plants with underground storage organs (geophytes) (Dean and Milton 1999). In contrast, the Mojave Desert is taller in average stature and structurally more diverse, with a diversity of deeper-rooted perennials present. Geophytes are notably few, making up less than 1 per cent of the flora (Rundel 1996).

In the Mediterranean, de Dios Miranda et al. (2010), studied an arid shrubland (242 mm mean annual precipitation) in Spain. They compared their results to those published from California (Jacobsen et al. 2007; 2008) and found that the shrubs within these winter-rainfall arid systems appeared to be similar in their physiological response to water stress. Shrubs within both these systems tended to rely on their ability either to avoid drought (leaf shedding) or to tolerate desiccation, and this differed from the response of evergreen MTV shrubs (see Chapter 6).

Figure 4.4 **(Opposite) Arid shrub communities.** At the arid margins of mediterranean-type climate regions, there are many diverse shrub communities, including (b), the Mojave Desert, California (a), the succulent karoo, South Africa (b), and thorn scrub, Chile (c). These communities often have open canopies and may include some succulent species. (See Plate 8)

Source: Photos a and b from R. Brandon Pratt; photo c from Anna L. Jacobsen.

4.3 Woodlands and forests

Evergreen and winter-deciduous forests and woodlands are present in all MTC regions (Case Study 8). Broadly distinguished by the cover of their tree components, the woodland canopies are generally sufficiently sparse (anywhere between 10 and 60 per cent tree cover; Davis et al. 2016) to allow understorey shrubs, herbs, and grasses to coexist, whereas forests tend to be more densely spaced with substantial (>60 per cent) tree canopy cover. Woodlands and forests generally tend to occur along the moister end of aridity gradients in MTC regions. In California, for example, the distribution of forest types is strongly linked to temperature and rainfall gradients associated with distance from the Pacific Ocean and elevation (North et al. 2016). This pattern is, however, not apparent in South Africa, where forests are limited in extent and confined to fire-protected refugia or moist sites, and above-ground biomass shows little variation across rainfall gradients (Mucina and Rutherford 2006).

Extensive evergreen and deciduous oak woodlands occur in California (including winter-deciduous blue oak *Quercus douglasii* and valley oak *Q. lobata,* evergreen California live oak *Q. agrifolia,* and semi-evergreen Engelmann oak *Q. engelmannii*) and, to a lesser extent, in the Mediterranean (including evergreen kermes oak *Q. coccifera,* holm oak *Q. ilex,* and cork oak *Q. suber,* and deciduous Pyrenean oak *Q. pyrenaica* and Portuguese oak *Q. faginea*). These woodlands may form a complex mosaic with shrubland communities, but are also associated with an understorey of alien-dominated annual grass species in California. In Australia, eucalypts (*Eucalyptus, Corymbia,* and *Angophora* spp. within the Myrtaceae) are the dominant tree species in woodlands and forests, generally associated with a shrubby understorey on infertile sandy or rocky soils and a mixture of grasses and shrubs on deeper, more fertile, soils. In Chile, the key tree species in woodlands and forests is *Nothofagus* (including evergreen Dombey's beech *N. dombeyi* in Valdivian rainforest and deciduous *Nothofagus* in higher-elevation montane forests).

Forests with substantial or closed canopies are most abundant in California (22 per cent total vegetated area), followed by the Mediterranean (15 per cent) and Chile (13 per cent) (Figure 4.2). Forest replaces oak woodlands at elevations above 2000 m in Southern California and coniferous forests occur at higher elevations, with this transition happening at lower elevations towards the north. Species composition can vary over short distances associated with water availability (e.g. *Pinus ponderosa* generally occurs on lower elevations and equator-facing slopes while *Abies concolor* occurs on higher elevations and pole-facing slopes) and topography or microclimate (e.g. lodgepole pine *Pinus contorta,* western white pine *Pinus monticola,* and red fir *Abies magnifica* occur where cold air drainage retains snow pack) (North et al. 2016).

In the Mediterranean, forests and woodlands are best represented in France, Spain, and Italy, whereas the southern and eastern Mediterranean countries have far lower forest and woodland cover, many of which have been exploited for fuel wood and timber, damaged by wildfire, converted to cropland, or impacted by urbanization (Zdruli 2014). Large-scale deforestation over the past two to three millennia (also see Chapter 8) has seen progressive replacement of broad-leaved forests (dominated, for example, by deciduous downy oak *Quercus pubescens*) by open woodlands and sclerophyllous shrublands of varying physiognomy and composition (e.g. holm oak *Q. ilex*; kermes oak *Q. coccifera*) in the south and west, while these processes occurred over much longer timeframes in the eastern parts of the basin (Blondel et al. 2010). Resultant agrosilvopastural savannas (dehesas of Spain, montados of Portugal, and Argan *Argania spinosa* 'forests' of southwestern Morocco) are diverse, heterogeneous, and today regarded as iconic plant assemblages of the Mediterranean. Ironically these anthropogenically maintained systems are under increasing threat, in many areas being replaced by Australian *Eucalyptus* (Blondel et al. 2010). Similar agrosilvopastural savannas occur in Chile (espinal; see Chapter 8). Regardless of deforestation trends, large, old, and venerable trees, such as the iconic coast redwood *Sequoia sempervirens* of California, are associated with forests in all MTC regions (Case Study 8).

Case Study 8 Large old (venerable) trees of fire-prone mediterranean-type climate regions

Grant W. Wardell-Johnson, Department of Environment and Agriculture, Curtin University, Perth, Australia

Comparative analysis of forests in mediterranean-type climate (MTC) regions is limited. Here, primary forests of climatically similar MTC regions, particularly areas with greater than 65 per cent winter rainfall are considered (Aschmann 1973; Ackerly et al. 2014). These environments are all fire-prone, and promote a distinctive suite of structural and functional features. Thus, MTC seasonality enhances propensity for fire by supporting high productivity (i.e. fuel build-up) and therefore ideal conditions to support fire (Bond et al. 2005). Frequency of destructive fires decreases, as does sclerophylly, in higher rainfall areas. Interactions between fire as an evolutionary pressure (Keeley et al. 2011) and the tree form are a hallmark of forests in fire-prone MTC regions, where many species are able to form large old (henceforth venerable) trees (see Lindenmayer et al. 2012). The regions considered contain the largest and some of the oldest organisms on the planet (Waring and Franklin 1979; Tng et al. 2012).

Contrasting pattern and process is evident in the forests of the southwestern Australian MTC region. In higher fertility sites in the highest rainfall zone (>1100 mm mean annual rainfall [MAR]), destructive fires occur infrequently. Dominant shade-intolerant, fire-regenerating species (e.g. *Eucalyptus diversicolor, E. jacksonii*—Case Study Figure 8a) grow

Continued

Case Study 8 (*Continued*)

rapidly under intense intraspecific and interspecific competition (Wardell-Johnson 2000). In more sclerophyllous environments, rapid growth commences only after site capture and fire insurance through development of a lignotuber. This strategy of persistence is exemplified by *Eucalyptus marginata* (Case Study Figure 8f), a dominant species (see Wardell-Johnson et al. 2015) in infertile, sclerophyllous forest with >600 mm MAR. In these environments, accumulation of nutrient-poor biomass fuels fire, while repeated fires exacerbate nutrient poverty and enhance flammability (Orians and Milewski 2007).

Eucalypts are naturally largely endemic to Australia, with over 800 species. Many giant eucalypts (i.e. those reaching 70 m in height; Tng et al. 2012), including the tallest angiosperm (*E. regnans*—99.6 m) occur in mesic and relatively fertile habitats (Wardell-Johnson et al. 2017a). All are shade-intolerant, fire regenerators of rapid height growth in the first 100 years of life, with senescence after c. 500 years (Wood et al. 2010). All have capacity to resprout epicormic shoots from deeply embedded buds; an ancient adaptation in the genus (Crisp et al. 2011). The two species of giant eucalypts in south-west Australia (*Eucalyptus diversicolor, E. jacksonii*) occur in the most isolated of MTC forests, and are separated from similar forest by 1500 km of arid lands (Wardell-Johnson et al. 1997). Unlike forests of similar MAR, but less seasonality in southeastern Australia, these forests no longer include members of Antarctic affinity (the resprouting shrub *Podocarpus drouynianus* is exceptional) owing to local extinction during drier periods of the late Pliocene. Nevertheless, this area, like other Australian mesic areas, harbours numerous species of tall (>50 m) eucalypts.

The growth habits of eucalypts enabling great height differ from other taxa capable of growing into venerable trees in other MTC regions. These species are primarily conifers capable of achieving great height that regenerate in tree-fall gaps and grow relatively slowly in areas where return intervals of destructive fires are very long. For example, *Sequoia sempervirens* (Cupressaceae)—the tallest tree (115.5 m) is a dominant in high-rainfall, near-coastal habitat in the Californian MTC region. This species has thick insulating bark and is a vigorous resprouter both basally and from epicormics, features it shares with the second-tallest conifer, coast Douglas fir (*Pseudotsuga menziesii* var. *menziesii* (Pinaceae) (99.4 m), occurring over a larger area in the same MTC region. Individuals of *Sequoia* can live for over 2000 years, and coast Douglas fir for more than 1000 years, the latter exhibiting greater morphological plasticity. Large size provides a buffer against nutrient and moisture stress.

For both of these species, long intervals between destructive fires and storms enable dominance over co-occurring hardwoods with more limited stature and life spans (Waring and Franklin 1979). This feature is shared with *Fitzroya cupressoides* (Cupressaceae—the tallest tree in South America at >60 m, and second oldest overall—3622 years). In California the family Cupressaceae also includes the largest tree (*Sequoiadendron giganteum*: 1486.9 m³ trunk volume; Case Study Figure 8b), a dominant long-lived fire-resistant tree with very thick bark on the western slopes of the Sierra Nevada Mountains of California, part of the high-rainfall California MTC region. Unlike most giant trees of the Pacific North-West, members of the Cupressaceae are derived from subtropical, xeric-adapted elements.

Among conifers, the Cupressaceae of worldwide distribution is disproportionately represented in MTC regions. Thus *Cypressus sempervirens, Juniperus thurifera* (Case Study Figure 8d) and *Tetraclinus articulata* are all capable of great age or size, occurring in the drier, more typical Mediterranean MTC. However, in the Mediterranean, only isolated, relictual stands or single venerable trees now provide evidence of capacity to attain great age or stature. This is because forested life zones of this region have been modified by upwards of 10,000 years of agriculture, pastoralism, and resource extraction. Thus, forest at low altitudes has been replaced by plantations of short-lived pines or anthropogenic communities dominated by dwarf prickly shrubs, while those at higher elevations have been heavily exploited. For example, in montane forests scattered, ancient trees of *Juniperus* and *Abies* occur above 3000 m in the Atlas Mountains and to 2600 m in the Taurus Range of southern Turkey (Blondel et al. 2010). There are also many other conifers (e.g. species of *Cedrus, Abies, Pinus* in the Pinaceae) able to form forests of venerable trees in the Mediterranean. In addition, the indigenous (Mesogean) component of the angiosperm flora that differentiated in the Mediterranean in the Oligocene includes many taxa able to occur as venerable trees. These are all sclerophyllous trees that are fire-resistant and vertebrate dispersed (e.g. *Olea europaea*—Case Study Figure 8e, *Quercus suber, Quercus ilex*).

In Chile the dominant MTC region temperate taxa are largely Antarctic elements (e.g. *Nothofagus*) and, like boreal elements of the northern hemisphere, not well fire-adapted. The higher rainfall areas of the Chilean MTC region are in general not as fire-prone as other MTC regions. *Fitzroya* and those species of *Nothofagus* able to achieve venerable status occur on the margins of the high rainfall MTC in relatively non-fire-prone areas. An important exception is the thick-barked, fire-resistant taxon *Araucaria araucana* (Araucariaceae, Case Study Figure 8c) occurring in the relatively fire-prone Pacific coastal range (e.g. Nahuelbuta) as well as the foothills of the Andes. In the Andes, its distribution coincides with areas of volcanic activity. In both situations this species occurs with both evergreen and deciduous species of *Nothofagus*. This fire-resistant taxon can achieve great age and size in relatively fire-prone areas of the Chilean MTC region.

Thus, many species capable of growing into venerable trees occur in more frequently burnt habitats in four of the five MTC regions. This is because one response to fire has favoured reducing nutrient losses by investment in longlasting organs (Mucina and Wardell-Johnson 2011). Hence persistence of individual trees in fire-prone landscapes requires adaptations to cope with low nutrients (Mucina and Wardell-Johnson 2011), avoid damage (Pausas 2015), and escape the fire trap (Bond 2008). Therefore one important adaptive strategy for trees in fire-prone environments includes the capacity for individual persistence, sometimes allowing trees of great age or size. However, taxonomic affinity and biogeographic history are also important enablers of venerable trees in fire-prone MTC regions.

Exceptionally among MTC regions, South Africa is without indigenous forest or venerable trees. Lineages from which the Cape flora are derived include Gondwanan, Antarctic, boreal, and African stocks (Goldblatt 1978), with the forest being largely African and the fynbos Gondwanan. Neither contributes plants capable of great age or longevity. Thus, forests are restricted to less seasonal, high rainfall and protected areas to the east. Here the divergent southern afromontane and southern coastal forests both attain their

Continued

Case Study 8 (*Continued*)

Case Study 8 Figure. Forests and venerable trees in fire-prone mediterranean-type climate ecosystems: a) *Eucalyptus jacksonii* and *E. diversicolor* (white-barked tree in background) in wet sclerophyll forest (mean annual rainfall [MAR] 1200 mm) in southwestern Australia. Note a relatively dense understorey and signs of a fire 6 years prior to the photograph; b) *Sequoiadendron giganteum* on the western slopes of the Sierra Nevadas', California. Note the large hollow-butt from previous fires; c) *Araucaria araucana* among granite outcrops in the coastal range (Cordillera de Nahuelbuta), Chile; d) *Juniperus thurifera* at about 1500 m elevation in southwestern Spain; e) very old specimen of *Olea europaea* in southwestern Spain; f) *Eucalyptus marginata* in dry sclerophyll forest (MAR 1000 mm) in southwestern Australia. Note the relatively open understorey.

Source: Photos from Grant W. Wardell-Johnson.

greatest richness and structural diversity, and they become increasingly impoverished westwards (Bergh et al. 2014). The southern afromontane forest is a tall temperate rainforest, notable for venerable trees of the Antarctic family Podocarpaceae (Podocarpus, Afrocarpus) sometimes exceeding 40 m in height. In the absence of fire (generally an unlikely scenario—though see Poulsen and Hoffman 2015), southern afromontane forest potentially occurs throughout the South Africa MTC region (Coetsee et al. 2015).

Venerable trees are particularly well represented in the forests (or former forests) of MTC regions. They provide disproportionately important ecosystem functions wherever they occur (Lindenmayer and Laurance 2016) and are significant structural and functional elements of these regions. However, they are also highly vulnerable to changes in temperature and water regimes and to extreme events triggered by anthropogenic global warming (Bennett et al. 2015; Lindenmayer and Laurance 2016; Wardell-Johnson et al. 2017a). Though they are highly resilient to the range of fire regimes that they have endured evolutionarily, new fire regimes suggest likely large-scale loss as a result of interactions with climate disruption (Wardell-Johnson et al. 2017a,b; Bowman et al. 2014). Prognosis for venerable MTC region trees and associated forests is therefore bleak. Increased protection and valorization of individual venerable trees and old growth forests is therefore required. The alternative is landscapes that are even more fire-prone and difficult to manage as a result of interactions with people and new climatic regimes.

4.4 Grasslands

Towards the arid end of climatic gradients mediterranean-type shrublands give way to grassland communities in California, Chile, the Mediterranean, and, to a small extent, Australia and South Africa. In California, native grasslands may once have been far more extensive, occurring on deep soils in large areas of California's Central Valley (but see Hamilton 1997). These diverse communities are comprised of perennial grasses (e.g. *Bromus carinatus, Elymus glaucus, Stipa cernua, Stipa pulchra, Poa secunda*) adapted to cool season growth, annual grasses, and a wide diversity of other herbaceous species.

Closely associated with these grasslands are **vernal pool communities**, shallow-basin seasonal wetlands that, while occurring in non-MTC zones, also feature in other MTC systems (Keeley and Zedler 1998), including Chile (Deil et al. 2007). California is the only region with such an extensive endemic flora to vernal pools, although the drivers for this are uncertain (this is an area ripe for comparative research across MTEs). Over 100 vascular plant species are restricted to Californian vernal pools, which host a variety of specialized endemics, mostly (80 per cent) annuals that have evolved from diverse genera in adjacent terrestrial grassland communities (Keeley and Zedler 1998). Drought, overgrazing, agriculture, and invasion by exotic annuals have contributed to the degradation and loss of these diverse grassland and vernal communities that today are restricted to disjunct, isolated fragments.

Grasslands also occur as alternative states to shrublands and woodlands, particularly after habitat type conversion (Chapter 8). The grasslands that occur in California and Chile today are dominated by non-native grasses and herbs, and are not only a consequence of native grassland degradation but also the result of habitat type conversion of areas formerly dominated by shrublands and woodlands (Figure 4.5; see Plate 8; Gulmon 1977;

Figure 4.5   **Grassland in a mediterranean-type climate region.** Grasslands are highly seasonal communities, with abundant growth and flowering of species during the winter (a) and into the late spring (b), but little growth or flowering during the summer and fall (c). In California (all photos shown), grasslands are diverse and contain a mix of annual forbs and grasses, geophytes, and perennial bunch grasses. These communities are also heavily invaded by annual grasses, primarily from the Mediterranean. (See Plate 9)

Source: Photos a and c from R. Brandon Pratt; photo b from Anna L. Jacobsen.

Hamilton 1997; Deil et al. 2007). In Chile, grasslands within the coastal mountains contain many native species, as well as a large number of introduced species, including many Eurasian introduced grass species, and grasslands also extend into degraded shrublands (Gulmon 1977; Ovalle et al. 2006; Deil et al. 2007).

Grassland and herbaceous communities across the Mediterranean have also been subject to long-term anthropogenic impacts but, in contrast to Californian and Chilean grasslands, appear to be relatively well adapted to grazing because of their long history of human association (Sternberg et al. 2000). Spatial heterogeneity in soil properties leads to high turnover at small spatial scales in grassland communities. For example, in south France, shallow sandy soils are codominated by a variety of perennial graminoids (e.g. *Festuca christiani-bernardii, Stipa pennata,* or *Carex humilis*), whereas communities on deeper clay soils are dominated by *Bromus erectus* and various subdominant grasses, forbs, and rosettes (e.g. *Carex flacca, Poa badensis, Lotus corniculatus, Potentilla neumanniana,* or *Hieracium pilosella*) (Barkaoui et al. 2013). In continually grazed grasslands on deep, granite-derived soils in southwest Spain, winter annuals dominate, including *Agrostis pourretii, Bromus hordeaceus, Chamaemelum fuscatum, C. mixtum, Crepis capillaris, Echium plantagineum, Medicago polymorpha, Molineriella minuta, Spergularia rubra, Trifolium glomeratum, T. subterraneum,* and *Vulpia geniculata* (Fernández Alés et al. 1993). In the more arid eastern Mediterranean, grassland communities also tend to be dominated by annual species that are structurally stable, resistant to grazing, and subject to temporal variation in precipitation patterns (Sternberg et al. 2000; Bar-Massada and Hadar 2017). In the absence of disturbance (grazing, fire) these grasslands degrade to communities dominated by a few species of tall perennial grasses (e.g. *Hordeum bulbosum*), forbs (e.g. *Echinops* spp.), or tall large-seeded annual grasses (*Triticum, Hordeum,* and *Avena* spp.) (Noy-Meir 1995). It is no accident that many of the world's cosmopolitan herbaceous weeds, especially grasses, originated in the Mediterranean, where ruderal traits have been selected for. These have been exported to other mediterranean regions where habitat type conversion to alien-dominated grasslands of Mediterranean origin has occurred (Chapter 8).

Grasslands are relatively rare in Australia and South Africa. In these MTC regions, grasslands occur primarily in interior arid regions where agriculture and/or rising water tables have caused a replacement of native perennial vegetation by non-native pastures, mostly originating from the Mediterranean region. Native perennial grasses such as *Rytidosperma* spp. (wallaby grasses) that once sustained Australia's early wool export industry still occur in high-rainfall areas of southeast Australia while towards the arid end of the gradient in Western Australia, *Spinifex* grassland co-occurs with *Eucalyptus* spp. In South Africa, native grasslands can occur in sandy areas and are more prevalent in the eastern portions of the MTC region.

4.5 Riparian communities

The term **riparian** refers to the section of a river bank influenced by elevated water tables, flooding, or the ability of soils to hold water (Naiman and Décamps 1997). Riparian vegetation is supported by these moist conditions, and tends to be distinctly different in species composition from neighbouring terrestrial communities. This is a particular feature of MTC regions, where these matrix communities are more xeric shrublands, woodlands, or grasslands. Most information on riparian zones is based upon research carried out in cooler, wetter temperate environments and on riparian zones adjoining forested uplands. Riparian communities are not typically included in discussions of mediterranean-type communities, yet the unique conditions linked to flashy flow regimes, an abundance of water in winter, summer low flows, and fire hold several implications for riparian vegetation processes in these systems, such as seedling establishment, plant growth, survival, and population and community dynamics (Figure 4.6). MTC region riparian communities are more resilient to fire (recovery 1–4 years) than their non-MTC counterparts (5–10 years) (Verkaik et al. 2013).

Gasith and Resh (1999) provided the first comparative review of MTC region river ecosystems, pointing out that they are 'ecological reflections' of MTC, being fluvial systems that are physically, chemically, and biologically moulded by successive, seasonally predictable flooding and drying events

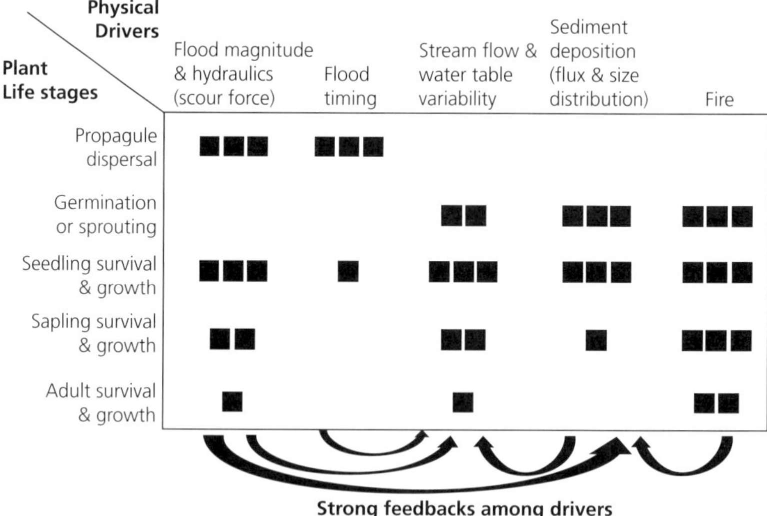

Figure 4.6 **Influential drivers for riparian plants.** Relative magnitude (number of symbols) of physical influences on woody riparian plant life stages in mediterranean-type climate regions. Number of symbols indicates magnitude of influence and arrows indicate direction of drivers. (Modified from Stella et al. 2013.)

that can vary in intensity both seasonally and annually. In 2013, a special issue of the journal *Hydrobiologica* synthesized information available on MTC region river ecology since this seminal work. In their synthesis overview, Bonada and Resh (2013) highlighted that, in all MTC regions, rivers are hotspots for biodiversity, but no two mediterranean rivers are the same for the general characteristics considered (Figure 4.7). Like their terrestrial counterparts, riparian vegetation has been described to various levels of detail across regions and this is summarized by Stella et al. (2013). Riparian zones in each MTC region are represented by distinctly regional flora, with some commonalities; for example, Australia and Chile share the genus *Muehlenbeckia*, South Africa and Chile share *Podocarpus* and *Maytenus*, South Africa and the Mediterranean share *Erica, Ilex*, and *Olea*, California and Chile share *Baccharis* and *Salix*, California and the Mediterranean share *Populus* and *Salix*, and the Mediterranean and Chile share *Coriaria*. There are also within-region differences, particularly in the Mediterranean, where different genera dominate the east and the west and where glacial refugia hotspots contain important reservoirs of plant diversity (Stella et al. 2013). Floristically, riparian communities of California and the Mediterranean are the most similar (Figure 4.7); they are dominated by winter-deciduous taxa and juxtapose starkly with adjacent evergreen upland MTV.

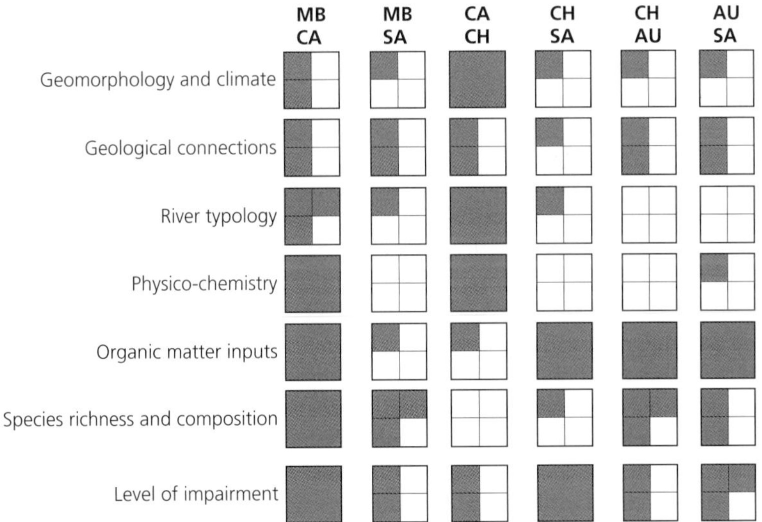

Figure 4.7 **Mediterranean-type climate (MTC) region river characteristics.** Pairwise regional comparisons of MTC region river characteristics, showing relative similarity among mediterranean rivers (indicated by number of black squares, with each large square corresponding to 100%, divided into 25% units) of the Mediterranean Basin (MB), California (CA), South Africa (SA), Chile (CH), and Australia (AU, southwestern and south Australia combined). (Modified from Bonada and Resh 2013.)

In contrast, winter-deciduous riparian taxa do not feature in South Africa and Australia.

4.6 Vegetation dynamics: patchiness in space and time

So far, we have described some of the similarities and many of the differences in the occurrence, structure, and composition of MTC communities between regions. Another major difference is tied to post-fire succession and timing of peak diversity. Fire-dependent reproduction, such as heat and/or smoke-stimulated seed germination (Dixon et al. 1995; Brown et al. 2003; Moreira et al. 2010; Hall et al. 2017) and fire-simulated flowering (Le Maitre & Brown 1992; Lamont & Downes 2011; Keeley et al. 2012b), is common in all MTC regions except for Chile. In Californian and Mediterranean shrublands, the diversity of vascular plants dramatically increases after fire when these fire-stimulated species emerge after long periods of inter-fire dormancy. This diversity tends to decline over time as these ecosystems recover after fire when the overstorey canopy closes up and the fire-stimulated species return to their dormant state. In California, the post-fire community is distinctive in that a significant number of these species (notably Hydrophyllaceae and Polemoniaceae) are confined only to the post-fire environment; i.e. they are **pyroendemics** (fire endemics) (Keeley et al. 2006). The trend of declining post-fire diversity is also apparent in the Mediterranean (e.g. Pausas et al. 1999); however, pyroendemics are not a feature of this region's post-fire communities, as these species also occur in other disturbed sites across the landscape. In general, however, the diversity of ephemerals and short-lived species in the Mediterranean is exceptionally high, peaking in the east where a large regional pool of species exists due to the confluence of floras from Europe, Asia, and Africa (Naveh and Whittaker 1979; Keeley and Fotheringham 2003). Indeed, apart from fynbos, grazed Mediterranean pastures are amongst the richest (and yet vastly different) temperate plant communities known.

Similar post-fire diversity levels (typically between 9 and 15 spp./m^2; Keeley et al. 2012b) occur in South Africa and Australia, but, unlike their northern hemisphere counterparts, this diversity does not decline markedly as the systems recover. In these more nutrient-limited systems, high diversity is maintained in mature post-fire stands. Some fynbos species (mostly annuals, fire ephemerals, and geophytes) do complete their reproductive cycles in the immediate post-fire environment and then become dormant until the next fire, but most remain present in mature stands even though they might be visually less obvious (Le Maitre and Brown 1992; Van Wilgen and Forsyth 1992; Goldblatt and Manning 2000). Finally, the absence of a modern predictable natural source of fire in Chilean matorral has been linked to an absence of post-fire ephemerals in this system (Montenegro et al. 2004), although herbaceous perennials and geophytes are relatively abundant in mature matorral (Montenegro et al. 1978).

Literature cited

Ackerly DD, Stock WD, Slingsby JA. 2014. Geography, climate, and biogeography; the history and future of Mediterranean-type ecosystems. Pages 361–75 in Allsopp N, Colville JF, Verboom AG, eds. Fynbos: ecology, evolution, and conservation of a megadiverse region. Oxford University Press, Oxford.

Alvarado-Cárdenas LO, Martínez-Meyer E, Feria TP, Eguiarte LE, Hernández HM, Midgley G, Olson ME. 2013. To converge or not to converge in environmental space: testing for similar environments between analogous succulent plants of North America and Africa. Annals of Botany 111: 1125–38.

Araya YN, Silvertown J, Gowing DJ, McConway KJ, Linder H, Midgley G. 2011. A fundamental, eco-hydrological basis for niche segregation in plant communities. New Phytologist 189: 253–8.

Armesto JJ, Arroyo MTK, Hinojosa LF. 2007. The mediterranean environment of Central Chile. Pages 184–99 in Veblen TT, Young KR, Orme AR, eds. The physical geography of South America. Oxford University Press, Oxford.

Aschmann H. 1973. Distribution and peculiarity of Mediterranean ecosystems. Pages 11–19 in di Castri F, Mooney HA, eds. Mediterranean type ecosystems. Springer, Berlin.

Barbour MG, Keeler-Wolf T, Schoenherr AA (eds). 2007. Terrestrial vegetation of California. University of California Press, Los Angeles, California. 713 pp.

Barkaoui K, Bernard-Verdier M, Navas ML. 2013. Questioning the reliability of the point intercept method for assessing community functional structure in low-productive and highly diverse mediterranean grasslands. Folia Geobotanica 48: 393–414.

Bar-Massada A, Hadar L. 2017. Grazing and temporal turnover in herbaceous communities in a Mediterranean landscape. Journal of Vegetation Science 28(2): 270–80.

Beard JS. 1979. Western Australia: an atlas of human endeavour. Dept Land Administration and Ministry of Justice, Perth. p. 60.

Beard JS. 1983. Ecological control of the vegetation of Southwestern Australia: moisture versus nutrients. Pages 66–73 in Kruger FJ, Mitchell DT, Jarvis JUM, eds. Mediterranean-type ecosystems: the role of nutrients. Springer-Verlag, Berlin, Germany.

Bennett AC, McDowell NG, Allen CD, Anderson-Teixeira KJ. 2015. Larger trees suffer most during drought in forests worldwide. Nature Plants 1, 15139 10.1038/nplants.2015.139.

Bergh NG, Verboom GA, Rouget M, Cowling RM. 2014. Vegetation types of the Greater Cape Floristic region. Pages 1–25 in Allsopp N, Colville JF, Verboom GA, eds. Fynbos: ecology, evolution, and conservation of a megadiverse region. Oxford University Press, Oxford.

Blondel J, Aronson J, Bodiou J-Y, Boeuf G. 2010. The Mediterranean region: biological diversity in space and time (2nd ed.). Oxford University Press, Oxford.

Bonada N, Resh VH. 2013. Mediterranean-climate streams and rivers: geographically separated but ecologically comparable freshwater systems. Hydrobiologia 719: 1–29.

Bond WJ. 2008. What limits trees in C4 grasslands and savannas? Annual Review of Ecology, Evolution and Systematics 39: 641–59.

Bond WJ, Woodward FI, Midgley GF. 2005. The global distribution of ecosystems in a world without fire. New Phytologist 165: 525–38.

Born J, Linder HP, Desmet P. 2007. The Greater Cape Floristic region. Journal of Biogeography 34: 147–62.

Bowman DMJS, Murphy BP, Neyland DLJ, Williamson GJ, Prior LD. 2014. Abrupt fire regime change may cause landscape-wide loss of mature obligate seeder forests. Global Change Biology 20: 1008–15.

Bradshaw PL, Cowling RM. 2014. Landscapes, rock types and climate of the Greater Cape Floristic region. Pages 1–25 in Allsopp N, Colville JF, Verboom AG, eds. Fynbos: ecology, evolution, and conservation of a megadiverse region. Oxford University Press, Oxford.

Brown NAC, Van Staden J, Daws MI, Johnson T. 2003. Patterns in the seed germination response to smoke in plants from the Cape Floristic Region, South Africa. South African Journal of Botany 69: 514–25.

Carlson JE, Holsinger KE, Prunier R. 2011. Plant responses to climate in the Cape Floristic Region of South Africa: evidence for adaptive differentiation in the Proteaceae. Evolution 65: 108–24.

Coetsee C, Bond WJ, Wigley BJ. 2015. Forest and fynbos are alternative states on the same nutrient poor geological substrate. South African Journal of Botany 101: 57–65.

Cornwell WK, Ackerly DD. 2009. Community assembly and shifts in plant trait distributions across an environmental gradient in coastal California. Ecological Monographs 79: 109–26.

Cowling RM. 1990. Diversity components in a species-rich area of the Cape Floristic Region. Journal of Vegetation Science 1: 699–710.

Cowling RM, Holmes PM. 1992. Endemism and speciation in a lowland flora from the Cape Floristic Region. Biological Journal of the Linnean Society 47: 367–83.

Cowling RM, Gxaba T. 1990. Effects of a fynbos overstorey shrub on understorey community structure: implications for the maintenance of community-wide species richness. South African Journal of Ecology 1: 1–7.

Cowling RM, Rundel PW, Lamont BB, Arroyo MK, Arianoutsou M. 1996. Plant diversity in Mediterranean-climate regions. Trends in Ecology & Evolution 11: 362–6.

Cowling RM, Potts AJ, Bradshaw PL, Colville J, Arianoutsou M, Ferrier S, Forest F, Fyllas NM, Hopper SD, Ojeda F, Proches S. 2015. Variation in plant diversity in mediterranean-climate ecosystems: the role of climatic and topographical stability. Journal of Biogeography 42: 552–64.

Crisp MD, Burrows GE, Cook LG, Thornhill AH, Bowman DM. 2011. Flammable biomes dominated by eucalypts originated at the Cretaceous-Paleogene boundary. Nature Communications 2: 193.

Dallman PR. 1998. Plant life in the world's Mediterranean climates: California, Chile, South Africa, Australia, and the Mediterranean basin. University of California Press, Oakland.

Davis FW, Baldocchi DD, Tyler CM. 2016. Oak woodlands. Pages 509–34 in Mooney HA, Zavaleta E, eds. Ecosystems of California. University of California Press, Oakland.

Dean WRJ, Milton SJ. 1999. The Karoo: ecological patterns and processes. Cambridge University Press, Cambridge.

de Dios Miranda J, Padilla FM, Martínez-Vilalta J, Pugnaire FI. 2010. Woody species of a semi-arid community are only moderately resistant to cavitation. Functional Plant Biology 37: 828–39.

Deil U, Alvarez M, Paulini I. 2007. Native and non-native species in annual grassland vegetation in Mediterranean Chile. Phytocoenologia 37: 769–84.

Dixon KW, Roche S, Pate JS. 1995. The promotive effect of smoke derived from burnt native vegetation on seed germination of Western Australian plants. Oecologia 101: 185–92.

Ellis AG, Verboom GA, van der Niet T, Johnson SD, Linder HP. 2014. Speciation and extinction in the Greater Cape Floristic Region. Pages 119–41 in Allsopp N, Colville JF, Verboom AG, eds. Fynbos: ecology, evolution, and conservation of a megadiverse region. Oxford University Press, Oxford.

Esler KJ, Rundel PW. 1999. Comparative patterns of phenology and growth form diversity in two winter rainfall deserts: the Succulent Karoo and Mojave Desert ecosystems. Plant Ecology 142: 97–104.

Esler KJ, Rundel PW, Cowling RM. 1999. The Succulent Karoo in a global context: plant structural and functional comparison with North American winter rainfall deserts. Pages 303–13 in Dean WRJ, Milton SJ, eds. The Karoo: ecological patterns and processes. Cambridge University Press, Cambridge.

Esler KJ, Pierce SM, de Villiers C. 2014. Fynbos: ecology and management. Briza Publications, Pretoria.

Esler KJ, von Staden L, Midgley GF. 2015. Determinants of the fynbos/Succulent Karoo biome boundary: insights from a reciprocal transplant experiment. South African Journal of Botany 101: 120–8.

Fernández Alés R, Laffarga JM, Ortega F. 1993. Strategies in Mediterranean grassland annuals in relation to stress and disturbance. Journal of Vegetation Science 4: 313–22.

Francioli SE, Muñoz AM, eds. 2009. Parque nacional la campana: origen de una reserva de la biosfera en Chile central. Taller La Era, Viña del Mar, Chile.

Gasith A, Resh VH. 1999. Streams in Mediterranean climate regions: abiotic influences and biotic responses to predictable seasonal events. Annual Review of Ecology and Systematics 30: 51–81.

Goldblatt P. 1978. An analysis of the flora of southern Africa: its characteristics, relationships, and origins. Annals of the Missouri Botanical Garden 65: 369–436.

Goldblatt P, Manning J. 2000. Cape plants. A conspectus of the Cape flora of South Africa. Strelitzia 9. National Botanical Institute, Pretoria.

Groom PK, Lamont BB. 2015. Plant life of southwestern Australia. De Gruyter Open Ltd, Warsaw/Berlin.

Gulmon SL. 1977. A comparative study of the grassland of California and Chile. Flora 166: 261–78.

Hall SA, Newton RJ, Holmes PM, Gaertner M, Esler KJ. 2017. Heat and smoke pre-treatment of seeds to improve restoration of an endangered Mediterranean climate vegetation type. Austral Ecology 42: 354–66.

Hamilton JG. 1997. Changing perceptions of pre-European grasslands in California. Madrono 44: 311–33.

Jacobsen AL, Pratt RB, Davis SD, Ewers FW. 2007. Cavitation resistance and seasonal hydraulics differ among three arid Californian plant communities. Plant, Cell and Environment 30: 1599–609.

Jacobsen AL, Pratt RB, Davis SD, Ewers FW. 2008. Comparative community physiology: non-convergence in water relations among three semi-arid shrub communities. New Phytologist 180: 100–13.

Jacobsen AL, Pratt RB, Moe LM, Ewers FW. 2009. Plant community water use and invasibility of semi-arid shrublands by woody species in southern California. Madroño 56: 213–20.

Keeley JE. 2000. Chaparral. Pages 203–53 in Barbour MG, Billings WD, eds. North American terrestrial vegetation. Cambridge University Press, Cambridge.

Keeley JE, Davis FW. 2007. Chaparral. Pages 339–66 in Barbour MG, Keeler-Wolf T, Schoenherr AA, eds. Terrestrial vegetation of California. University of California Press, Los Angeles, California.

Keeley JE, Fotheringham CJ. 2003. Species–area relationships in Mediterranean-climate plant communities. Journal of Biogeography 30: 1629–57.

Keeley JE, Zedler PH. 1998. Characterization and global distribution of vernal pools. Pages 1–14 in Witham CW, Bauder ET, Belk D, Ferren Jr WR, Ornduff R, eds. Ecology, conservation, and management of vernal pool ecosystems—Proceedings from a 1996 Conference. California Native Plant Society, Sacramento, CA.

Keeley JE, Fotheringham CJ, Baer-Keeley M. 2006. Demographic patterns of post-fire regeneration in mediterranean-climate shrublands of California. Ecological Monographs 76: 235–55.

Keeley JE, Pausis JG, Rundel PW, Bond WJ, Bradstock RA. 2011. Fire as an evolutionary pressure shaping plant traits. Trends in Plant Science 16: 406–11.

Keeley JE, Fotheringham CJ, Rundel PW. 2012a. Postfire chaparral regeneration under mediterranean and non-mediterranean climates. Madroño 59: 109–27.

Keeley JE, Bradstock RJ, Bond WA, Pausas JG, Rundel PW. 2012b. Fire in Mediterranean ecosystems: ecology, evolution and management. Cambridge University Press, Cambridge. 515 p.

Killick DJB. 1979. African mountain heathlands. Pages 97–116 in Specht RL, ed. Heathlands and related shrublands of the world. Elsevier, Amsterdam.

Lambers H. 2014. Plant life on the sandplains in southwest Australia: a global biodiversity hotspot. The University of Western Australia Publishing, Crawley, Australia. 332 p.

Lamont BB, Keith D. 2017. Heathlands and associated shrublands. Pages 339–68 in Keith D, ed. Vegetation of Australia. 3rd ed. Cambridge University Press, Cambridge.

Lamont BB, Downes KS. 2011. Fire-stimulated flowering among resprouters and geophytes in Australia and South Africa. Plant Ecology 212: 2111–25.

Latimer AM, Silander JA, Rebelo AG, Midgley GF. 2009. Experimental biogeography: the role of environmental gradients in high geographic diversity in Cape Proteaceae. Oecologia 160: 151–62.

Le Houerou HN. 2004. An agro-bioclimatic classification of arid and semiarid lands in the isoclimatic Mediterranean zones. Arid Land Research and Management 18: 301–46.

Le Maitre DC, Brown PJ. 1992. Life cycles and fire-stimulated flowering in geophytes. Pages 145–60 in van Wilgen BW, Richardson DM, Kruger FJ, van Hensbergen HJ, eds. Fire in South African mountain fynbos: ecosystem, community and species response at Swartboskloof. Springer-Verlag, Berlin.

Lindenmayer DB, Laurance WF. 2016. The ecology, distribution, conservation and management of large old trees. Biological Review 92: 1434–58.

Lindenmayer DB, Laurance WF, Franklin JF. 2012. Global decline in large old trees. Science 338: 1305–6.

Loidi J, Biurrun I, Campos JA, García-Mijangos I, Herrera M. 2010. A biogeographical analysis of the European Atlantic lowland heathlands. Journal of Vegetation Science 21: 832–42.

Marais KE, Pratt RB, Jacobs SM, Jacobsen AL, Esler KJ. 2014. Postfire regeneration of resprouting mountain fynbos shrubs: differentiating obligate resprouters and facultative seeders. Plant Ecology 215: 195–208.

Mitchell AA, Wilcox DG. 1994. Arid shrubland plants of Western Australia. University of Western Australia Press, Nedlands, Australia.

Montenegro G, Rivera O, Bas F. 1978. Herbaceous vegetation in the Chilean matorral. Oecologia 36: 237–44.

Montenegro GL, Ginocchio RO, Segura AL, Keeley JE, Gomez M. 2004. Fire regimes and vegetation responses in two Mediterranean-climate regions. Revista Chilena de Historia Natural 77: 455–64.

Mooney HA, Parsons DJ. 1973. Structure and function of the California chaparral—an example from San Dimas. Pages 83–112 in di Castri F, Mooney HA, eds. Mediterranean-type ecosystems: origin and structure. Springer-Verlag, New York, USA.

Mooney HA, Kummerow J, Johnson AW, Parsons DJ, Keeley S, Hoffmann A, Hays RI, Giliberto J, Chu C. 1977. The producers—their resources and adaptive responses. Pages 85–43 in Mooney HA, ed. Convergent evolution in Chile and California. Dowden, Hutchinson & Ross, Inc. Stroudsburg, Pennsylvania, USA.

Moreira B, Tormo J, Estrelles E, Pausas JG. 2010. Disentangling the role of heat and smoke as germination cues in Mediterranean Basin flora. Annals of Botany 105: 627–35.

Moreno G, Pulido F. 2008. The functioning, management and persistence of dehesas. Pages 127–60 in Rigueiro-Rodriguez A, McAdam J, Mosquera-Losada MR, eds. Agroforestry in Europe. Current Status and Future Prospects. Springer, Netherlands.

Morton SR, Smith DS, Dickman CR, Dunkerley DL, Friedel MH, McAllister RR, Reid JR, Roshier DA, Smith MA, Walsh FJ, Wardle GM. 2011. A fresh framework for the ecology of arid Australia. Journal of Arid Environments 75: 313–29.

Mucina L, Rutherford MC. (eds) 2006. The vegetation of South Africa, Lesotho and Swaziland. Strelitzia 19. South African National Biodiversity Institute, Pretoria.

Mucina L, Wardell-Johnson GW. 2011. Landscape age and soil fertility, climatic stability, and fire regime predictability: beyond the OCBIL framework. Plant and Soil 341: 1–23.

Naiman RJ, Decamps H. 1997. The ecology of interfaces: riparian zones. Annual Review of Ecology and Systematics 28: 621–58.

Naveh Z, Whittaker RH. 1979. Structural and floristic diversity of shrublands and woodlands in northern Israel and other Mediterranean areas. Plant Ecology 41: 171–90.

North M, Collins B, Safford H, Stephenson N. 2016. Montane forests. Pages 553–77 in Mooney HA, Zavelta E, eds. Ecosystems of California. University of California Press, Berkeley, California.

Noy-Meir I. 1995. Interactive effects of fire and grazing on structure and diversity of Mediterranean grasslands. Journal of Vegetation Science 6: 701–10.

Ojeda F, Arroyo J, Marañón T. 1995. Biodiversity components and conservation of Mediterranean heathlands in southern Spain. Biological Conservation 72: 61–72.

Ojeda F, Marañón T, Arroyo J. 1996. Patterns of ecological, chorological and taxonomic diversity at both sides of the Strait of Gibraltar. Journal of Vegetation Science 7: 63–72.

Ojeda F, Marañón T, Arroyo, J. 2000. Plant biodiversity in the Aljibe Mountains (S. Spain): a comprehensive account. Biodiversity Conservation 9: 1323–43.

Orians GH, Milewski AV. 2007. Ecology of Australia: the effects of nutrient-poor soils and intense fires. Biological Review 82: 393–423.

Ovalle C, Del Pozo A, Casado MA, Acosta B, de Miguel JM. 2006. Consequences of landscape heterogeneity on grassland diversity and productivity in the Espinal agroforestry system of central Chile. Landscape Ecology 21: 585–94.

Pausas JG. 2015. Bark thickness and fire regime. Functional Ecology 29: 315–27.

Pausas JG, Carbó E, Caturla RN, Gil JM, Vallejo R. 1999. Post-fire regeneration patterns in the eastern Iberian Peninsula. Acta Oecologica 20: 499–508.

Pirie MD, Oliver EG, de Kuppler AM, Gehrke B, Le Maitre NC, Kandziora M, Bellstedt DU. 2016. The biodiversity hotspot as evolutionary hot-bed: spectacular radiation of Erica in the Cape Floristic Region. BMC Evolutionary Biology 16: 764.

Porqueddu C, Ates S, Louhaichi M, Kyriazopoulos AP, Moreno G, Pozo A, Ovalle C, Ewing MA, Nichols PGH. 2016. Grasslands in 'Old World' and 'New World' Mediterranean-climate zones: past trends, current status and future research priorities. Grass and Forage Science 71: 1–35.

Potts AJ, Midgley JJ, Child MF, Larsen C, Hempson T. 2011. Coexistence theory in the Cape Floristic Region: revisiting an example of leaf niches in the Proteaceae. Austral Ecology 36: 212–19.

Poulsen ZC, Hoffman MT. 2015. Changes in the distribution of indigenous forest in Table Mountain National Park during the 20th Century. South African Journal of Botany 101: 49–56.

Rebelo AG, Boucher C, Helme N, Mucina L, Rutherford MC. 2006. Fynbos biome. Pages 52–219 in Mucina L, Rutherford MC, eds. The vegetation of South Africa, Lesotho and Swaziland. Strelitzia 19. South African National Biodiversity Institute, Pretoria.

Richards MB, Cowling RM, Stock WD. 1997. Soil factors and competition as determinants of the distribution of six fynbos Proteaceae species. Oikos 79: 394–406.

Rundel PW. 1981. The matorral zone of central Chile. Pages 271–308 in di Castri F, Goodall DW, Specht RL, eds. Mediterranean-type shrublands. Elsevier, Amsterdam.

Rundel PW. 1996. Monocotyledonous geophytes in the California flora. Madroño 43: 354–68.

Rundel PW. 2007. Sage scrub. Pages 208–28 in Barbour MG, Keeler-Wolf T, Schoenherr AA, eds. Terrestrial vegetation of California. University of California Press, Berkeley.

Rundel PW, Arroyo MT, Cowling RM, Keeley JE, Lamont BB, Vargas P. 2016. Mediterranean biomes: evolution of their vegetation, floras, and climate. Annual Review of Ecology, Evolution, and Systematics 47: 383–407.

Schimper AFW. 1903. Plant-geography upon a physiological basis. (Transl. by WR Fisher). Clarendon Press, Oxford.

Silvertown J, Araya YN, Linder HP, Gowing DJ. 2012. Experimental investigation of the origin of fynbos plant community structure after fire. Annals of Botany 126: 276–84

Specht RL. 1979. Heathlands and related shrublands of the world. In: Specht RL, ed. Ecosystems of the world. Elsevier Scientific, New York.

Specht RL, Moll EJ. 1983. Mediterranean-type heathlands and sclerophyllous shrublands of the world: an overview. Pages 41–65 in Kruger FJ, Mitchell DT, Jarvis JUM, eds. Mediterranean-type ecosystems. Springer, Berlin, Heidelberg.

Specht RL, Montenegro G, Dettmann ME. 2015. Structure and alpha biodiversity of major plant communities in Chile, a distant Gondwanan relation. Journal of Environment and Ecology 6: 21–47.

Stella JC, Rodríguez-González PM, Dufour S, Bendix J. 2013. Riparian vegetation research in Mediterranean-climate regions: common patterns, ecological processes, and considerations for management. Hydrobiologia 719: 291–315.

Sternberg M, Gutman M, Perevolotsky A, Ungar ED, Kigel J. 2000. Vegetation response to grazing management in a Mediterranean herbaceous community: a functional group approach. Journal of Applied Ecology 37: 224–37.

Stock WD, Verboom GA. 2012. Phylogenetic ecology of foliar N and P concentrations and N: P ratios across mediterranean-type ecosystems. Global Ecology and Biogeography 21: 1147–56.

Thuiller W, Lavorel S, Midgley GU, Lavergne S, Rebelo T. 2004. Relating plant traits and species distributions along bioclimatic gradients for 88 Leucadendron taxa. Ecology 85: 1688–99.

Tng DYP, Williamson GJ, Jordan GJ, Bowman DM. 2012. Giant eucalypts—globally unique fire-adapted rain-forest trees? New Phytologist 196: 1001–14.

Tomaselli R. 1981. Main physiognomic types and geographic distribution of shrub systems related to mediterranean climates. Pages 95–106 in di Castri F, Goodall DW, Specht RL, eds. Ecosystems of the world 11: mediterranean-type shrublands. Elsevier, Amsterdam.

Underwood EC, Klausmeyer KR, Cox RL, Busby SM, Morrison SA, Shaw MR. 2009. Expanding the global network of protected areas to save the imperiled Mediterranean biome. Conservation Biology 23: 43–52.

Van Wilgen BW, Forsyth GG. 1992. Regeneration strategies in fynbos plants and their influence on the stability of community boundaries after fire. Pages 54–80 in van Wilgen BW, Richardson DM, Kruger FJ, van Hensbergen HJ, eds. Fire in South African mountain fynbos: ecosystem, community and species response at Swartboskloof. Springer-Verlag, Berlin.

Verboom AG, Linder HP, Forest F, Hoffmann V, Bergh NG, Cowling RM. 2014. Cenozoic assembly of the Greater Cape Flora. Pages 93–118 in Allsopp N, Colville JF, Verboom AG, eds. Fynbos: ecology, evolution, and conservation of a megadiverse region. Oxford University Press, Oxford.

Verkaik I, Rieradevall M, Cooper SD, Melack JM, Dudley TL, Prat N. 2013. Fire as a disturbance in Mediterranean climate streams. Hydrobiologia 19: 353–82.

Wardell-Johnson G. 2000. Responses of locally endemic and regionally distributed eucalypts to moderate and high intensity fire in the Tingle Mosaic, south-western Australia. Austral Ecology 25: 409–21.

Wardell-Johnson G, Williams J, Hill K, Cummings R. 1997. Evolutionary biogeography and contemporary distribution of eucalypts. Pages 92–128 in Williams J, Woinowski J, eds. Eucalypt ecology: individuals to ecosystems. Cambridge University Press, Cambridge.

Wardell-Johnson GW, Calver M, Burrows N, Di Virgillio G. 2015. Integrating rehabilitation, restoration and conservation for a sustainable jarrah forest future during climate disruption. Pacific Conservation Biology 21: 175–85.

Wardell-Johnson G, Neldner J, Balmer J. 2017a. Wet sclerophyll forests. Pages 281–313 in Keith DA, ed. Vegetation of Australia. 3rd edn. Cambridge University Press, Cambridge.

Wardell-Johnson G, Crellin L, Napier C, Meigs G, Stevenson A, Ing Wong S. 2017b. Has canopy height and biomass recovered 78 years after an intense fire in south-western Australia's red tingle (*Eucalyptus jacksonii*) forests? International Journal of Wildland Fire 26: 148–55.

Waring RH, Franklin JF. 1979. Evergreen coniferous forests of the Pacific Northwest. Science 204: 1380–6.

Warming E. 1909. Oecology of plants. An introduction to the study of plant communities. Oxford University Press, London.

Whittaker RH. 1960. Vegetation of the Siskiyou mountains, Oregon and California. Ecological Monographs 30: 279–338.

Wood SW, Allen KJ, Hua Q, Bowman DM. 2010. Age and growth of a fire prone Tasmanian temperate old-growth forest stand dominated by *Eucalyptus regnans*, the world's tallest angiosperm. Forest Ecology and Management 260: 438–47.

Yates MJ, Anthony Verboom G, Rebelo AG, Cramer MD. 2010. Ecophysiological significance of leaf size variation in Proteaceae from the Cape Floristic Region. Functional Ecology 24: 485–92.

Zdruli P. 2014. Land resources of the Mediterranean: status, pressures, trends and impacts on future regional development. Land Degradation & Development 25: 373–84.

Choreography: Life in Motion

5 Evolution and Diversity

Abstract

As mediterranean-type climate (MTC) regions emerged and expanded, species from the regional pool colonized and persisted in these new climate regions. In general, taxa were derived from a few types of historical 'geoflora' communities: temperate, subtropical and tropical, and semi-arid or arid. Some of the taxa within modern mediterranean-type vegetation represent relatively ancient relict taxa that pre-date the emergence of mediterranean-type drivers. Other lineages underwent subsequent speciation, resulting in the evolution of new MTC region-specific taxa, including the production of many new species through evolutionary radiations. Low extinction rates associated with historically stable climate and limited recent geological activity might explain the high diversity found in some MTC regions, while in regions with more topographical variation the ability of species to move across elevation gradients has been suggested also to have allowed species to be buffered from climatic changes that may otherwise have led to extinctions.

5.1 Origins

Prior to the onset of the key drivers of mediterranean-type ecosystems (MTEs) other types of ecosystems and plant communities existed in the regions that would become MTEs. As mediterranean-type climate (MTC) regions expanded and other drivers developed, such as predictable fire regimes, species that already existed in the region migrated into these regions and/or persisted in the face of newly emerging conditions (Figure 5.1). Over time, other species from the regional pool may also have been able to move into MTEs. Some of these lineages underwent subsequent speciation, resulting in the evolution of new MTE-specific taxa, and other taxa went extinct. As has already been introduced in Chapter 4, these community assembly

The Biology of Mediterranean-Type Ecosystems. Karen J. Esler, Anna L. Jacobsen, and R. Brandon Pratt,
Oxford University Press (2018). © Karen J. Esler, Anna L. Jacobsen, and R. Brandon Pratt 2018.
DOI 10.1093/oso/9780198739135.001.0001

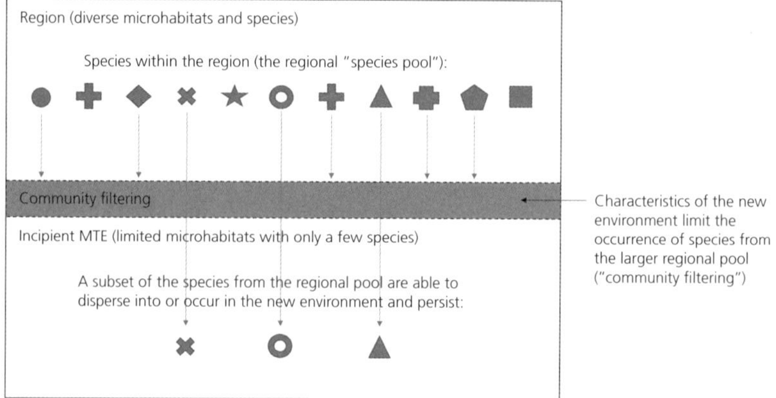

Figure 5.1   **Species pools and filters.** Early mediterranean-type ecosystems (MTEs) were populated by species that filtered into newly emerging habitats from adjacent communities. The species that first occurred within developing MTEs were likely species from the regional species pool that were able to persist in these changing and emerging environments. These species represented just a fraction of the species that occurred near the incipient MTEs and which would have been able to disperse into the community or which existed in that area prior to changes in the environment. The characteristics of the developing MTEs (climate, fire, etc.) acted as a filter that allowed for the persistence of only species with certain traits.

processes resulted in the formation of suites of species co-occurring within different MTE communities.

Some of the taxa within modern mediterranean-type vegetation (MTV) represent relatively ancient taxa that pre-date the emergence of mediterranean-type drivers and which may have changed very little over time. These taxa may be traced back to several different types of communities that existed prior to the emergence of MTEs (Figure 5.2). In general, MTE taxa may be derived from a few types of historical communities: 1) communities that contained temperate forest taxa; 2) communities that contained subtropical and tropical taxa; and 3) semi-arid or arid interior communities that may have existed near emerging MTEs.

Pre-mediterranean taxa are discussed as belonging to different 'geofloras'. Geofloras are identified and named based on their biogeographical origin, the timing of their origination (based on geological times; see Figure 2.1), and their floristic composition. There may be several overlapping geoflora units within an individual region, based on differences in how they have been defined by different authors. Additionally, because of periods of past climatic change, geoflora types are sometimes difficult to interpret. Modern elements of past geofloras are often identified through a combination of phylogenetic relationships and fossil records. For instance, if a plant family has numerous taxa that exist within rainforests and only a few related taxa

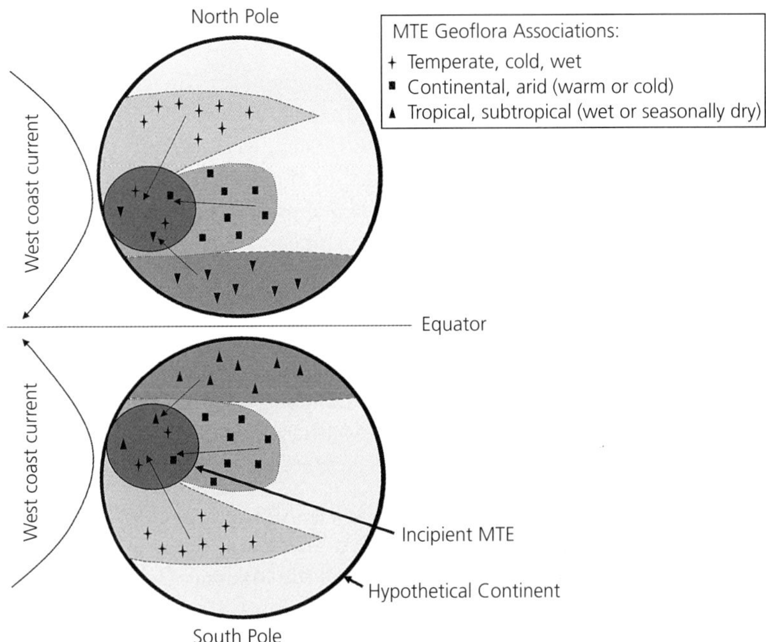

Figure 5.2 **Origin of taxa that existed in and colonized incipient mediterranean-type ecosystems (MTEs).** Incipient MTEs (shown in dark grey) were inhabited by species that were related to or the same as the taxa that existed in extant adjoining communities (other shades of grey, with taxa indicated with differing shapes). For most MTEs, pre-mediterranean taxa included those from temperate, subtropical, and tropical communities, and limited contributions from continental arid communities. (This figure is based loosely on the 'hypothetical continent' precipitation regimes presented in Thrower and Bradbury 1973.)

within MTEs, that family may be interpreted as deriving from a tropical geoflora. Alternatively, if fossils from a taxon are identified as being part of a past temperate flora and that taxon is now found in MTEs, that taxon may be interpreted as deriving from a formerly temperate geoflora.

The idea of geofloras as describing communities of origin for different MTEs has been developed to differing extents for each of the different regions. In brief, for each region, some elements of the modern floras are affiliated with the following pre-mediterranean origins:

- In the Mediterranean, taxa have been associated with several different historical groups, and names for these differ depending on their biogeographical origin as well as their timing of origination:

  - The Laurasian geoflora (Quézel 1978) includes two groups of taxa. The first group represents temperate species that have northern origins, including many species with circumpolar or circumboreal distributions and affiliations (Quézel 1985). This is roughly analogous to the Holarctic group as discussed in Blondel et al. (2010). This group includes evergreen and deciduous tree species. Example taxa include *Juglans*, *Platanus*, *Pinus*, *Rhododendron*, and *Rhamnus*. There is also a group of subtropical and arid species, the Madrean-Tethyan flora, that include representatives such as *Quercus*, *Rhus*, *Olea*, *Nerium*, and *Laurus* (Kovar-Eder et al. 2006; Rodríguez-Sánchez et al. 2009).

- The Irano-Turanian taxa represent species from the cold arid steppes to the east of the Mediterranean (Quézel 1985; Blondel et al. 2010), and includes some arid-adapted shrub elements, such as *Salsola* and *Artemisia*, as well as deciduous tree species, including *Prunus* and *Sorbus*.
- The Sahelian taxa (Quézel 1985), identified as the Saharo-Arabian component by Blondel et al. (2010), represent southern warm arid species as well as some dry tropical components. These elements are particularly well represented in Mediterranean Africa (Quézel 1978) and include many Chenopodiaceae species.

- In California, there is considerable overlap in the origin of pre-mediterranean elements of the taxa with the Mediterranean, since both contain ancient elements of the Madrean-Tethyan geoflora:

  - Many species within California have been linked to the ancient Madrean-Tethyan flora (Axelrod 1975; Valiente-Banuet et al. 1998), which represents floral connections from the Mesophytic when North America, Europe, and Asia were connected and formed the super-continent of Laurasia (see Figures 2.1 and 2.2). Elements of this geoflora would have shared a continent with the Mediterranean during this period. Axelrod (1975) includes *Arbutus*, *Pinus*, *Platanus*, *Quercus*, *Prunus*, *Rhamnus*, and *Rhus* within this group, as already described above for the Mediterranean as part of the Laurasian floral group.

  - When classified using geofloras based upon the more recent Palaeogene, many of these same floristic elements are divided into Arcto-Tertiary taxa and Madro-Tertiary taxa (Raven and Axelrod 1978). The Arcto-Tertiary geoflora represents cold-adapted and mesic northern forest taxa, while the Madro-Tertiary geoflora represents sclerophyllous species from warmer tropical and arid environments.

  - A smaller proportion of the flora of California is derived from taxa that have origins in warm and arid adjacent communities. About 14% of the flora of the California Floristic Province have affinities to adjacent desert-affiliated taxa, including some annuals and subshrubs (Raven and Axelrod 1978). These taxa colonized the MTEs more recently than those mentioned above.

  - Relatively few species appear to have entered the California flora from South America, although the North American and South American continents are currently connected. There are more taxa that have moved from North America to South America than in the opposite direction (Raven and Axelrod 1978).

- In Chile, the geofloras represent ancient connections to Gondwanan and the other southern hemisphere continents, more recent exchange following the connection of North and South America, and a strong connection to the South American tropics (see also Case Study 6 in Chapter 3):

  - Elements of the Gondwanan geoflora are present in Chile as well as the other southern hemisphere MTEs (Rundel et al. 2016). In Chile, examples of mesic Gondwanan flora elements include *Nothofagus* and *Araucaria* (Arroyo et al. 1995). There are also some arid elements of the flora that have Gondwanan-associated distributions, including *Prosopis* and *Acacia*, which are shared between South America and Africa (Solbrig et al. 1977). These two elements of the flora that have Gondwanan connections are sometimes separated into southern and Antarctic geofloras.

  - The neotropical taxa are derived from the tropical forests of South America. There are strong connections between the MTC region of Chile and the tropical forests found to the east of the Andes. Taxa of neotropical rainforest affinity include *Aextoxicon*, *Myrceugenia*, *Quillaja*, and *Colliguaja*. The derivation of taxa within Chile from these rainforest taxa likely reflects the initial uplift that cut off tropical species along the western coast while also preventing colonization by arid-adapted continental species (Arroyo et al. 1995). Several of these genera also have disjunct distributions, with related taxa occurring in Brazil and Chile, with the uplifted Andes in between.

  - There are some desert elements represented within the Chilean MTE flora, but these are relatively limited (Arroyo et al. 1995).

  - More recent introduction of North American species, following the joining of North and South America, occurred (Solbrig et al. 1977). This has included mostly herbaceous lineages (Arroyo et al. 1995), as well as a few shrub components, such as *Sambucus* (Solbrig et al. 1977).

- In Australia, there are clear geoflora connections to Gondwana, with the flora largely developing in isolation following the separation of the supercontinent. The continent also went through several periods of drying:

  - The Gondwanan elements of the geoflora represent families such as the Proteaceae, Myrtaceae, and Restoniaceae, which are conspicuous members of the modern floras of both Australia and South Africa. Climate was much wetter and milder during the period of ancient connection between these landmasses (Dodson and Kershaw 1995), with subsequent drying and the appearance of a more arid-adapted flora (Mucina et al. 2014). This geoflora included both tropical and temperate floristic elements.

  - The arid geoflora arrived in Australia following its separation from other landmasses and moved into the MTEs following the onset of a drying climate. These elements include the families Chenopodiaceae and Asteraceae (Crisp et al. 2004).

- In South Africa, once again, there are clear Gondwanan connections of the flora. Additionally, periods of drying, when combined with the geography of the African continent, which lacks landmass to the south to allow for polar migration during dry periods, meant that many mesic and temperate taxa were lost from this region over evolutionary time (Goldblatt 1978).

  - As already mentioned above for Chile and Australia, there are many taxa that are derived from the Gondwanan geoflora. In several studies, this is referred to as the Cape component of the geoflora because of the strong association with the occurrence of these families to the present day Cape Floristic Region (Lubke et al. 1986; Galley and Linder 2006).

  - The tropical and subtropical geoflora includes many taxa that are today represented in the Afromontane forests and thickets, as well as some of the woody elements of the karoo, such as *Euclea* (Goldblatt 1978; Lubke et al. 1986) and also *Searsia* (formerly *Rhus*) (Galley and Linder 2006).

  - The arid geoflora likely developed during periods of drying from African taxa and moved into dry regions more recently (Cowling 1983; Galley and Linder 2006). Today, many of these families represent highly diverse elements of the Succulent Karoo.

Taxa from these historical plant communities that were able to colonize and persist in emerging MTEs had characteristics that enabled them to respond to different types of general disturbance, including deep root systems, sclerophyllous and evergreen leaves, and long lifespan (Peñuelas et al. 2001), as well as reduced flowers and large animal-dispersed seeds (Herrera 1992; Groom and Lamont 2015). These traits may be considered pre-adaptations (see Chapter 3) to some of the disturbance-related drivers present in MTEs. Taxa that later would evolve within MTEs (neo-mediterranean species) often do not have these same traits. Rather, they display many of the characters that were previously discussed in Chapter 3, including being smaller shrubs that do not resprout and have soil-stored seeds (Chapter 6). One of the reasons that this suite of traits has been identified as being adaptations to MTE drivers is that they, for some lineages, evolved within MTC regions and were not present prior to the onset of MTC conditions (see Chapter 3).

In some cases, the origins and evolutionary history of taxa within MTEs continue to have an influence on the functional responses of some species. For instance, a study that examined plant drought response in Spain separated plants into pre-mediterranean and neo-mediterranean species (Peñuelas et al. 2001). The pre-mediterranean flora represent taxa that pre-date the emergence of MTEs and appear to have changed little from their pre-mediterranean ancestors. In this study, this included *Pistacia*, *Olea*, *Quercus*, and *Pinus*, and they were compared to species that were identified as emerging following the establishment of mediterranean conditions, including *Lavandula, Genista,* and *Erica.* In response to drought, the neo-mediterranean species were heavily impacted, but exhibited a stronger recovery after the drought was relieved when compared to the pre-mediterranean taxa. A separate study also found this same difference between pre-mediterranean and neo-mediterranean taxa drought response and recovery (Gratani and Varone 2004). However, in interpreting these studies there are assumptions associated with the categorization of modern species as pre- or neo-mediterranean (Ackerly 2009). It is important to evaluate the time of trait origin for specific relevant traits of interest rather than the time of taxa origin (Ackerly 2009). Also, since all of these species are present in MTC regions, they all have traits that enable continued persistence within these regions.

Ancient differences linked to the origins of lineages may also be related to the assemblies of species that are found today among different communities within MTC regions (Chapter 4). The environmental factors that drive changes in communities and their species composition across the landscape select for characters that may be associated with specific lineages (Herrera 1992). This results in communities potentially being composed of a higher proportion of specific lineages than neutral assembly processes would predict. This process has been called 'lineage sorting' (Vrba and Gould 1986) and it results in species co-occurring that share certain traits that are phylogenetically conserved and that share common origins. An example of this is the thicket communities of South Africa, which contain a high proportion of species that can be linked to tropical origins (Linder et al. 1992).

5.2 Macroevolutionary patterns

5.2.1 Divergence

From the initial colonization of incipient MTC regions by existing taxa, selection for traits that were adaptive in these new environments resulted in changes in species characters over time, **anagenesis**. Sometimes, these changes lead to the reproductive isolation of populations and the development of new species, **cladogenesis**. Other taxa were able to track suitable environmental conditions and microhabitats and appear to have changed very little over time (i.e. evolutionary stasis).

Within modern MTC regions, there are many examples of taxa that appear to be little changed over time. These **relict species** often exist in isolated mesic pockets within broader modern MTEs and have distributions that are much reduced compared to past distributions, as inferred from the fossil record or the distances between modern disjunct populations. Examples

Figure 5.3   **Relict species currently have limited distributions.** Within mediterranean-type ecosystems there are many examples of relict species that formerly had larger ranges, but now are quite restricted. Some of these species have changed very little over time, including *Drimys winteri* (a), *Sequoiadendron giganteum* (b); inset shows the cone of this species), and *Podocarpus latifolius* (c). (See Plate 10)

Source: Photos for panels a and b from Anna L. Jacobsen and photo for panel c from Stuart Hall.

of these relatively unchanged species include *Drimys winteri* (Winteraceae; Chile), *Podocarpus latifolius* (Podocarpaceae; South Africa) and *Sequoiadendron giganteum* (Cupressaceae; California), which is extant today in a limited range but is known from fossils and previously had a much larger distribution than its current range (Raven and Axelrod 1995; Figure 5.3; see Plate 9).

Other taxa changed over time and also diverged to produce new species. The production of many new species from a common ancestor is called **evolutionary radiation** (sometimes used synonymously with the term adaptive radiation). A high proportion of endemic species within MTC regions, discussed in Chapter 3, is largely a result of high numbers of neo-mediterranean species. Approximately a quarter of the species within Mediterranean Africa have evolved as new species within the region (1037 of 4034 total species) (Quézel 1978). In some of the other MTC regions, this proportion is even higher. This pattern is apparent within many lineages within these regions that currently display high levels of diversity. A large portion of the diversity within South Africa and Australia is caused by high levels of speciation within just a few families and genera that have radiated and are endemic to these regions (Linder 2003; Hopper and Gioia 2004).

Table 5.1 Number of species within the top ten most diverse genera. The top ten most speciose genera for the mediterranean-type ecosystems (MTEs) of Australia, California, Chile, and South Africa are included, along with the number of taxa within each genus that occur within the MTE. For Chile, 11 species are included in the list because of several genera with the same reported number of species. Sources include Hopper and Gioia (2004); Raven and Axelrod (1995); Baldwin et al. (2012); Moreira-Muñoz (2011); Goldblatt and Manning (2002)

| Australia | | California | | Chile | | South Africa | |
Genus (family)	Taxa (#)	Genus (family)	Taxa (#)	Genus (family)	Taxa (#)	Genus (family)	Taxa (#)
Acacia (Mimosaceae)	502	*Eriogonum* (Polygonaceae)	218	*Senecio* (Asteraceae)	224	*Erica* (Ericaceae)	657
Eucalyptus (Myrtaceae)	362	*Astragalus* (Fabaceae)	147	*Adesmia* (Fabaceae)	130	*Aspalathus* (Fabaceae)	272
Grevillea (Proteaceae)	229	*Carex* (Cyperaceae)	143	*Viola* (Violaceae)	72	*Pelargonium* (Geraniaceae)	148
Melaleuca (Myrtaceae)	185	*Phacelia* (Boraginaceae)	107	*Carex* (Cyperaceae)	67	*Agathosma* (Rutaceae)	143
Stylidium (Stylidiaceae)	170	*Lupinus* (Fabaceae)	104	*Calceolaria* (Calceolariaceae)	60	*Phylica* (Rhamnaceae)	133
Leucopogon (Ericaceae)	165	*Arctostaphylos* (Ericaceae)	95	*Haplopappus* (Asteraceae)	54	*Lampranthus* (Aizoaceae)	124
Caladenia (Orchidaceae)	162	*Mimulus* (Phrymaceae)	83	*Oxalis* (Oxalidaceae)	53	*Oxalis* (Oxalidaceae)	119
Verticordia (Myrtaceae)	138	*Cryptantha* (Boraginaceae)	76	*Astragalus* (Fabaceae)	49	*Moraea* (Iridaceae)	115
Dryandra (Proteaceae)	136	*Penstemon* (Plantaginaceae)	73	*Nolana* (Solanaceae)	44	*Cliffortia* (Rosaceae)	114
Hakea (Proteaceae)	105	*Ceanothus* (Rhamnaceae)	62	*Solanum* (Solanaceae)	44	*Senecio* (Asteraceae)	110
				Valeriana (Valerianaceae)	44		

One indicator that MTC regions have been centres for evolutionary radiations is the prevalence of large numbers of closely related species within each of the five MTC regions. Within each region, there are a limited number of genera that make up a large portion of the floral diversity (Tables 5.1 and 5.2). Even among the most species-rich genera, the top few genera within each region contain many more species than the other genera. Raven and Axelrod (1995) calculated that the ten largest genera accounted for 15.2 per cent of the species within the California Floristic Province and 17.5 per cent of the species within the South African Cape flora. It is also interesting to note that many of these diverse genera represent herbaceous plant diversity, with the exception of Australia.

Many studies have examined the traits that may have undergone divergent selection as part of the radiations within diverse MTE genera. These traits

Table 5.2 Diverse genera of the Mediterranean. Genera from different Mediterranean regions that have been reported as being particularly diverse are included; there is not a centralized listing for the species of the entire Mediterranean mediterranean-type ecosystem. All genera are included that were reported in the top ten most diverse genera of at least two of the included studies. Taxa numbers for the mediterranean portion of Northern Africa represent only the endemic species within each genus. For all regions, the absence of taxa numbers in the table indicates that that genus is not reported as being particularly diverse, although there may still be representatives of that genus present in the region. Sources include Quézel (1978); Blondel and Aronson (1999); Thompson (2005); Georghiou and Delipetrou (2010); Bacchetta et al. (2012); Cueto et al. (2014)

Mediterranean					
	Africa	Turkey	Greece	Sardinia, Italy	Eastern Andalusia, Spain
Genus (family)	Taxa (#)	Taxa (#)	Taxa (#)	Taxa (#)	Taxa (#)
Silene (Caryophyllaceae)	48	9	119	6	45
Centaurea (Asteraceae)	25	13	122	6	48
Astragalus (Fabaceae)	18	7		6	30
Galium (Rubiaceae)	7	9			32
Verbascum (Scrophylariaceae)	5	6	72		
Ranunculus (Ranunculaceae)	6	6			31
Limonium (Plumbaginaceae)	13			33	
Ophrys (Orchidaceae)			90	6	
Teucrium (Lamiaceae)	25				40
Vicia (Fabaceae)		10			32
Ononis (Fabaceae)	16		33		
Alyssum (Brassicaceae)	6	12			
Salvia (Lamiaceae)	7	11			
Anchusa (Boraginaceae)		6		6	
Stachys (Lamiaceae)	11				52

differ for different taxa. Within *Cistus* in the Mediterranean, it appears that selection for differences in leaf morphology was important (Guzmán et al. 2009). For some genera, floral selection and interactions with pollinators appear to have been important in speciation, such as *Ophrys* (the 'bee orchids'; Stebbins 1970) and *Salvia* (Claßen-Bockhoff et al. 2004) in the Mediterranean and *Gladiolus* (Goldblatt et al. 1998) and *Disa* (Johnson et al. 1998) in South Africa. Response to fire and selection for different modes of post-fire regeneration were important in the evolutionary radiations of taxa such as *Arctostaphylos* (Figure 5.4; see Plate 10) and *Ceanothus* in California (Wells 1969; Raven and Axelrod, 1978), as well as taxa from other MTEs (Pausas and Keeley 2014; He et al. 2016). Finally, changes in climate, acidification, and the creation of topographical and edaphic microhabitats has likely been important in the evolution of diversity within some lineages (Crisp et al. 2004; Burge et al. 2011; Cowling et al. 2015).

Figure 5.4 **Resprouting and non-resprouting species from the same genus.** *Arctostaphylos* species in California include many species that resprout following fire as well as many species that do not resprout following fire. Photos in panels a, c, d, and f are of the resprouting species *A. glandulosa* and photos in panels b, e, and g are of the non-resprouting species *A. glauca*. While these species differ in their post-fire resprouting ability, other features, such as their flowers, leaves, and fruit, are not very divergent. (See Plate 11)

Source: Photos c, d, e, and f from Anna L. Jacobsen and photos a, b, and g from R. Brandon Pratt.

5.2.2 Convergence

An important topic of research within MTC regions has been the examination of convergent evolution (discussed in Chapters 1 and 2). This topic is mentioned again here, as it relates to evolution and patterns of diversity that are apparent within MTC regions. Convergence may be examined across several different levels of organization, including at the ecosystem, community, and species levels. At the specific level, a pattern of evolutionary convergence would be demonstrated by the evolution of similar characters in response to developing MTC conditions within the different regions. That is, taxa within each of the different regions would display changes in their characters over time, with the end result being that, from whatever their different historical characters may have been, they are now more similar to one another than they used to be. At this level, a classic example has been the presence of species with very similar leaf types across the different regions. For instance, similar needle-like leaves are found in *Adenostoma* (Rosaceae; California), *Erica* (Ericaceae; Mediterranean), and *Stoebe* (Asteraceae; South Africa), and similar broad sclerophyllous leaves are found in *Heteromeles* (Rosaceae; California), *Lithraea* (Anacardiaceae; Chile), *Quercus* (Fagaceae; Mediterranean), and *Protea* (Proteaceae; South Africa) (these examples are from Cody and Mooney 1978; Chapter 6).

At broader levels, convergent evolution has been examined by looking across MTC regions for similarities in community structure, function, and/or resource use. This is often based on an analysis of the frequency of certain traits outside of MTC regions compared to within them. For instance, MTC region communities contain more species that depend on fire for successful seedling recruitment compared to non-MTE species (Carrington and Keeley 1999). These comparisons can be complicated depending on how broadly they are examined, the types of communities within MTC regions that are being compared, and how sites are matched (Case Study 9). For instance, a study

---

Case Study 9 Mediterranean heathland and fynbos: a neglected example of convergence between mediterranean climate regions

Fernando Ojeda, Departamento de Biología-IVAGRO, Universidad de Cádiz, Campus Río San Pedro, Spain

Mediterranean-type ecosystems (MTEs) are recognized worldwide by high levels of plant diversity and endemism. Many studies have explored the processes that underlie this biodiversity and have documented sound examples of evolutionary convergence across mediterranean-type climate (MTC) regions. For instance, similar mediterranean heathland-like formations are found in both the South African (fynbos) and Australian (kwongan) MTC regions. They harbour large numbers of endemic species from a limited number of lineages (i.e. low taxonomic distinctness), mainly edaphic endemics associated with a shrub habit and pyrogenic (fire-stimulated) germination (Cowling et al. 1994). This remarkable example of convergence is often explained by the shared presence of highly infertile, acid soils and a similar regime of recurrent fires (Cowling et al. 1994; 1996), together with a stable topography through the Cenozoic and stable climatic conditions at least since the late Pliocene (Cowling et al. 2015). Richer, non-acidic soils, pronounced climate changes during the Pleistocene, and a lesser influence of fire, in terms of longer fire-return intervals, may account for different biodiversity patterns in MTEs of California, Chile, and the Mediterranean Basin (Cowling et al. 1996). In these three MTC regions, most endemism is not edaphic but orographic and largely associated with the herbaceous habit (Cowling et al. 1996), and pyrogenic germination is not as prevalent (Pausas et al. 2004). However, importantly, the patterns above are usually based on comparisons of just one type of sclerophyllous shrubland as representative of each region (e.g. Cowling et al. 1996; Dallman 1998; Keeley and Fotheringham 2003): fynbos in the South African Cape, kwongan in southwestern Australia, matorral in Chile, chaparral in California, and garrigue in the Mediterranean Basin.

The Mediterranean Basin is the largest of the five MTC regions and, although garrigue-like shrublands are nearly ubiquitous (Dallman 1998), heathlands are frequent in its western end (western Iberian Peninsula and northwestern tip of Africa; Rivas-Martínez 1979; Loidi et al. 2007). Heathlands are shrubland formations generally lacking a tree layer and dominated by evergreen, fine-leaved dwarf shrubs in the Ericaceae family. They are globally widespread but mostly restricted to nutrient-poor, acid soils under temperate climatic conditions. This Mediterranean heathland, known as herriza in the Strait of Gibraltar region (Ojeda 2011), is associated with a mild mediterranean climate and the relatively high occurrence of acidic, infertile soils, otherwise scarce in the Mediterranean Basin (see Böhner et al. 2008). It has often been regarded as interdigitations of the European Atlantic heathland with 'xerophytic' Mediterranean shrublands (Rivas-Martínez 1979; Loidi et al. 2007) and has been largely ignored in comparisons of MTEs across MTC regions.

The existence of remarkable floristic and ecological differences between this Mediterranean heathland and garrigue-like shrublands suggests it is a unique vegetation type within the Mediterranean Basin. Apart from biogeographical patterns that contrast with garrigue-type shrublands (Arroyo and Marañón 1990), endemism richness is higher and taxonomic distinctness lower in the Mediterranean heathland (Ojeda et al. 1995; 2000). Unlike some other Mediterranean shrublands, endemism is edaphic (related to highly acidic, nutrient-poor, sandstone soils; Ojeda et al. 1995; 2000) and markedly associated with the shrub lifeform (Ojeda et al. 2000; 2001) in this community type.

Ojeda et al. (2001) analysed the similarities in biodiversity patterns between the Mediterranean heathland from the southwestern Mediterranean Basin and fynbos from the South African Cape and highlighted them as a sound example of convergence between the two MTC regions (Case Study 9 Figure). Their shared association with (i) acidic, infertile soils, (ii) a somewhat buffered Pleistocene, and (iii) recurrent fires could explain the similarities in biodiversity patterns (Ojeda et al. 2001). However, similarities in patterns are not paralleled by similarities in levels, as endemism richness is much higher

Continued

Case Study 9 (*Continued*)

Case Study 9 Figure. Convergence between heathland communities. Two images of heathland-type communities illustrate similar community structure, both of which contain many dominant dwarf shrubs in the Ericaceae. The image on the top (a) is from a Mediterranean heathland located near the Strait of Gibraltar (sierra del Niño, Spain). The image on the bottom (b) is of fynbos shrub communities in the SW Cape Region (Klein River Mountains, South Africa).

Source: Photos from Fernando Ojeda.

and taxonomic distinctness lower in fynbos (Ojeda et al. 2001). Higher winter-rainfall reliability (Cowling et al. 2005) and higher topographical and climatic stability since the Pliocene (Cowling et al. 2015) in the South African MTC region might determine a dramatic asymmetry in diversification-extinction rates (e.g. Segarra-Moragues et al. 2013;

Onstein et al. 2015) and thus explain overall differences in diversity levels between the southwestern Mediterranean heathland and the South African Cape fynbos (Ojeda et al. 2001).

Although there is clear evidence of the evolutionary influence of fire in the Mediterranean Basin (Keeley et al. 2012), this influence is generally regarded as lower than in other MTC regions, such as the South African Cape and southwestern Australia (Cowling et al. 1996; Pausas et al. 2004; Keeley et al. 2012). However, regarding its ecological relationship with fire, the Mediterranean heathland is also more similar to heathland-like formations from other MTC regions (e.g. fynbos) than to other shrublands from the western Mediterranean. It is more flammable than neighbouring garrigue-like shrublands (Ojeda et al. 2001) and, unlike them, pyrogenic germination (P+) is prevalent in both resprouter (R+) and non-sprouter (R-) herriza species, suggesting that, at a landscape scale, fire has played a stronger role in its evolution (Ojeda et al. 2001; Ojeda et al. 2010).

Similarities between the Mediterranean heathland and fynbos highlight the importance of comparing across the many communities present within each MTE, rather than just the dominant or 'type' communities that are often cited but potentially not representative of some of the most informative comparisons that could be made. It would be valuable in the future to examine heathland-type communities in other MTC regions for comparison. For instance, shrubland communities on nutrient-poor, serpentine soil patches in the Californian Coast Ranges are highly diverse, including many Ericaceae species (*Arctostaphylos*), with many edaphic (serpentine) endemics (e.g. Kruckeberg 1984; Safford and Harrison 2004). Thus, there may be additional opportunities to explore patterns of heathland convergence across MTEs.

of plant hydraulic traits found that species from a California chaparral community displayed similar traits to those from a South African fynbos community, but that species from other types of MTC region communities within California displayed different traits (Figure 5.5). Given the diversity of community types that occur within MTC areas (Chapter 4) it is likely that future studies of this question will benefit from a broader sampling of MTC communities than have been included to date. As scientific study broadens and increases, there are many opportunities to evaluate patterns of convergence across MTC communities (Case Study 9).

Evolutionary convergence among MTC regions is particularly interesting given the relatively isolated nature of the different regions from one another, both now and since the early emergence of these climate regions. Although there are some ancient connections between the floras of some regions, as described above, the origins for the vast majority of taxa within MTC regions pre-date the emergence of modern MTCs. This provides unique opportunities to evaluate evolutionary patterns with five independent 'replicates' of emergent MTC with different taxa and different geomorphological and climatic histories. This makes MTC regions unique natural laboratories for evolutionary study.

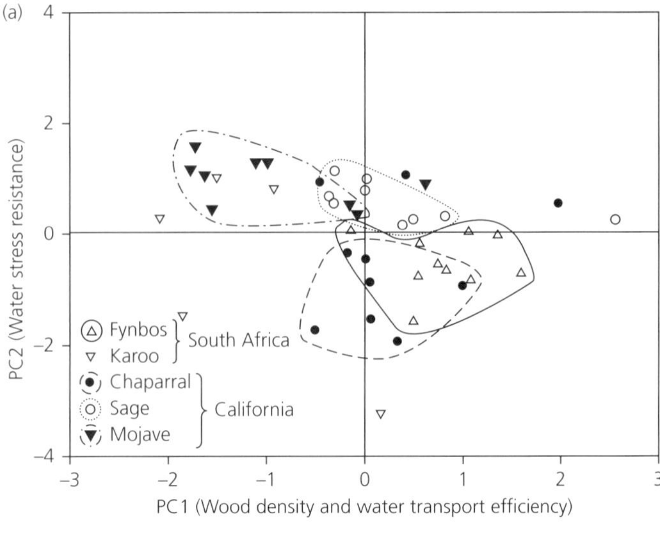

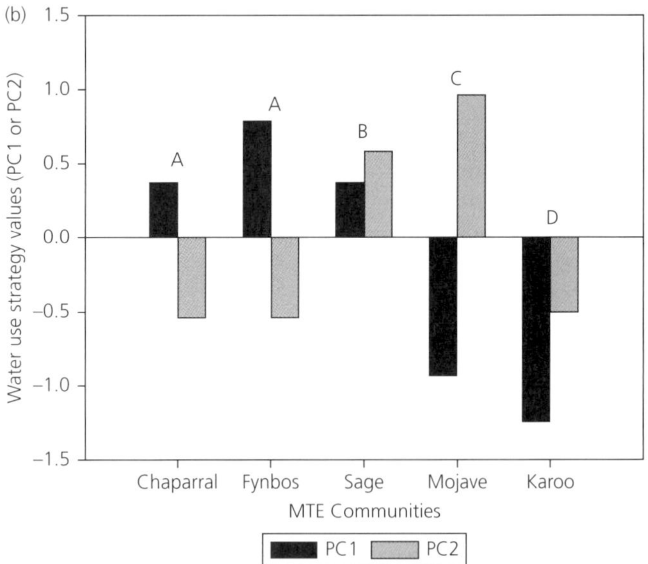

Figure 5.5 **Convergence among mediterranean-type ecosystem (MTE) communities.** Convergence between the hydraulic strategies of species from different mediterranean-type climate (MTC) region communities from two MTC regions, California and South Africa, were examined. Multiple hydraulic traits were combined into two principal components (PC1 and PC2); see Jacobsen et al. 2009 for additional details. In panel a, each symbol represents a different species within each community. In panel b, each bar represents the mean principal component values for each community and the letters above the bars indicate differences between communities; communities with the same letter are not different, while communities with different letters are different. Convergence was evident between two communities, California chaparral and South Africa fynbos, but other types of MTC region communities contained species that displayed significantly different sets of hydraulic traits.

5.2.3 Extinction

Both of the evolutionary topics discussed above, divergence and convergence, deal with how species change over time. An important additional topic is the loss of species, i.e. species extinction. While some taxa are ancient and have been around for seemingly long periods of time and potentially radiated, other taxa exist for relatively brief periods. Understanding both rates of speciation and rates of extinction is important when examining the evolution of diversity, because high levels of diversity can result from high speciation rates and average extinction rates, low extinction rates and average speciation rates, or both low extinction and high speciation rates. The influence of extinction on MTE biodiversity is discussed below.

5.3 Diversity

MTC regions have high biodiversity. There are numerous factors that may contribute to the accumulation of species through time, including both high speciation rates and low extinction rates. Numerous studies have examined the influence of these different factors on diversity within MTC regions. The question of what drives high MTC region diversity has been framed as whether or not MTC regions are 'cradles' or 'museums' of biodiversity (Stebbins 1974); that is, do they foster the formation of many new species (cradles) or do they maintain many ancient species (museums)? Studies that have examined plant diversity patterns within individual MTC areas have found that both high speciation and low extinction have been important contributors to high diversity, but that low extinction and the maintenance of ancient and relict taxa has been a particularly important determinant of modern levels of diversity (reviewed in Rundel et al. 2016).

Regarding MTC regions as 'cradles,' there is evidence to support the diversification of some lineages within the regions; however, when compared across regions, the high diversity of MTC areas cannot be explained by their levels of diversification alone. Some examples of highly diverse lineages have already been discussed above in the context of evolutionary divergence and the presence of very speciose genera within each of the MTC regions (Tables 5.1 and 5.2). These exceptional cases are balanced by other genera that have very few representatives that have not diverged, sometimes even over very long periods of time. For example, in a study comparing rates of diversification in South African and Australian Proteaceae, the results suggested that the MTC regions were centres of diversification for some genera within this plant family, but that there were also lineages within the Proteaceae that had not diverged and had been maintained over long periods of time (Sauquet et al. 2009). Taken as a whole, diversification rates within MTC regions are not different than those outside of them (reviewed in Rundel et al. 2016). Among

MTC regions, the highest speciation rates have been recorded in the Mediterranean Basin (>1 species per million years in herbaceous and subshrub clades; Verdú and Pausas 2013), intermediate rates in the Cape and California, and the lowest rates in Chile and southwest Australia (Linder 2008; Madriňán et al. 2013). These differences in speciation rates are not consistent with present-day diversity differences across MTC regions (see Chapter 2, Case Study 4).

High species diversity within MTC regions appears to be related to low rates of extinction (Rundel et al. 2016; Onstein and Linder 2016). Extinctions are difficult to quantify because not all species are captured in the fossil record, but extinction rates can be inferred by evaluating the difference between divergence rates and diversity. An additional source of information on extinction rates is the prevalence of ancient taxa. In Australia, high diversity has been linked to some recently evolved taxa as well as the maintenance of relict species (Hopper 1979; Hopper and Gioia 2004). The same has been reported for South Africa, with a large number of species having radiated in a few lineages as well as the presence of many relict taxa (Goldblatt 1978; Linder 2003). Chile has many relict species and communities (Hinojosa et al. 2006; Núñez-Ávila and Armesto 2006; Fleury et al. 2015; see Figure 3.5), as does California (Stebbins and Major 1965; Lancaster and Kay 2013). Conservation of relict taxa is an important component of conservation planning to maintain diversity within MTC communities (Wardell-Johnson and Horwitz 1996; Médail and Quézel 1999; Garreaud et al. 2008; Schut et al. 2014).

The dual role of both divergence and low extinction as contributors to MTC region diversity is well illustrated by the diversity of **refugia** within the Mediterranean. Refugia are small regions that have been buffered from extreme climatic changes and disturbances when compared to the surrounding area and they often contain relict species. In the Mediterranean, refugia tend to be at mid-elevation, providing the opportunity for both upslope and downslope migration if conditions change, and they also tend to be in canyons that are less arid than surrounding areas (Médail and Diadema 2009). Refugia contain both higher numbers of Tertiary taxa and also evidence of recent speciation within some lineages (Tzedakis 2009). The combination of both new and old species within these refugia makes them centres of regional endemism. Hotspots of biodiversity within the Mediterranean are strongly associated with refugia and the co-occurrence of new and ancient taxa (Médail and Quézel 1997; Médail and Diadema 2009; Tzedakis 2009; Hewitt 2011).

5.4 Drivers of diversity

There are numerous proposed explanations for why MTEs may have high diversity (Case Study 10). One of the most commonly cited factors influencing high present diversity is climatic stability, which permitted the

Case Study 10 The origin of the mediterranean biome

Philip W. Rundel, Department of Ecology and Evolutionary Biology, University of California, Los Angeles (UCLA), Los Angeles, California, USA

Mediterranean-type ecosystems (MTEs) have an important place in the biodiversity of plant species. This significance leads to a question of when and how these ecosystems evolved and the factors that have led to the evolution of their remarkable plant diversity. It was once widely held that the novel mediterranean-type climate (MTC) regime of winter rainfall and summer drought that links the five MTE biomes of the world first appeared about 4–3 million years ago (Ma) in the Pliocene (e.g. Suc 1984; Axelrod 1989). However, there is emerging evidence that the onset of a proto-MTC occurred much earlier, at least in the mid-Miocene (Rundel et al. 2016). A key event in the onset of an MTC was the Middle Miocene Climate Optimum 17–14 Ma, associated with global cooling and growth of the East Antarctic ice sheet. Atmospheric and oceanic circumpolar circulation intensified during this period, resulting in increased strength of Hadley cell circulation and seasonal movement of the subtropical high-pressure centres, thereby promoting conditions favourable for MTC formation. More speculatively, there may have been periods of MTC formation even earlier, contemporaneously with Antarctic glaciation in the Oligocene. Although the global patterns of atmospheric circulation that determine MTC are clear, the seasonal intensity of MTC have likely varied through time. Changing offshore ocean currents influence the nature of MTC regimes, with cold currents intensifying summer drought and warm currents increasing summer rainfall.

The rich modern floras of southwest Australia and the South African Cape Region include many lineages that have ancient origins in evergreen sclerophyll shrublands on oligotrophic soils in the Upper Cretaceous–early Cenozoic (Lamont and He 2012). Much of the contemporary diversity of these climatically buffered, fire-prone, and nutrient-poor landscapes is related to the persistence and diversification of these old lineages (Cowling et al. 2015). Supplementing the diversity of the South African Cape has been the juxtaposition of an ancient and topographically heterogeneous landscape associated with the Cape Fold Belt and a relatively young lowland landscape that has provided novel habitats for the diversification of geophytes and leaf succulents, with the Iridaceae and Aizoaceae as good examples.

Although palaeoendemic taxa that pre-date an MTC regime are present in the younger and more dynamic landscapes of the Mediterranean, California, and Chile, these form only a small part of their modern species diversity. More recent immigration and diversification account for a large portion of modern species richness in their floras. For the Mediterranean and California, and, to a lesser extent, southwest Australia and South Africa, the novel climate seasonality and predictable crown fire regimes associated with the development of an MTC in the Miocene allowed colonization from a broad regional species pool. This colonization was followed by diversification stimulated by strategies of post-fire persistence. Consistent with this hypothesis is the breadth of independent lineages in disparate plant families that developed key life-history strategies to cope with fire. Central Chile, which lacks a history of fire since the Miocene uplift of the Andes, is an exception and shows no evidence of fire-stimulated diversification (Keeley et al. 2012).

Continued

Case Study 10 (*Continued*)

The pattern of colonization through permeable habitat boundaries open to recruitment from other biomes, followed by diversification, suggests a relative absence of phylogenetic niche conservatism in the assembly of the modern floras. This manner of floristic assembly is in contrast to that present in the Palaeogene lineages of the MTC regions of Australia and South Africa (Rundel et al. 2016).

There are many other ecological factors not unique to MTEs that have promoted speciation in these five regions. These include adaptations to oligotrophic soils in southwest Australia and the South African Cape Region, and diverse spatial patterns of climatic, topographic, and edaphic heterogeneity during the Pliocene and Quaternary in all five MTE regions, but most notably in the younger three. Despite differing tectonic and climatic histories, evidence for the moderate climate regimes of these regions can be seen in their relict palaeoendemic lineages. In the same manner, the key to diversity for many MTE lineages has been in their ability to speciate and persist at fine spatial scales, with low rates of extinction of the adapted lineages (Rundel et al. 2016).

accumulation of species over time. While all five MTC regions are incredibly diverse (see Table 3.2), South Africa stands out as being uniquely diverse compared to the other regions; low extinction rates associated with historically stable climate and limited recent geological activity might explain the high diversity found there relative to other MTC regions (Cowling et al. 2015). The same factors may also be important in driving high diversity in Australia (Cook et al. 2015). In California and the Mediterranean, which have greater variation in topography than either South Africa or Australia, the ability of species to move across elevation gradients has been suggested also to have allowed species to be buffered from climatic changes that may otherwise have led to extinctions (Médail and Diadema 2009; Lancaster and Kay 2013).

South Africa and Australia MTC regions have often been contrasted with the other three MTC regions in studies examining drivers of diversity (see Chapter 2, Case Study 4). These two regions have been the most climatically stable and did not experience the recent geological uplift or glaciation of the other regions. They also contain more extensive nutrient-poor soils when compared to other MTEs, especially Chile (see Chapter 7, Table 7.2; Stock and Verboom 2012). Plant diversity may be influenced by soil nutrients, with higher diversity associated with low soil mineral nutrition (Laliberté et al. 2013). The influence of these features on diversity has been developed into a framework that contrasts 'old, climatically buffered, infertile landscapes' (OCBILs) to 'young, often disturbed, fertile landscapes'

(YODFELs) (Hopper 2009); however, given the high diversity of all MTC regions this may be a more useful framework for comparing MTC regions to non-MTC ones.

Within MTC regions, communities are not all equally diverse and some of the same patterns that influence high diversity of these regions generally may also impact and explain differences between communities within the regions. Across MTC regions, species richness is linked to small ranges and limited distributions of species and many locally endemic species (Greuter 1994). Similar to the general pattern described above, climatic stability also seems to drive diversity within South Africa and Australia MTC areas, with a pronounced west–east gradient of declining diversity within these two regions and this pattern attributed to greater climatic stability (and therefore low extinction rates) in the west (Cowling and Lombard 2002; Sniderman et al. 2013).

At a local scale of 0.1 ha or less, a wide range of theories has been invoked to explain diversity and coexistence within MTC regions, particularly in MTV communities. These can be related to the theoretical framework proposed by MacArthur (1972) to explain **patterns in species richness** and include:

- High diversity because of species specialization. Species are differentiated spatially along structural or growth-form niche axes or spatially/temporally along a regeneration niche axis, with more species because each species is more specialized (e.g. Richardson et al. 1995; Vlok and Yeaton 2000; Pauw 2006; 2013).
- High diversity because of spatial variation in resource availability that allows for niche separation among a larger number of species (e.g. hydrological niche separation; Silvertown et al. 2012).
- High diversity because of disturbance-associated variation in resource availability. Disturbances such as fire or grazing maintain diversity because spatial and temporal variation increases the resource spectrum and lessens the role of competition in shaping community structure (e.g. continuum of sprouting responses to disturbances of varying severity; Bond and Midgley 2001).
- High diversity because of increased interactions with other species. Interactions (direct or indirect) with neighbours or neighbourhood effects maintain diversity because each species is able to overlap in its niche with other species and neighbours more (e.g. Bond and Midgley 1995; Nottebrock et al. 2017).
- High diversity because of temporal variation in niche availability. Space is allocated randomly to propagules of different species in a lottery process, with different post-fire conditions favouring different species at different times (e.g. Bond et al. 1995; Lamont and Witkowski 1995; Laurie and Cowling 1995).

Thus patterns of MTV diversity are believed to be a result of high levels of species persistence (Hubbell's 2001 **neutral theory**) coupled with interacting drivers of speciation driven by geologic (poor soils and low topographical diversity), climatic (drought), and ecological (fire, novel herbivores, and pollinators) factors (**local determinism**) (Rundel et al. 2016).

Literature cited

Ackerly DD. 2009. Evolution, origin and age of lineages in the Californian and Mediterranean floras. Journal of Biogeography 36: 1221–33.

Arroyo J, Marañón T. 1990. Community ecology and distributional spectra of Mediterranean shrublands and heathlands in southern Spain. Journal of Biogeography 17: 163–76.

Arroyo MTK, Cavieres L, Marticorena C, Muñoz-Schick M. 1995. Convergence in the Mediterranean floras in central Chile and California: insights from comparative biogeography. Pages 43–88 in Arroyo MTK, Zedler PH, Fox MD, eds. Ecology and biogeography of Mediterranean ecosystems in Chile, California, and Australia. Springer, New York, USA.

Axelrod DI. 1975. Evolution and biogeography of Madrean-Tethyan sclerophyll vegetation. Annals of the Missouri Botanical Garden 62: 280–334.

Axelrod DI. 1989. Age and origin of chaparral. Pages 7–19 in Keeley SC, ed. The California chaparral: paradigms reexamined. Natural History Museum of Los Angeles County, Los Angeles, California, USA.

Bacchetta G, Fenu G, Mattana E. 2012. A checklist of the exclusive vascular flora of Sardinia with priority rankings for conservation. Anales del Jardín Botánico de Madrid 69: 81–9.

Baldwin BG, Goldman DH, Keil DJ, Patterson R, Rosatti TJ, Wilken DH, eds. 2012. The Jepson manual: vascular plants of California. University of California Press, USA.

Blondel J, Aronson J. 1999. Biology and wildlife of the Mediterranean region. Oxford University Press, USA.

Blondel J, Aronson J, Bodiou J-Y, Boeuf G. 2010. The Mediterranean region: biological diversity in space and time. 2nd edn. Oxford University Press, Oxford, UK. 376 pp.

Böhner J, Blaschke T, Montanarella L. 2008. SAGA-seconds out. Hamburger Beiträge zur Physischen Geographie und Landschaftsökologie 19: 113.

Bond WJ, Midgley JJ. 1995. Kill thy neighbour: an individualistic argument for the evolution of flammability. Oikos 73: 79–85.

Bond WJ, Midgley JJ. 2001. Ecology of sprouting in woody plants: the persistence niche. Trends in Ecology & Evolution 16: 45–51.

Bond WJ, Maze K, Desmet P. 1995. Fire life histories and the seeds of chaos. Ecoscience 2: 252–60.

Burge DO, Erwin DM, Islam MB, Kellermann J, Kembel SW, Wilken DH, Manos PS. 2011. Diversification of *Ceanothus* (Rhamnaceae) in the California Floristic Province. International Journal of Plant Sciences 172: 1137–64.

Carrington ME, Keeley JE. 1999. Comparison of post-fire seedling establishment between scrub communities in mediterranean and non-mediterranean climate ecosystems. Journal of Ecology 87: 1025–36.

Claßen-Bockhoff R, Speck T, Tweraser E, Wester P, Thimm S, Reith M. 2004. The staminal lever mechanism in *Salvia* L. (Lamiaceae): a key innovation for adaptive radiation? Organisms Diversity & Evolution 4: 189–205.

Cody ML, Mooney HA. 1978. Convergence versus nonconvergence in Mediterranean-climate ecosystems. Annual Review of Ecology, Evolution, and Systematics 9: 265–321.

Cook LG, Hardy NB, Crisp MD. 2015. Three explanations for biodiversity hotspots: small range size, geographical overlap and time for species accumulation. An Australian case study. New Phytologist 207: 390–400.

Cowling RM. 1983. Phytochorology and vegetation history in the south-eastern Cape, South Africa. Journal of Biogeography 10: 393–419.

Cowling RM, Lombard AT. 2002. Heterogeneity, speciation/extinction history and climate: explaining regional plant diversity patterns in the Cape Floristic Region. Diversity and Distributions 8: 163–79.

Cowling RM, Witkowski ETF, Milewski AV, Newbey KR. 1994. Taxonomic, edaphic and biological aspects of narrow endemism on matched sites in mediterranean South Africa and Australia. Journal of Biogeography 21: 651–64.

Cowling RM, Rundel PW, Lamont BB, Arroyo MK, Arianoutsou M. 1996. Plant diversity in mediterranean climate regions. Trends in Ecology in Evolution 11: 362–6.

Cowling RM, Ojeda F, Lamont BB, Rundel PW, Lechmere-Oertel R. 2005. Rainfall reliability, a neglected factor in explaining convergence and divergence of plant traits in fire-prone mediterranean-climate ecosystems. Global Ecology and Biogeography 14: 509–19.

Cowling RM, Potts AJ, Bradshaw P, Colville J, Arianoutsou M, Ferrier S, Forest F, Fylas N, Hopper SD, Ojeda F, Proches S, Smith RJ, Rundel PW, Vassilakis E, Zutta BR. 2015. Variation in plant diversity in mediterranean climate ecosystems: the role of climatic and topographic stability. Journal of Biogeography 42: 552–64.

Crisp M, Cook L, Steane D. 2004. Radiation of the Australian flora: what can comparisons of molecular phylogenies across multiple taxa tell us about the evolution of diversity in present–day communities? Philosophical Transactions of the Royal Society of London B: Biological Sciences 359: 1551–71.

Cueto M, Blanca López G, Salazar C, Cabezudo B. 2014. Diversity and ecological characteristics of the vascular flora in the western mediterranean (Eastern Andalusia, Spain). Acta Botanica Malacitana 39: 81–97.

Dallman PR. 1998. Plant life in the world's mediterranean climates. University of California, Los Angeles, USA.

Dodson JR, Kershaw AP. 1995. Evolution and history of Mediterranean vegetation types in Australia. Pages 21–40 in Arroyo MTK, Zedler PH, Fox MD, eds. Ecology and biogeography of Mediterranean ecosystems in Chile, California, and Australia. Springer, New York, USA.

Fleury M, Marcelo W, Vásquez RA, González LA, Bustamante RO. 2015. Recruitment dynamics of the relict palm, *Jubaea chilensis*: intricate and pervasive effects of invasive herbivores and nurse shrubs in central Chile. PLOS one 10(7): p.e0133559.

Galley C, Linder HP. 2006. Geographical affinities of the Cape flora, South Africa. Journal of Biogeography 33: 236–50.

Garreaud R, Barichivich J, Christie DA, Maldonado A. 2008. Interannual variability of the coastal fog at Fray Jorge relict forests in semiarid Chile. Journal of Geophysical Research: Biogeosciences 113: G04011.

Georghiou K, Delipetrou P. 2010. Patterns and traits of the endemic plants of Greece. Botanical Journal of the Linnean Society 162: 130–422.

Goldblatt P. 1978. An analysis of the flora of southern Africa: its characteristics, relationships, and origins. Annals of the Missouri Botanical Garden 65: 369–436.

Goldblatt P, Manning JC. 2002. Plant diversity of the Cape region of southern Africa. Annals of the Missouri Botanical Garden 89: 281–302.

Goldblatt P, Manning JC, Bernhardt P. 1998. Adaptive radiation of bee-pollinated *Gladiolus* species (Iridaceae) in southern Africa. Annals of the Missouri Botanical Garden 85: 492–517.

Gratani L, Varone L. 2004. Adaptive photosynthetic strategies of the Mediterranean maquis species according to their origin. Photosynthetica 42: 551–8.

Greuter W. 1994. Extinctions in mediterranean areas. Philosophical Transactions of the Royal Society of London B: Biological Sciences 344: 41–6.

Groom PK, Lamont BB. 2015. Plant lift of southwestern Australia. De Gruyter Open, Warsaw/Berlin.

Guzmán B, Lledó MD, Vargas P. 2009. Adaptive radiation in mediterranean *Cistus* (Cistaceae). PLoS One 4(7): p.e6362.

He T, Lamont BB, Manning J. 2016. A Cretaceous origin for fire adaptations in the Cape flora. Scientific Reports 6: 34880.

Herrera CM. 1992. Historical effects and sorting processes as explanations for contemporary ecological patterns: character syndromes in Mediterranean woody plants. The American Naturalist 140: 421–46.

Hewitt GM. 2011. Mediterranean peninsulas: the evolution of hotspots. Pages 123–47 in Zachos FE, Habel JC, eds. Biodiversity hotspots. Springer, Berlin, Germany.

Hinojosa LF, Armesto JJ, Villagrán C. 2006. Are Chilean coastal forests pre-Pleistocene relicts? Evidence from foliar physiognomy, palaeoclimate, and phytogeography. Journal of Biogeography 33: 331–41.

Hopper SD. 1979. Biogeographical aspects of speciation in the southwest Australian flora. Annual Review of Ecology and Systematics 10: 399–422.

Hopper SD. 2009. OCBIL theory: towards an integrated understanding of the evolution, ecology and conservation of biodiversity on old, climatically buffered, infertile landscapes. Plant and Soil 322: 49–86.

Hopper SD, Gioia P. 2004. The southwest Australian floristic region: evolution and conservation of a global hot spot of biodiversity. Annual Review of Ecology, Evolution, and Systematics 35: 623–50.

Hubbell SP. 2001. The unified neutral theory of biodiversity and biogeography. Princeton University Press, Princeton, NJ, USA. 392 pp.

Jacobsen AL, Esler KJ, Pratt RB, Ewers FW. 2009. Water stress tolerance of shrubs in mediterranean-type climate regions: convergence of fynbos and succulent karoo communities with California shrub communities. American Journal of Botany 96: 1445–53.

Johnson S, Linder H, Steiner K. 1998. Phylogeny and radiation of pollination systems in *Disa* (Orchidaceae). American Journal of Botany 85: 402–411.

Keeley JE, Fotheringham CJ. 2003. Species–area relationships in mediterranean-climate plant communities. Journal of Biogeography 30: 1629–57.

Keeley JE, Bond WJ, Bradstock RA, Pausas JG, Rundel PW. 2012. Fire in Mediterranean ecosystems: ecology, evolution and management. Cambridge University Press, Cambridge, UK.

Kovar-Eder J, Kvaček Z, Martinetto E, Roiron P. 2006. Late Miocene to early Pliocene vegetation of southern Europe (7–4Ma) as reflected in the megafossil plant record. Palaeogeography, Palaeoclimatology, Palaeoecology 238: 321–39.

Kruckeberg AR. 1984. California serpentines: flora, vegetation, geology, soils, and management problems. University of California Press, Berkeley, California, USA.

Laliberté E, Grace JB, Huston MA, Lambers H, Teste FP, Turner BL, Wardle DA. 2013. How does pedogenesis drive plant diversity? Trends in Ecology & Evolution 28: 331–40.

Lamont BB, He T. 2012. Fire-adapted Gondwanan angiosperm floras evolved in the Cretaceous. BMC Evolutionary Biology 12: 223.

Lamont BB, Witkowski ET. 1995. A test for lottery recruitment among four Banksia species based on their demography and biological attributes. Oecologia 101: 299–308.

Lancaster LT, Kay KM. 2013. Origin and diversification of the California flora: re-examining classic hypotheses with molecular phylogenies. Evolution 67: 1041–54.

Laurie H, Cowling RM. 1995. Lottery coexistence models extended to plants with disjoint generations. Journal of Vegetation Science 6: 161–8.

Linder HP. 2003. The radiation of the Cape flora, southern Africa. Biological Reviews 78: 597–638.

Linder HP. 2008. Plant species radiations: where, when, why? Philosophical Transactions of the Royal Society B: Biological Sciences 363: 3097–105.

Linder HP, Meadows ME, Cowling RM. 1992. History of the Cape flora. Pages 113–34 in Cowling RM, ed. The ecology of fynbos: nutrients, fire and diversity. Oxford University Press, Cape Town, South Africa.

Loidi J, Biurrun I, Campos JA, García-Mijangos I, Herrera M. 2007. A survey of heath vegetation of the Iberian Peninsula and Northern Morocco: a biogeographical and bioclimatic approach. Phytocoenologia 37: 341–70.

Lubke RA, Everard DA, Jackson S. 1986. The biomes of the eastern Cape with emphasis on their conservation. Bothalia 16: 251–61.

MacArthur RH. 1972. Geographical ecology. Harper & Row, New York.

Madriñán S, Cortés AJ, Richardson JE. 2013. Páramo is the world's fastest evolving and coolest biodiversity hotspot. Frontiers in Genetics 4: 192.

Médail F, Diadema K. 2009. Glacial refugia influence plant diversity patterns in the Mediterranean Basin. Journal of Biogeography 36: 1333–45.

Médail F, Quézel P. 1997. Hot-spots analysis for conservation of plant biodiversity in the Mediterranean Basin. Annals of the Missouri Botanical Garden 84: 112–27.

Médail F, Quézel P. 1999. Biodiversity hotspots in the Mediterranean Basin: setting global conservation priorities. Conservation Biology 13: 1510–13.

Moreira-Muñoz A. 2011. Plant geography of Chile (Vol. 5). Springer Science & Business Media.

Mucina L, Laliberté E, Thiele KR, Dodson JR, Harvey J. 2014. Biogeography of kwongan: origins, diversity, endemism and vegetation patterns. Pages 35–79 in Lambers H, ed. Plant life on the sandplains in Southwest Australia, a global biodiversity hotspot. UWA Publishing, Crawley, Australia.

Nottebrock H, Schmid B, Treurnicht M, Pagel J, Esler KJ, Böhning-Gaese K, Schleuning M, Schurr FM. 2017. Coexistence of plant species in a biodiversity hotspot is stabilized by competition but not by seed predation. Oikos 126: 276–84.

Núñez-Ávila MC, Armesto JJ. 2006. Relict islands of the temperate rainforest tree *Aextoxicon punctatum* (Aextoxicaceae) in semi-arid Chile: genetic diversity and biogeographic history. Australian Journal of Botany 54: 733–43.

Ojeda F. 2011. Singularidad botánica de la herriza o brezal Mediterráneo del estrecho de Gibraltar. Migres 2: 17–23.

Ojeda F, Arroyo J, Marañón T. 1995. Biodiversity components and conservation of Mediterranean heathlands in southern Spain. Biological Conservation 72: 61–72.

Ojeda F, Marañón T, Arroyo J. 2000. Plant biodiversity in the Aljibe Mountains (S. Spain): a comprehensive account. Biodiversity and Conservation 9: 1323–43.

Ojeda F, Simmons MT, Arroyo J, Marañón T, Cowling RM. 2001. Biodiversity patterns in South African fynbos and Mediterranean heathland. Journal of Vegetation Science 12: 867–74.

Ojeda F, Pausas JG, Verdú M. 2010. Soil shapes community structure through fire. Oecologia 163: 729–35.

Onstein RE, Linder HP. 2016. Beyond climate: convergence in fast evolving sclerophylls in Cape and Australian Rhamnaceae predates the Mediterranean climate. Journal of Ecology 104: 665–77.

Onstein RE, Carter RJ, Xing Y, Richardson JE, Linder HP. 2015. Do mediterranean-type ecosystems have a common history?—Insights from the Buckthorn family (Rhamnaceae). Evolution 69: 756–71.

Pausas JG, Keeley JE. 2014. Evolutionary ecology of resprouting and seeding in fire-prone ecosystems. New Phytologist 204: 55–65.

Pausas JG, Bradstock RA, Keith DA, Keeley JE, GCTE Fire Network. 2004. Plant functional traits in relation to fire in crown-fire ecosystems. Ecology 85: 1085–100.

Pauw A. 2006. Floral syndromes accurately predict pollination by a specialized oil-collecting bee (*Rediviva peringueyi*, Melittidae) in a guild of South African orchids (Coryciinae). American Journal of Botany 93: 917–26.

Pauw A. 2013. Can pollination niches facilitate plant coexistence? Trends in Ecology & Evolution 28: 30–7.

Peñuelas J, Lloret F, Montoya R. 2001. Severe drought effects on Mediterranean woody flora in Spain. Forest Science 47: 214–18.

Quézel P. 1978. Analysis of the flora of Mediterranean and Saharan Africa. Annals of the Missouri Botanical Garden 65: 479–534.

Quézel P. 1985. Definition of the Mediterranean region and the origin of its flora. Pages 9–24 in Gomez-Campo C, ed. Plant conservation in the Mediterranean area. Geobotany 7, W. Junk, Dordrecht.

Raven PH, Axelrod DI. 1978. Origin and relationships of the California flora. University of California Publications in Botany, Vol. 72. University of California Press, USA.

Raven PH, Axelrod DI. 1995. Origin and relationships of the California flora. California Native Plant Society, Sacramento, California, USA. 134 pp.

Richardson DM, Cowling RM, Lamont BB, Hensbergen H. 1995. Coexistence of Banksia species in southwestern Australia: the role of regional and local processes. Journal of Vegetation Science 6: 329–42.

Rivas-Martínez S. 1979. Brezales y jarales de Europa occidental. (Revisión fitosociológica de las clases Calluno-Ulicetea y Cisto-Lavanduletea.) Lazaroa 1: 5–128.

Rodríguez-Sánchez F, Guzmán B, Valido A, Vargas P, Arroyo J. 2009. Late Neogene history of the laurel tree (Laurus L., Lauraceae) based on phylogeographical analyses of Mediterranean and Macaronesian populations. Journal of Biogeography 36: 1270–81.

Rundel PW, Arroyo MT, Cowling RM, Keeley JE, Lamont BB, Vargas P. 2016. Mediterranean biomes: evolution of their vegetation, floras, and climate. Annual Review of Ecology, Evolution, and Systematics 47: 383–407.

Safford HD, Harrison S. 2004. Fire effects on plant diversity in serpentine vs. sandstone chaparral. Ecology 85: 539–48.

Sauquet H, Weston PH, Anderson CL, Barker NP, Cantrill DJ, Mast AR, Savolainen V. 2009. Contrasted patterns of hyperdiversification in Mediterranean hotspots. Proceedings of the National Academy of Sciences 106: 221–5.

Schut AG, Wardell-Johnson GW, Yates CJ, Keppel G, Baran I, Franklin SE, Hopper SD, Van Niel KP, Mucina L, Byrne, M. 2014. Rapid characterisation of vegetation structure to predict refugia and climate change impacts across a global biodiversity hotspot. PLoS One 9(1): p.e82778.

Segarra-Moragues JG, Torres-Díaz C, Ojeda F. 2013. Are woody seeder plants more prone than resprouter to population genetic differentiation in Mediterranean-type ecosystems? Evolutionary Ecology 27: 117–31.

Silvertown J, Araya YN, Linder HP, Gowing DJ. 2012. Experimental investigation of the origin of fynbos plant community structure after fire. Annals of Botany 126: 276–84.

Sniderman JK, Jordan GJ, Cowling RM. 2013. Fossil evidence for a hyperdiverse sclerophyll flora under a non–mediterranean-type climate. Proceedings of the National Academy of Sciences 110: 3423–8.

Solbrig O, Cody ML, Fuentes ER, Glanz W, Hunt JH, Moldenke AR. 1977. The origin of the biota. Pages 13–26 in Mooney, ed. Convergent evolution in Chile and California: mediterranean climate ecosystems. Dowden, Hutchinson & Ross, Inc.

Stebbins GL. 1970. Adaptive radiation of reproductive characteristics in angiosperms, I: pollination mechanisms. Annual Review of Ecology and Systematics 1: 307–26.

Stebbins GL. 1974. Flowering plants: evolution above the species level. The Belknap Press of Harvard University Press, Cambridge, MA.

Stebbins GL, Major J. 1965. Endemism and speciation in the California flora. Ecological Monographs 35: 1–35.

Stock WD, Verboom GA. 2012. Phylogenetic ecology of foliar N and P concentrations and N: P ratios across mediterranean-type ecosystems. Global Ecology and Biogeography 21: 1147–56.

Suc J-P. 1984. Origin and evolution of the Mediterranean vegetation and climate of Europe. Nature 307: 429–32.

Thompson JD. 2005. Plant evolution in the Mediterranean. Oxford University Press, Oxford.

Thrower NJ, Bradbury DE. 1973. The physiography of the Mediterranean lands with special emphasis on California and Chile. Pages 37–52 in di Castri F, Mooney HA, eds. Mediterranean type ecosystems. Springer, Berlin, Germany.

Tzedakis PC. 2009. Museums and cradles of Mediterranean biodiversity. Journal of Biogeography 36: 1033–4.

Valiente-Banuet A, Flores-Hernández N, Verdú M, Dávila P. 1998. The chaparral vegetation in Mexico under nonmediterranean climate: the convergence and Madrean-Tethyan hypotheses reconsidered. American Journal of Botany 85: 1398–408.

Verdú M, Pausas JG. 2013. Syndrome-driven diversification in a mediterranean ecosystem. Evolution 67: 1756–66.

Vlok JH, Yeaton RI. 2000. The effect of short fire cycles on the cover and density of understorey sprouting species in South African mountain fynbos. Diversity and Distributions 6: 233–42.

Vrba ES, Gould SJ. 1986. The hierarchical expansion of sorting and selection: sorting and selection cannot be equated. Paleobiology 12: 217–28.

Wardell-Johnson G, Horwitz P. 1996. Conserving biodiversity and the recognition of heterogeneity in ancient landscapes: a case study from south-western Australia. Forest Ecology and Management 85: 219–38.

Wells PV. 1969. The relation between mode of reproduction and extent of speciation in woody genera of the California chaparral. Evolution 23: 264–7.

Plate 1 Landscape photos from each of the five mediterranean-type ecosystem (MTE) regions (a–e) (see p. 5).

Plate 2 Convergent evolution of a succulent plant form in different deserts (a–d) (see p. 6).

Plate 3 Conspicuous common northern hemisphere and southern hemisphere taxa (see p. 26).

Plate 4 Examples of plants that re-establish through resprouting following fire (see p. 45).

Plate 5 Examples of plants that re-establish through seeding following fire (see p. 46).

Plate 6 Plant physical defence structures related to herbivory (see p. 94).

Plate 7 Flowers may be suited for generalist or specialist pollinators (see p. 96).

Plate 8 Arid shrub communities (see p. 126).

Plate 9 Grassland in a mediterranean-type climate region (see p. 134).

Plate 10 Relict species currently have limited distributions (see p. 156).

Plate 11 Resprouting and non-resprouting species from the same genus (see p. 159).

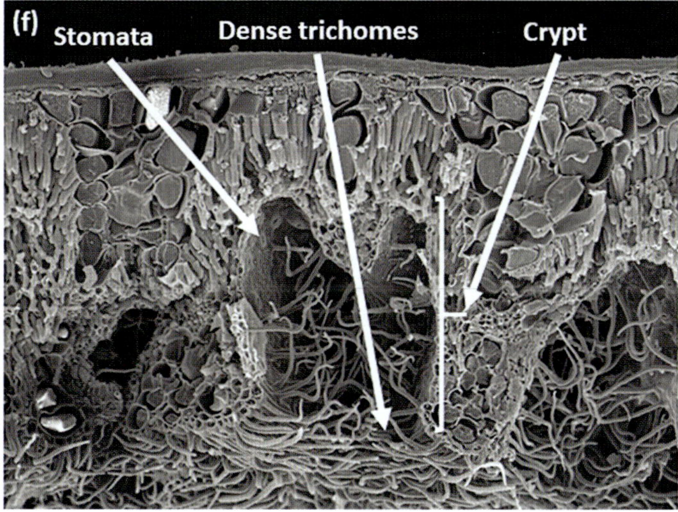

Plate 12 Adaptive leaf traits (see p. 184).

Plate 13 Carnivorous and parasitic plants (see p. 230).

Plate 14 Mediterranean region traditional land use (a, b) (see p. 246).

Plate 15 A degraded Chilean landscape (see p. 257).

Plate 16 Invasive alien plants (see p. 264).

Plate 17 Pathogen impacts on mediterranean-type climate region plants (see p. 269).

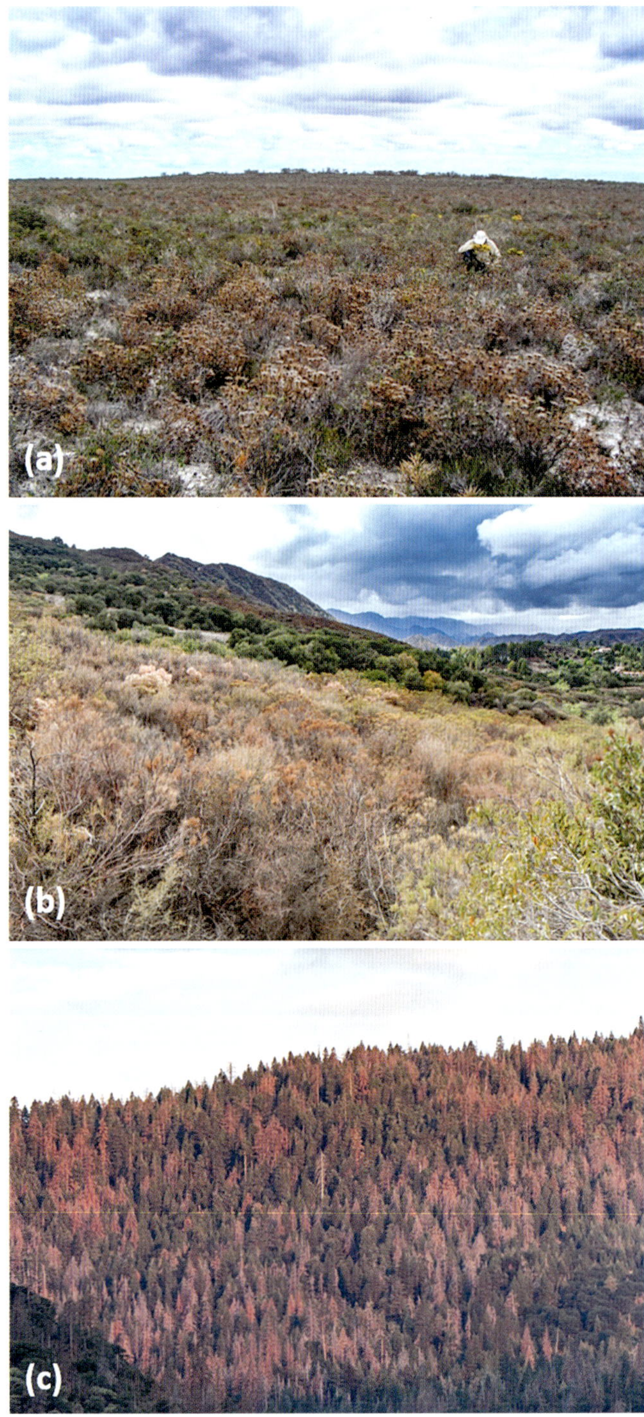

Plate 18 Drought-induced mortality (see p. 274).

6 Form and Function of Mediterranean Shrublands

Abstract

The archetypal shrub type that dominates most of the regions that experience mediterranean-type climate (MTC) is an evergreen shrub with thick and leathery leaves (sclerophyllous). The occurrence of large stands of such shrubs in all MTC regions led early biogeographers to hypothesize that the MTC selects for this growth form and leaf type and that this had led to convergent evolution (see Chapters 1 and 2). This hypothesis has received considerable research interest and continues to be examined. In this chapter we consider the structure and physiology of these archetypal MTC region shrub species and examine evidence for convergent evolution in their structure and function. We also assess the key adaptive traits that enable the shrub species that compose mediterranean-type vegetation (MTV) communities to thrive in MTC regions.

6.1 Structure and physiology

6.1.1 Sclerophyllous leaves

Evergreen thick and leathery leaves (i.e. sclerophyllous leaves) have been the focus of considerable study in the context of being adaptive to a mediterranean-type climate (MTC). This is because of the striking similarity in leaf structure between the dominant shrubs of the five MTC regions and because of the importance of leaves to plant physiology and fitness and ecosystem processes (Chapter 7). Leaves perform multiple functions, chief of which is the acquisition of CO_2 for energy and growth (photosynthesis). This energy produced by plant leaves represents the chief input into these systems that all other, non-photosynthetic, organisms depend on. Moreover, their

The Biology of Mediterranean-Type Ecosystems. Karen J. Esler, Anna L. Jacobsen, and R. Brandon Pratt,
Oxford University Press (2018). © Karen J. Esler, Anna L. Jacobsen, and R. Brandon Pratt 2018.
DOI 10.1093/oso/9780198739135.001.0001

decomposition is a key factor affecting nutrient cycling of these ecosystems (Chapter 7).

A leaf is described as **sclerophyllous** if it has a texture that is hard, and a sclerophyll is a plant that has such leaves. Other commonly used terms to describe leaves that are sclerophyllous are dense, tough, thick, and leathery (coriaceous). It has been quantified as **specific leaf area** (SLA; leaf area/dry mass) or 1/SLA, which is called **leaf mass area** (LMA). These are easily measured traits, which is one reason why they are popular measures, and they are directly related to sclerophylly. Loveless (1961) devised a sclerophylly index that estimates the proportion of cell wall to intracellular content (crude fibre dry mass/crude protein dry mass). The sclerophylly index and LMA are correlated to one another when analysed across a broad range of angiosperm taxa (data analysed from Read and Sanson 2003; however, see Groom and Lamont 1999) and when analysed for MTC region species only (Figure 6.1). Both estimates are also correlated with a range of leaf mechanical properties and subjective assessment of sclerophylly by experts (Read and Sanson 2003). There is no one best way to measure sclerophylly, but in this chapter we will use LMA as a proxy for the degree to which leaves are sclerophyllous (Groom and Lamont 1999).

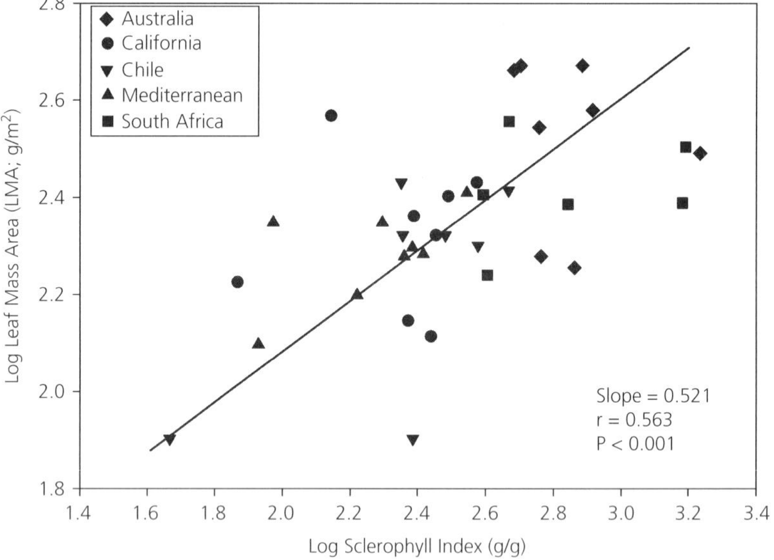

Figure 6.1   **Sclerophylly and leaf structure.** Relationship between leaf mass area and sclerophylly index (data are log transformed values). Statistics and best fit line are for standardized major axis (SMA) regression. Data are only from evergreen shrub species from regions with a mediterranean-type climate (Mooney 1977; Specht 1988; von Willert et al. 1989).

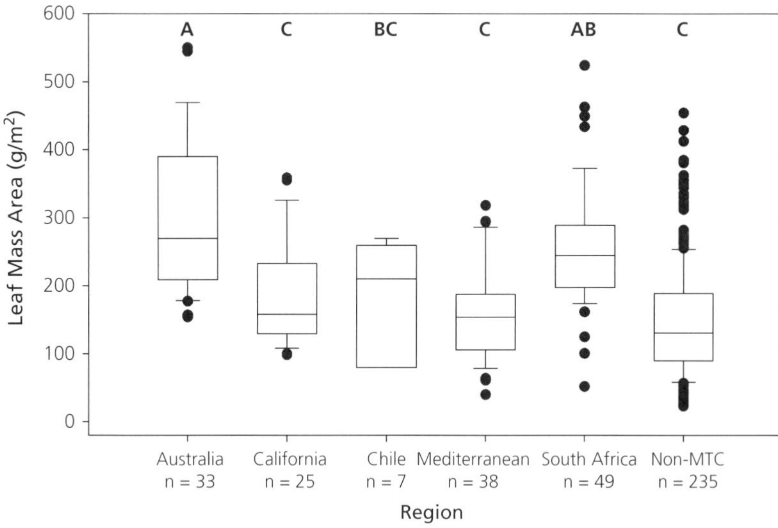

Figure 6.2   **Leaf structure by region.** Box and whisker plots of leaf mass per area (LMA) of ever-green shrubs from the five mediterranean-type climate (MTC) regions and other regions of the globe (non-MTC data from Wright et al. 2004 and MTC data from the same and Griffin 1978; Hassiotou et al. 2009; Mooney 1977; Pratt, unpublished; Specht 1988). The sample size (n) is shown beneath each region. The box and whisker plots depict the median (line in the box), the lower and upper horizontal lines of the box denote the 25% and 75% quartiles (half of the values fall within the box). The 'whiskers' are the lines outside of the box and they extend to the lowest values in the upper and lower limits. Values beyond the upper and lower whiskers are extreme values or outliers. The regions that have unique letters at the top of the figure are significantly different. There were many duplicate samples from the same species and, in general, the values were similar. In these cases only one value was used to represent the species to avoid pseudor-eplication. For the Australian MTC data, mean values were taken for *Hakea* and *Banksia* spp. because these taxa were overrepresented in the dataset and thus dominated. This did not change the final result, which is that Australia has the highest LMA of any MTC region. In five cases, extreme values were removed of needle-leaved taxa that had extreme LMA values >550 (all were from Australia with one from the MTC region and four from the non-MTC). The systems included in the non-MTC region category come from a wide range of biomes and from all over the globe.

When trying to understand if sclerophyllous leaves and their associated traits are uniquely adapted to the MTC we need to compare the leaf traits across the MTC regions and also to other systems (Figure 6.2). It is also important to consider the selective advantage of a given leaf trait. In the strictest sense, for a trait to represent an **adaptation** it needs to have been selected for and evolved in response to a particular set of pressures (Gould and Lewontin 1979), in this case the MTC. The clearest evidence for adaptation of a trait would be one with clear functional benefit in the context of the MTC and that is unique to mediterranean-type vegetation (MTV). Such a trait could be examined in the context of the fossil record and phylogeny to determine if the trait arose around the same time as the onset of the MTC. If so, this would be strong evidence for this trait being an adaptation to MTC. This, coupled with the traits' prevalence among the vegetation across all five MTC regions, would then make a strong case for convergent evolution (Chapter 2).

The striking dominance of sclerophylly among MTV in all five regions led some to conclude that this leaf type was an adaptation to MTC (Case Study 2). A strong case can be made that evergreen sclerophyllous leaves are adaptive (functionally well suited) to an MTC. One of the strongest arguments is that evergreen leaves are beneficial in the context of carbon gain because they can opportunistically respond to unpredictable rain events that occur in the autumn and spring (Mooney and Dunn 1970). Moreover, because winters are mild, some photosynthesis is possible for species that are moderate to deeply rooted at all times of the year, except during the driest and hottest part of the summer/fall. Drought-deciduous taxa with shallow roots commonly dominate in areas that are more arid than the MTC. In these taxa, new leaves have to be regrown to take advantage of a rainfall event leading to costs in the form of time lost to be productive and resources spent growing new leaves. In this way, species with evergreen leaves gain a competitive advantage in an MTC. For plants with evergreen leaves sclerophylly is important because longer-lived leaves need protection against the environment, including herbivores. Indeed, leaf longevity is strongly and positively correlated with LMA (Figure 6.3).

Sclerophylly may also be important for resisting the strains that accompany tissue dehydration during the summer dry period. The way plants extract water from soil is by lowering the water pressure at the point of transpiration in their leaves. The pressure is lowered substantially below the reference of pure water at atmospheric pressure, leading to large negative pressures. These negative pressures are transmitted to a continuum of water connected by cohesion (hydrogen bonding among water molecules) and located inside the cells that transport water called vessels. This places the water in the vessels under tension, which provides the motive force for moving water in the vascular system from the soil to the leaves. This mechanism of plant water transport is called the **cohesion-tension model**. If the soil is holding onto the water more tightly, as happens when the soil dries, the plant has to generate more negative pressures and greater tensions. These tensions place considerable strain on the vessels and cells connected to them. To avoid imploding, vessels need to have thick walls and surrounding tissues need to be reinforced with thick walls as well (Brodribb and Holbrook 2005). Beyond the negative pressures that develop in the vessels, negative pressures may also develop in individual cells (negative turgor pressure). In this case, having small cells with thick cell walls is advantageous to avoid deformation and collapse of cells when tissues are dehydrated (Oertli et al. 1990). Lastly, thick cell walls may lead to stiffer cells (higher elastic modulus). This, combined with accumulation of solutes in cells, may allow cells to avoid harmful levels

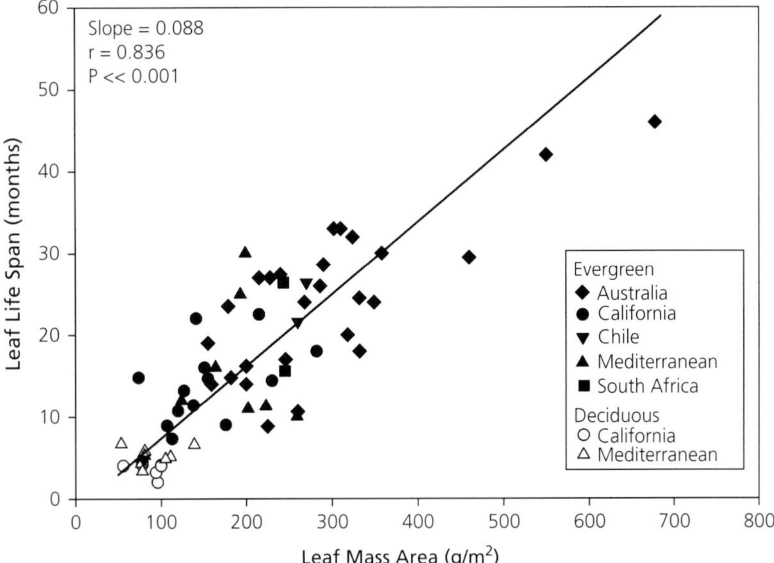

Figure 6.3 **Leaf longevity and sclerophylly.** Leaf longevity related to an estimate of sclerophylly (leaf mass per area) for evergreen and deciduous shrub species (each data point is the mean for a species). Statistics and best fit line are for standardized major axis regression. Data were taken from multiple sources (Midgley and Enright 2000; Wright et al. 2004; Mediavilla et al. 2008).

of dehydration (Bartlett et al. 2012). In summary, thick cell walls and smaller cells are likely associated with sclerophylly and facilitate a greater capacity for extracting water from drying soil.

While it is easy to make the argument that the evergreen sclerophyllous leaf habit is adaptive to the MTC, it does not appear that these characters represent unique adaptations to MTC. Evergreen sclerophyllous shrublands also occur in areas with a summer wet climate. Two examples of this are shrublands in parts of central Arizona and in the Tehuacán Valley region of Mexico (Case Study 7). Beyond the MTC regions, systems that are moderately water-limited with a degree of unpredictability in rainfall may favour sclerophylls. Another complicating factor is that sclerophylly is a trait found among species adapted to nutrient-poor soils and not only water-limited systems (McDonald et al. 2003). This relationship arises because sclerophyllous leaves are better protected from herbivory and therefore nutrient losses (Coley and Barone 1996). Having leaves that live longer also leads to greater nutrient use efficiency, as the leaves and their nutrients stay on the plant and photosynthesize for a longer timeframe and the loss of nutrients associated with senescence and leaf fall are minimized. Therefore, evergreen sclerophyllous species can be found abundantly in non-MTC communities (Sniderman et al. 2013; Turner 1994; Wright and Cannon 2001).

Did leaves that are sclerophyllous evolve from ancestors that were non-sclerophylls in response to the MTC; that is, are they adaptations? One important way to examine this question is to look at the evolutionary history of sclerophylly among MTV taxa. In some cases, there is evidence that taxa evolved higher LMA than their closest ancestors in non-MTC environments (e.g. *Heteromeles arbutifolia* in chaparral; Ackerly 2004a), whereas many taxa have not (Ackerly 2004a; Verdú et al. 2003).

Overall, there is little support for the evergreen sclerophyllous leaf habit being a unique adaptation to MTC for most species. However, the dominance of species with this leaf habit, and our understanding of this trait, leaves little doubt that this trait is adaptive to species that inhabit an MTC. As the MTC expanded in the five regions, some non-sclerophyllous taxa came under pressure to evolve them while those taxa that were already sclerophyllous were well suited to thrive in these systems. These results suggest that sclerophyllous taxa were 'pre-adapted' or exapted (Gould and Vrba 1982) to the MTC when it expanded millions of years ago (see Chapter 2, Case Study 2). Many taxa likely evolved the sclerophyllous habit in response to water limitations in areas that were not strictly MTC but that experienced seasonal water deficits, or in areas with nutrient-impoverished soils.

Because some MTC regions (Australia and South Africa) have both an MTC and soils that are generally impoverished in nutrients (Chapter 7), this may affect sclerophylly. We can examine this by comparing LMA across the MTC regions (Figure 6.2). In this analysis, Australia and South Africa have significantly greater LMA than the other MTC regions and evergreen shrubs from other regions of the world. The MTV from California, Chile, and the Mediterranean is not significantly different from evergreen shrubs from non-MTC vegetation types. This analysis suggests that soils impoverished in nutrients may drive evergreen shrubs toward greater sclerophylly. This would be advantageous, as greater sclerophylly would facilitate greater longevity, and the fewer times leaves need to be dropped means less loss of nutrients over time. This analysis also supports the statement above that the MTC does not uniquely contain evergreen shrubs with high sclerophylly when compared to evergreen shrubs from a diverse range of other systems (tundra, tropical, temperate, and desert). Nevertheless, there is evidence that aridity within South Africa and Australia MTC regions can contribute to greater sclerophylly, so the role of moisture is not to be fully discounted (Lamont et al. 2002).

6.1.2 Leaf traits other than sclerophylly

There are numerous other traits of leaves that are adaptive in the context of an MTC. In all five MTC regions, leaves of shrubs range in size (leaf area) from very small to medium sized (Orshan et al. 1984). Compared to other more mesic systems, the median leaf area for these shrubs is smaller and the maximum value is only about 40 cm^2. Compared to more arid systems the leaf size of shrubs in MTC regions is larger. In some cases taxa in MTC regions evolved smaller leaves, presumably in response to the MTC, while other taxa had small leaves prior to the onset of the MTC (Case Study 2, Ackerly 2004a).

A special small leaf type common among shrubs of MTV are **ericoid leaves**, so named because they are common among some taxa in the heather family (Ericaceae; see Chapter 5, Case Study 9). Ericoid leaves are typically small and are needle-like because the margins fold inward on the underside of the leaf and meet or nearly so at the centre axis (Figure 6.4a, b; see Plate 11). Other species have non-ericoid needle-shaped leaves; for example, chamise (*Adenostoma fasciculatum*), which is a common chaparral shrub throughout most of California (Figure 6.4c; see Plate 11). Small leaves are better at avoiding overheating because they have greater boundary layer conductance (the conductive capacity of the still layer of air that surrounds leaves) that allows them to better exchange heat with the air and track air temperature, thus avoiding overheating and excessive transpiration in bright sunlight. When water is available, transpiration serves to cool leaves effectively, but the MTC is water-limited and so evaporative cooling is limited (Yates et al. 2010). Small leaves may be linked to nutrient-poor soils and nutrient uptake by enhancing transpiration when water is available. Greater rates of transpiration would lead to greater uptake of soil water and the nutrients dissolved in that water (Yates et al. 2010; McDonald et al. 2003). Plants inhabiting the warmest and driest desert regions commonly have even smaller leaves than those found among MTV species (Cowling and Campbell 1980).

Evolutionary analyses have not widely supported leaf size as an adaptation to MTC. Ackerly (2004a) compared leaf size of chaparral taxa to their non-MTC ancestral state. Only one of the twelve lineages evolved reduced leaf size upon the evolutionary transition from a non-MTC to an MTC. This suggests that small leaf size was present in most of these lineages prior to the onset of the MTC in California. Thus, as with sclerophylly, small leaves represent an exaptation for many taxa that thrived in the MTC when it expanded millions of years ago.

One of the challenges in an MTC is that, when water is limited, sunlight is in excess. This excess can damage tissues by causing overheating and also from the production of free radicals that form when too much energy is absorbed. One way to mitigate the amount of energy absorbed is to display leaves in a way that reduces the surface area exposed to full sun. This can be achieved by angling leaves vertically so that leaf margins are facing the sky, which has the effect that, when the sun is at its midday peak, the smallest amount of leaf area is exposed to it (Comstock and Mahall 1985; Figure 6.4d; see Plate 11). Another way to reduce the energy absorbed is to cover leaves with reflective hairs or waxes (Figure 6.4e, f; see Plate 11). Biochemical adaptations, such as

Figure 6.4   Adaptive leaf traits. Small ericoid leaves are common among a range of taxa (a is an ericaceous *Phyllodoce* sp. from South Africa; b is a proteaceous *Grevillea* sp. from Australia). Chamise (*Adenostoma fasciculatum*), a common chaparral shrub throughout California, has small needle-like leaves (c). Hoary-leaf *Ceanothus* (*C. crassifolius*) angles its leaves skyward during the dry season and has white hairy undersurfaces of leaves that reflects light (d). Species differ in absorbance (Abs.) of light by leaves owing to the presence of hairs or degree and colour of waxes as shown between two co-occurring manzanita species (*Arctostaphylos glandulosa* on the left is greener and more absorptive and *A. glauca* is lighter in colour and absorbs less sunlight; e). Scanning electron micrograph of big-pod *Ceanothus megacarpus* showing stomatal crypts lined with stomata and dense trichomes (hairs; f). (See Plate 12)

Source: Photos a and f from Anna L. Jacobsen and others from R. Brandon Pratt.

shifts in key pigments such as chlorophylls and xanthophylls, can also allow for safe dissipation of absorbed radiation in excess of what is used in photosynthesis and for protection against free radicals. All of these types of leaf characters can be found among MTV, as well as more generally in dry and sunny environments.

One leaf feature found in some MTV sclerophyllous plants are **stomatal crypts**. These are indentations on the bottom of leaves that create a recess lined with stomata (Figure 6.4f; see Plate 11). In many species, the openings to the crypts are covered with a layer of hairs (trichomes), as is the case in Figure 6.4f (see Plate 11). Deep stomatal crypts are likely an adaptation to dry environments generally (Jordan et al. 2008). The crypts serve to create increased boundary layer resistance to water loss because the air in the crypt is humidified during transpiration and that same air is not well mixed by air currents being protected by the crypts, thus both the air inside of the leaf and the crypt air achieve high water vapor concentrations, slowing diffusion. At the same time, CO_2 diffusion will continue because photosynthetic CO_2 fixation in the leaf draws the CO_2 concentration down and ensures a gradient between the air immediately inside the leaf and the air in the crypt. In this way, crypts should increase water use efficiency (amount of water lost per CO_2 fixed); however, this explanation has not been well supported and other hypotheses have been proposed. One possibility is that crypts buffer leaves from large short-term changes in water loss by adding a fixed resistance in series with stomatal resistance to water loss (Hassiotou et al. 2009). Crypts are found among species that have thick leaves and in this context they have another important effect related to CO_2 uptake. Thick leaves create longer paths for CO_2 diffusion to the photosynthetic tissues of leaves. The crypts shorten this path by placing the CO_2 entry point, stomata, nearer to the top of the leaf where sunlight is being absorbed and energy generated to drive CO_2 fixation (Hassiotou et al. 2009). Supporting this, there is a strong positive correlation between crypt depth and leaf thickness in *Banksia* spp. (Lambers et al. 2014).

Species with stomatal crypts are widespread in MTV of Australia, California, and the Mediterranean. However, there is not a wide range of taxa that have crypts and they are most abundantly found in the *Banksia* spp. (clade *Cryptostomata*) of Australia and *Ceanothus* spp. subgenus *Cerastes* in California. In the Mediterranean, *Nerium oleander* is a widespread species with crypts and at least some *Phillyrea* and *Cistus* spp. have crypts (Rotondi et al. 2003). There are scant reports of crypts in South African and Chilean species. Ericoid leaves may function similarly to crypts in some ways. Also, species with stomata on both leaf surfaces (amphistomatous) may overcome diffusion limitations that accompany thick leaves, thus representing a different evolutionary solution than crypts to the same problem of diffusion limitation. Amphistomatous leaves are common in MTV species with thick leaves (e.g. *Arctostaphylos* spp. in California, *Hakea* spp. in Australia, and *Protea* spp. in South Africa).

6.2 Photosynthesis and growth

Photosynthesis is the process by which plants take up CO_2 from the atmosphere and use the energy from sunlight to fix this CO_2 and ultimately produce chemical energy (carbohydrates) or use the CO_2 as a building block to build new tissues. The energy and biomass that plants produce via photosynthesis is called primary production and it forms the basis of energy that fuels ecosystems.

The summer-dry and wet-winter cycle of the MTC has a strong impact on photosynthesis and growth. Because it is moist when it is cool and days are short, plants must be able to photosynthesize at these lower temperatures. Later in the spring, when temperatures are warmer and days are longer, and if water is still available, conditions may be optimal for photosynthesis and growth (Figure 6.5).

It has been hypothesized that the photosynthesis of the leaves of evergreen shrubs in MTV is relatively insensitive to temperature and that they can photosynthesize at or near maximum over a wide range of temperatures and are limited by only the hottest leaf temperatures. This hypothesis is logical given the generally mild climate of MTC regions and the evergreen leaves of

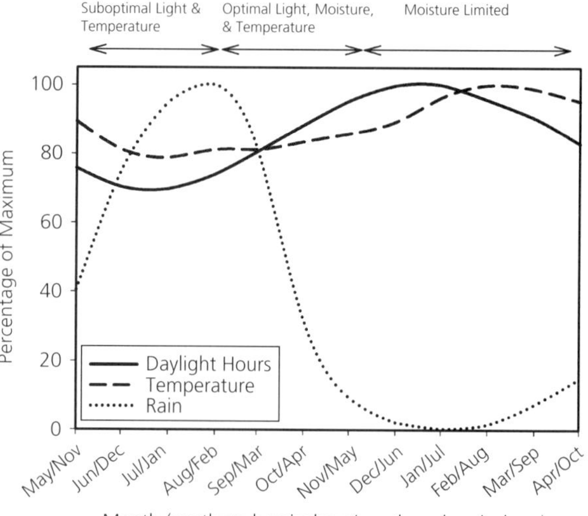

Figure 6.5　**Key aspects of the mediterranean-type climate that affect mediterranean-type vegetation species.** When water is available (fall, winter, and early spring), days are short and temperatures are cool, leading to suboptimal conditions for photosynthesis and growth. If water persists into the spring, warm temperatures and longer days can create optimal conditions. During the late spring and summer months water becomes limiting, and photosynthesis is greatly reduced and growth is commonly halted. (Redrawn from Parker et al. 2016.)

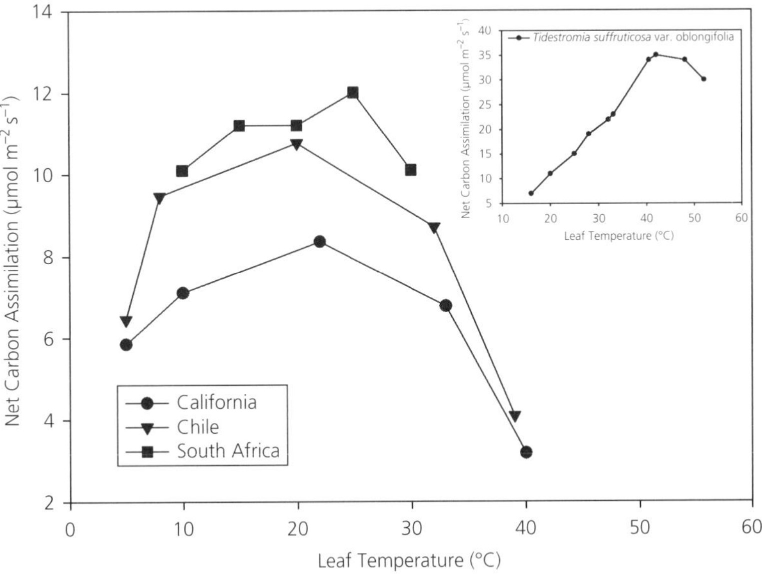

Figure 6.6   **Photosynthesis and temperature.** Photosynthesis (net carbon assimilation of individual leaves) measured across a range of temperatures for healthy plants. Data from California are for the shrub *Rhus ovata*, the data for Chile are the mean of four common matorral species (Oechel et al. 1981), and the data for South Africa are from *Protea repens* (Mooney et al. 1983). For comparison, the inset figure is from a C4 warm desert species that has a characteristically different photosynthesis response from the mediterranean-type vegetation (Bjorkman et al. 1975).

the dominant shrubs and the available data do support this hypothesis for the regions (Figure 6.6). By contrast, species adapted to hot environments show temperature limitations at mild temperatures and a much higher temperature optimum for photosynthesis (Figure 6.6 inset). There is evidence that species whose distributions span a broad range of temperatures exhibit photosynthetic adjustments to temperatures (Mooney et al. 1975). Since light provides the energy to turn CO_2 into chemical energy, short days during winter lead to an unavoidable reduction in photosynthesis (Figure 6.5); therefore, short days are more limiting than temperature to photosynthesis in these shrublands. Nevertheless, at colder sites at middle elevations or in cold air drainage basins photosynthesis may be depressed by cold winter temperatures (Flexas et al. 2014). Such sites appear to be suboptimal for evergreen sclerophylls from the standpoint of seasonal carbon gain of leaves (Flexas et al. 2014).

The most important factor limiting plant photosynthesis is lack of water. The summer-dry season depletes the shallower soil layers and this leads to dehydration of plant tissues. As this happens, the tiny cells in the leaves that function as valves (stomata) to control CO_2 uptake close (Figure 6.7a) and this restricts

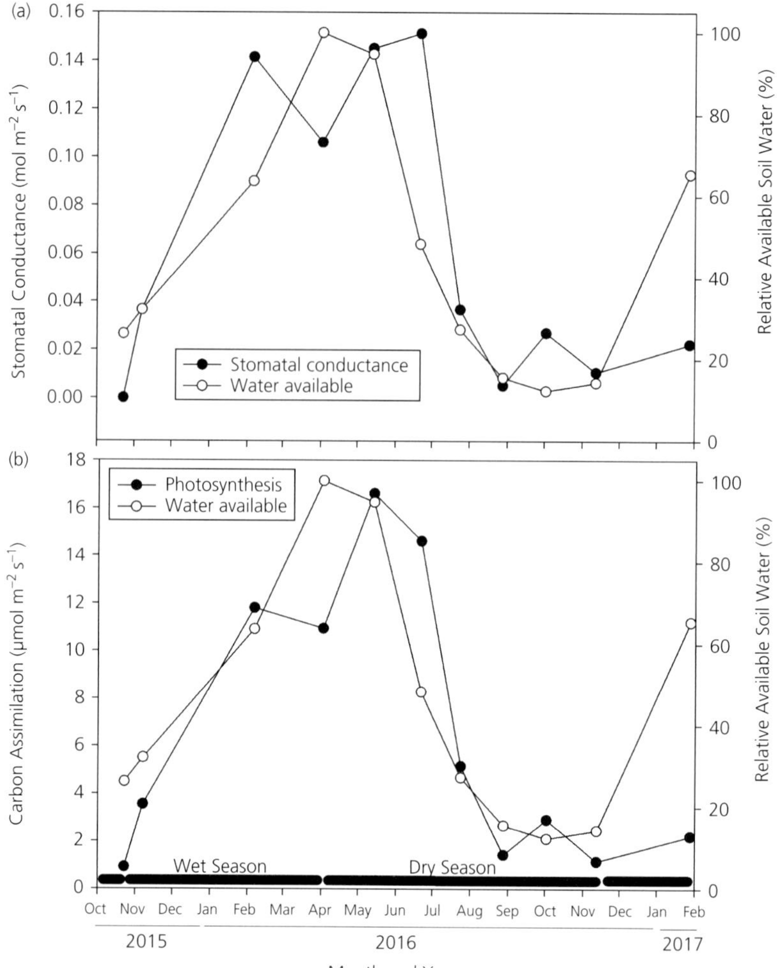

Figure 6.7   **Stomatal and photosynthetic response to available water.** Seasonal course of stomatal conductance (panel a) and photosynthesis (net carbon assimilation of individual leaves; panel b) for the chaparral shrub *Ceanothus vestitus* on the southern slope in the Tehachapi Mountains on Tejon Ranch (Pratt, unpublished data). The available soil moisture is the percentage of maximum predawn water potential of plants measured with a pressure chamber, which is a common method to estimate the soil water potential that a plant is able to access. Note, the months depicted are from a northern hemisphere mediterranean-type climate region, so Oct–Dec is fall, Jan–Mar is winter, Apr–June is spring, and July–Sept is summer.

CO_2 uptake (Figure 6.7b). The pattern of stomatal closure and photosynthetic limitation during the MTC dry season is well established for MTV species (Limousin et al. 2010). Plants have to expose their leaf tissue directly to the dry atmosphere to let CO_2 diffuse into leaves, thus they necessarily lose water from these moist tissues (transpiration) in what has been called the photosynthesis/transpiration compromise. At some point, CO_2 uptake fails to keep pace with CO_2 loss via respiration and the plant runs a carbon deficit. This deficit can be temporarily tolerated by drawing on energy stores (chiefly carbohydrates, such as starch and fructans). During droughts, when the carbon deficit is protracted, plants may suffer mortality due to carbohydrate depletion (Pratt et al. 2014; Pausas et al. 2016). This is a complicated process and carbohydrate depletion may at times be a proximate cause and not the ultimate cause of mortality (McDowell et al. 2008). For example, low levels of energy production and stores compromise plant defences against pathogens and can lead to mortality due to disease (Manion 1981). Desiccation owing to vascular failure to transport water, as discussed later in this chapter, is often a proximate or ultimate cause of mortality (Jacobsen et al. 2012).

Maximum photosynthetic rate of leaves (net carbon uptake), measured when plants are not experiencing stress, is an important trait that provides insight into key aspects of a plant's physiology and ecology (Figure 6.8). Specifically, leaves with higher photosynthetic rates generally have greater levels of nutrients (nitrogen and phosphorus) and reduced longevity and sclerophylly (LMA). These rates are commonly expressed on an area basis (Figure 6.8a), which provides insight into how the area of leaf tissue exposed to light is able to capture CO_2. Another way to express these values is on a per mass basis (Figure 6.8b), which provides economic information about how favourable of a return a leaf produces for a given resource investment (mass). The available data do not come from matched sites that would yield the most direct comparisons and there is scant data from Chile.

Comparing species across regions shows that Californian, Mediterranean, and Australian species have higher photosynthetic rates on a leaf area basis than the other regions, including non-MTC evergreen broadleaf shrubs (Figure 6.8a). When these rates are expressed on a per mass basis, the results slightly differ, with rates of Californian and Mediterranean species being greater than the other three MTC regions (Figure 6.8b). Australian plants are able to achieve high rates on an area basis, but they do this by increasing their sclerophylly (LMA; Figure 6.2), which translates into lower rates on a per mass basis (Figure 6.8b). This is possible because thicker and more sclerophyllous leaves have a greater volume that can be packed with more metabolic compounds per area. Nitrogen is a key nutrient that is a component of proteins and the nitrogen content of the leaves of Australian species fit this explanation; i.e. they are relatively higher on an area basis and lower on a per mass basis (Figure 6.9). South African species might be expected to follow the same pattern as Australian species, since the region generally has nutrient-impoverished soils, but they do not achieve the same high rates on an area basis. Nevertheless, on a mass basis they are very similar to Australian species and lower than Californian and Mediterranean species (Figure 6.8b; Mooney 1983).

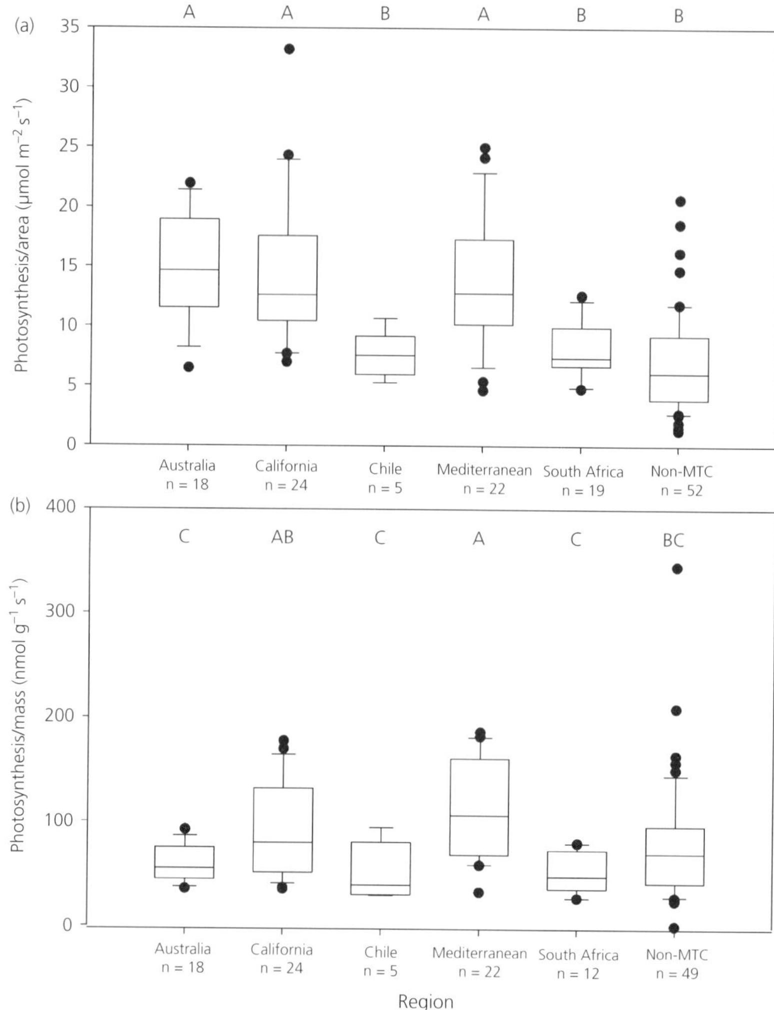

Figure 6.8 **Photosynthetic rates of mediterranean-type vegetation species.** Box and whisker plots (see Figure 6.2 for format) of net carbon assimilation (photosynthesis) on a per leaf area basis (a) and on a per mass basis (b) for C3 evergreen mediterranean-type climate (MTC) region species and a non-MTC group that represents evergreen broadleaf shrubs from non-MTC regions (data are from Field et al. 1983; Mooney 1977; Mooney et al. 1983; Oechel et al. 1981; Lawrence 1987; Pratt et al. 2012b; Pratt unpublished; von Willert et al. 1989; Wright et al. 2004). The sample size (n) is shown beneath each region. Regions within a panel that have unique letters are significantly different.

Mineral nutrition differences of soils of the different MTC regions are linked to leaf photosynthetic rates and foliar nutrient contents (Chapter 7). The species from California and Mediterranean regions have greater levels of foliar nitrogen content, particularly when expressed on a mass basis (Figure 6.9a, b). The nutrient explanation predicts that Chilean species should

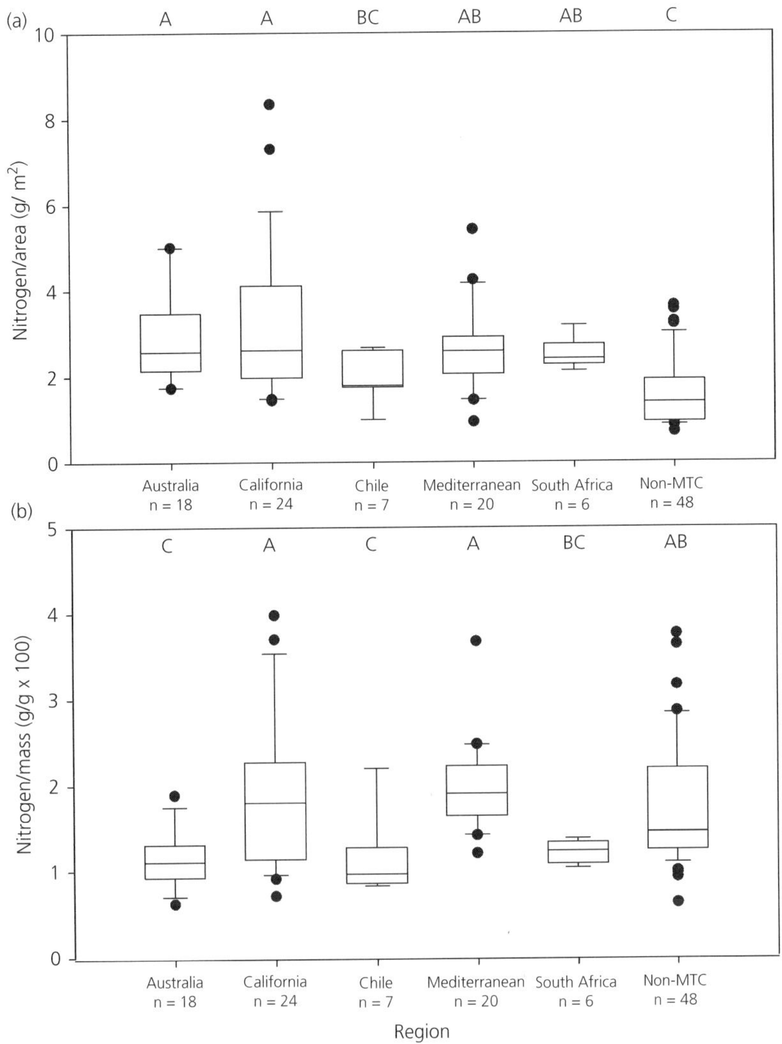

Figure 6.9 **Foliar nitrogen content of mediterranean-type vegetation species.** Box and whisker plots (see Figure 6.2 for format) of leaf nitrogen content on a per leaf area basis (panel a) and on a per mass basis (panel b) for C3 evergreen mediterranean-type climate (MTC) species and a non-MTC group that represents evergreen broadleaf shrubs from non-MTC regions (see Figure 6.8 for data sources). The sample size (n) is shown beneath each region. Regions within a panel that have unique letters are significantly different. Foliar nutrient content is also discussed in Chapter 7 and shown in Table 7.3; however, please note that the units and data used here differ from those in Table 7.3. Different data were used here because these data were limited to those that matched the data in Figure 6.8 to facilitate comparisons within this chapter.

be more like the Californian and Mediterranean ones, but they are more similar to Australian and South African species in both photosynthetic rates and nitrogen content. The small sample size for Chilean species means that this analysis is one that requires further examination.

The range of photosynthetic rates within regions, particularly on a mass basis (Figure 6.8b), appears to be greater in the Californian and Mediterranean species, which may be due to a greater range of environments, particularly in the context of mineral nutrition. There is a well established association that leaves with greater nitrogen contents are able to achieve greater maximum rates of photosynthesis (Field and Mooney 1986; Wright et al. 2004). This relationship is well supported by species from MTC regions and a sample of non-MTC evergreen shrubs (Figure 6.10; Field 1991). This suggests that available nutrients are a key factor affecting the maximum photosynthetic rates that MTV species are able to achieve.

The growth of shrubs is also limited in similar ways to photosynthesis, but growth is even more sensitive to water limitations than is photosynthesis.

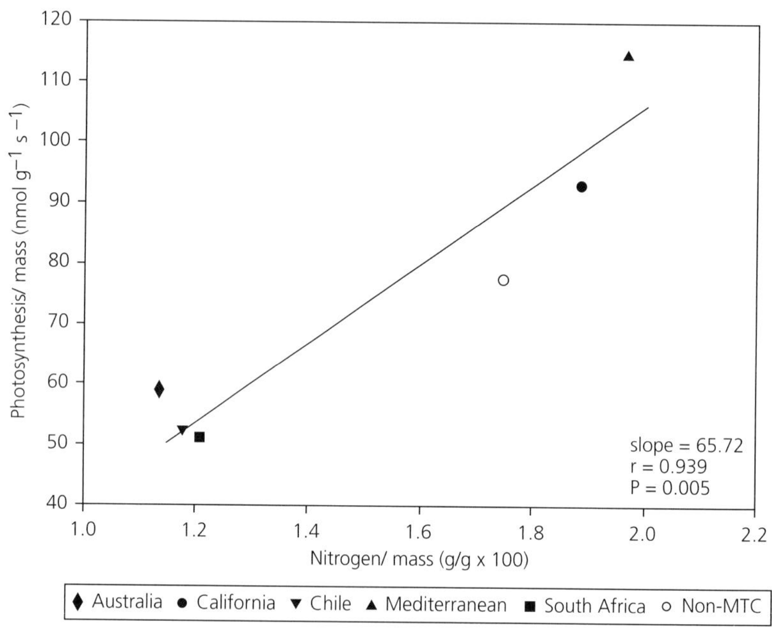

Figure 6.10   **Photosynthesis and foliar nitrogen content of mediterranean-type vegetation species.** Relationship between net carbon assimilation (photosynthesis) on a mass basis and leaf nitrogen content on a mass basis for C3 evergreen mediterranean-type climate (MTC) region species and a non-MTC group that represents evergreen broadleaf shrubs from non-MTC regions. Data points are means of the data from Figures 6.8 and 6.9. Best fit line and slope are from a standardized major axis regression, and *r* and *P* values are from correlation analysis.

Growth can be separated into two types: vegetative and reproductive. Vegetative growth is the production of the non-reproductive organs of the plants, roots, stems, and leaves, whereas reproductive growth is the production of flowers, fruits, and seeds. Vegetative growth of shoots may be under strong selection because of competition for light in dense stands of shrubs. Plants also require enough access to soil moisture to be able to expand cells and tissues. Because of this, the timing of vegetation growth may be relatively constrained and most species flush new growth in the spring, but there is a range from late winter to late spring (Kummerow 1983). By contrast, reproductive growth widely differs among co-occurring shrubs (Kummerow 1983). Competition for pollinators and fruit dispersers and avoidance of herbivores may be selective agents favouring different reproductive periods (Mooney et al. 1974).

Plants create more mass (biomass) chiefly by constructing their tissues out of organic carbon. This means they need carbon from CO_2 and photosynthesis or they have to use stored carbon to fuel their growth and that plant carbon balance influences growth amounts and the timing of vegetative and reproductive growth (Mooney et al. 1974). Another requirement is that their tissues need to be above a threshold level of hydration. This is because plant growth involves not only making more cells, but also enlarging these cells, and cell enlargement requires cells to be blown up like balloons using the positive pressure in living cells (turgor pressure). When plants are dehydrated this pressure can decline to zero, in which case cell expansion is impossible. For plants with shallow to moderate rooting depth, the turgor loss point is reached sometime during summer and this ends the growing season. Other plants that can access stable pools of moisture via deeper more extensive roots maintain turgor and can keep growing through the early part of the dry and hot summer (Montenegro et al. 1979; West et al. 2012; Mitchell et al. 2008). A larger percentage of species flower in the dry season in South Africa and Australia compared to Chile, the Mediterranean, or California (Figure 6.11). This may be related to the more reliable rainfall patterns found in Australia and South Africa (Cowling et al. 2005). Seasonal variations in the timing and amount of rainfall and variations in temperature leads to the amount and timing of growth being variable from year to year (Keeley 1987). During an intense drought growth may be minimal or even negative because of dieback of shoots (Venturas et al. 2016), and flowers or fruit may not be produced or may be aborted before reaching maturity (Kummerow et al. 1981; West et al. 2012). Root growth patterns are not well studied, with some notable exceptions (Montenegro et al. 1982; Kummerow et al. 1978).

Seasonality of growth can be affected by a number of things other than carbon balance and access to water. Cool temperatures may limit growth during the winter months as limited kinetic energy leads to sluggish enzymes and ridged cell walls. Kummerow (1983) argued that, after soil moisture,

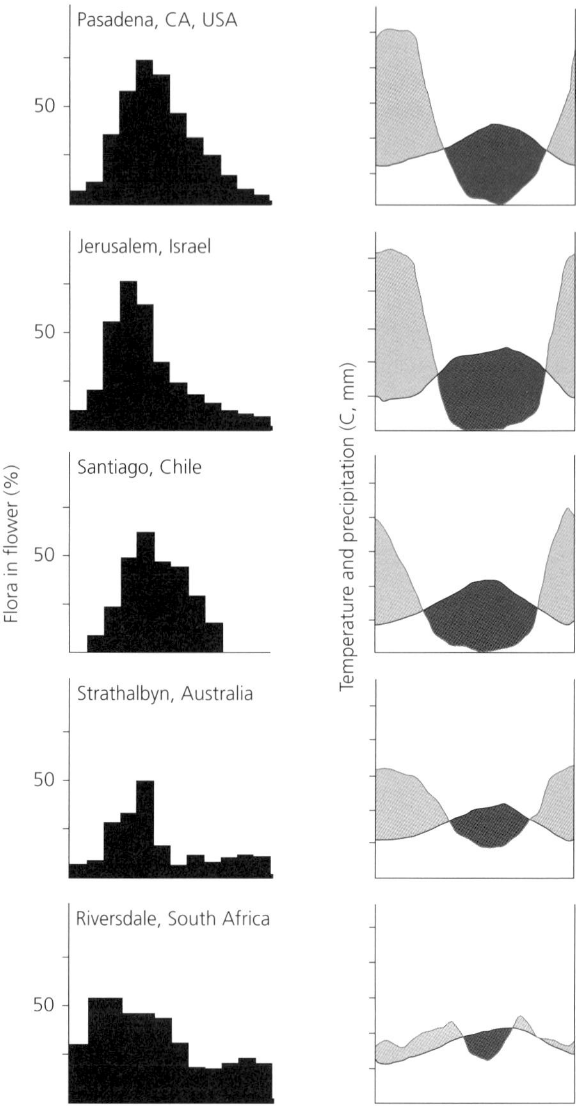

Figure 6.11   Flower times of shrubs. Flowering times of plants from areas representing all five mediterranean-type climate regions. Note that the months have been scaled so that northern and southern hemisphere seasons align. Climadiagrams for each site are included in the right column (see Figure 2.5 for an explanation of this diagram type). (Figure is redrawn from Mooney et al. 1974.)

cool temperatures were chief in limiting growth. Moreover, it may take some plants weeks or months after the fall dry season to recover to a fully hydrated state. Fine roots are strongly affected by depleted shallow soil moisture and need to be regrown (Kummerow 1983). In addition, an important part of this recovery for some species is likely the growth of new vascular tissues to support maximum rates of transpiration and photosynthesis (Pratt et al. 2005). Many species produce flower buds the growing season prior to their maturation (nascent flower buds), with flowering occurring in the subsequent growing season and typically in the winter (Hoffmann and Hoffmann 1976; Keeley 1987; Kummerow et al. 1981). By doing this, the amount of growth that needs to occur during the cool winter is reduced and flowering can rapidly commence during the winter rainy season. Interestingly, for species with this growth pattern, a drought during one year may prevent them from producing nascent flower buds and thus prevent them from flowering the following year even if that year is moist. By contrast, species that flower and fruit in the same season are more affected by the current year's conditions. Other factors, such as temperature and carbohydrate stores, play a role in amount of fruit produced on an annual basis (Keeley 1987).

Beyond cool temperatures, cold (subzero) temperatures are a factor affecting phenology and biogeography of species (Case Study 4). Temperature is not constant across the landscape in MTCs and this has important implications for plant function (Mooney et al. 1975). In some areas within all MTC regions, there are areas that are merely cool in the winter and there are also cold areas that experience hard frosts (minimum temperature is approximately less than –6°C). These cold areas are commonly found in mountain or hill basins (cold-air drainage basins) that gather the dense cold air that flows downslope by the pull of gravity after forming on surrounding high points by radiative cooling on clear nights. This leads to a temperature inversion where the coldest temperatures are at basin bottoms and the hills are warmer (Case Study 4). These conditions lead to abrupt freezing gradients with sites that commonly freeze in the winter being immediately adjacent to sites that virtually never experience freezing temperatures (Ewers et al. 2003). These gradients in freezing can be associated with changes in distribution of shrub species over very short distances (Ewers et al. 2003; Duker et al. 2015). The commonness of freezing sites will be reduced in areas with less topography, like much of Australia, and areas with coastal exposure, and broad differences may exist between the northern and southern hemispheres (Bannister 2007). Moreover, changing climates will likely affect these freezing gradients, which may lead to shifts in vegetation.

The harmful effects that freezing temperatures have on shrubs range from subtle to catastrophic. In some cases the cold temperatures may directly kill cells and lead to their death (Boorse et al. 1998; Duker et al. 2015). In other cases, the freezing and thawing of water in the xylem leads to the introduction of air into the xylem that causes blockages (emboli) in a process called freezing-induced cavitation. These emboli obstruct water flow and create deficits in foliar tissues in the days and weeks following the freeze-thaw event as water supplied to transpiring tissues cannot keep pace with water lost. Some plants can cope with this while for others this can be fatal in some contexts (Pratt et al. 2005). Interestingly, species differ in their vulnerability to freeze-thaw induced cavitation and the vulnerability is directly related to the diameter of the conduits (vessels) in the xylem that transport water. Wider conduits are more vulnerable to the formation of emboli when water in the xylem freezes and then thaws (Davis et al. 1999). Moreover, if a plant is dehydrated and experiences a freeze-thaw event it may lead to widespread and catastrophic levels of emboli (Davis et al. 2005).

6.3 Responding to limited water

The key factor that affects the physiology of shrubs in MTC regions is **drought**. The term drought has been used in different contexts. Commonly, a drought refers to below average annual precipitation; however, many authors use the term to refer to a period when water is limiting to organisms. In MTC regions, water is limiting every summer and fall during the rainless months when warm temperatures occur even when rainfall inputs are average or above average. More recently, in the context of climate change, droughts have been accompanied by warmer temperatures in what has been called **global change-type drought** or hot drought (Diffenbaugh et al. 2015). As climate changes, it is increasingly common for MTC regions to experience different types of **drought regimes** and understanding how MTV responds is an area of active research. Drought regimes describe the many factors that may contribute to atypical water deficit, including timing, duration, intensity, and temperature patterns. In many areas, rainfall totals are reaching record lows, leading to high-intensity droughts. Duration is an important context as well, and droughts may be short (1 year) or long term (multi-year). Intra-annual patterns are important as well, such as the number of consecutive months without rainfall or the co-occurrence of a drought with a heat wave. Moreover, timing of a drought is also an important component of a drought regime. If drought occurs during the wintertime, when freeze-thaw stress co-occurs, this will be significantly more stressful and cause greater strain on plants (Davis et al. 2005). Accuracy in predicting MTV response to drought will require consideration of different drought regimes. Moreover, the different MTC regions differ in their typical drought regime, which has implications for ecosystem convergence (Cowling et al. 2005).

The dominant shrubs have evolved different strategies for dealing with water limitations, ranging from **dehydration avoidance** to **dehydration tolerance** (Ackerly 2004b). Dehydration avoidance is a strategy whereby a plant avoids tissue dehydration even though water in the environment is limiting. It is important to think of dehydration avoidance and tolerance as a spectrum. Strict dehydration avoidance generally does not occur in MTV, but species do dramatically differ in the degree of dehydration that their tissues experience even when co-occurring (Jacobsen et al. 2007a; 2007b). One way to assess the degree to which a plant avoids or tolerates dehydration is to measure the water status of their tissues during the fall months before the winter rains. At this time, the available water is at a seasonal low and plants are maximally dehydrated. Water status can be measured in different ways, with the most common being to measure water potential of tissues, which provides a measure of the energy status of the water relative to pure water at standard conditions. For this parameter, more negative values indicate greater levels of dehydration. Water potentials have been measured across the five MTC regions. The range of values observed suggests that all five systems contain species that range from dehydration tolerators to dehydration avoiders (Table 6.1; Figure 6.12). A list of common taxa that fall along the dehydration tolerance to avoidance spectrum are listed in Table 6.1. One pattern that emerges is that South African and Australian MTC regions have greater numbers of species that are dehydration avoiders. This is partly due to the presence of so many species in the Proteaceae family, many of which avoid dehydration due to deep rooting (Higgins et al. 1987), and partly due to rainfall and hydrological factors that make these regions comparatively more mesic (Cowling et al. 2005).

When the shallow soil layers dry, dehydration avoidance can be achieved in a number of ways, including stomatal closure, leaf shedding, deep and extensive roots, and storage of water. Shallow roots risk leaking their water back into the soil as it dries. In some cases this may be tolerated or even advantageous as a means to promote the soil microbiota, nutrient uptake, and root longevity (Bauerle et al. 2008). Preventing water leakage can be accomplished by death of fine roots coupled to the presence of a thick water-impermeable wax layer of larger roots, thus it follows that the roots in the shallow soil layers generally die back each year (Kummerow et al. 1978). It could also be avoided by physical shrinkage of the roots creating an air gap between the roots and the soil, extensive cavitation that reduces flow through the xylem, or physiological regulation of root permeability. Dehydration-tolerant species likely keep more living and active roots in drier soil layers, which gives them an advantage in rapidly responding to unpredictable rain pulses. Other than roots, storage of water in woody shrubs is not enough to supply water to transpiring tissues for very long, thus storage cannot avoid tissue dehydration in the absence of stomatal closure. Dehydration-avoiding succulent plants are not well represented in MTC regions, but they are found in drier and warmer communities within MTC regions, such as the Succulent Karoo in South Africa, and in ecosystems that form ecotones with MTV, such as the summer-rainfall Sonoran Desert in California (Chapter 4). Leaf shedding is common to varying degrees, but evergreen plants retain significant leaf area year round, thus shedding leaves is limited. In summary, dehydration avoidance is achieved by some combination of deep and extensive roots, stomatal closure, and leaf shedding.

Table 6.1 Common taxa that differ in how they respond to water scarcity. Dehydration avoidance was defined as being any species that had minimum water potentials >–4 MPa; intermediate species had minimum water potentials between –4 and –6 MPa; and dehydration-tolerant species had minimum water potentials <–6 MPa

Region	Family	Dehydration-avoiders	Family	Intermediate	Family	Dehydration-tolerators
Australia[1,2,3,4]	Fabaceae Proteaceae	*Jacksonia floribunda* *Banksia attenuata* and *B. menziesii* *Hakea polyanthema* *Petrophile linearis* *Stirlingia latifolia*	Myrtaceae Proteaceae	*Eremaea pauciflora* *Melaleuca scabra* *Beaufortia micrantha* *Hakea* spp.	Dilleniaceae Fabaceae Ericaceae Casuarinaceae	*Hibbertia* spp. *Bossiaea eriocarpa* *Leucopogon conostephioides* *Allocasuarina campestris*
California[5,6]	Anacardiaceae Lauraceae	*Malosma laurina* *Umbellularia californica*	Fagaceae Rhamnaceae Rosaceae	*Quercus berberidifolia* *Ceanothus* subgenus *Ceanothus* *Heteromeles arbutifolia*	Ericacae Rhamnaceae Rosaceae	*Arctostaphylos* spp. *Ceanothus* spp. subgenus *Cerastes* *Cercocarpus betuloides*
Chile[7]	Anacardiaceae Quillajaceae	*Lithraea caustica* *Quillaja saponaria*	Lamiaceae	*Clinopodium chilense*	Euphorbioideae Lauraceae Rhamnaceae	*Colliguaja odorifera* *Cryptocarya alba* *Retanilla trinervia*
Mediterranean[8,9]	Anacardiaceae Aquifoliaceae Ericaceae	*Pistacia lentiscus* *Ilex aquifolium* *Arbutus unedo*	Cistaceae Fagaceae Lauraceae	*Cistus laurifolius* *Quercus* spp. *Laurus nobilis*	Cistaceae Oleaceae	*Cistus* spp. *Phillyrea latifolia*
South Africa[10,11]	Fabaceae Proteaceae Restionaceae	*Aspalathus hirta* *Protea* spp. *Leucadendron salignum* *Staberoha cernua*	Bruniaceae Ericaceae	*Brunia noduliflora* *Nebelia laevis* *Erica* spp.	Anacardiaceae Ericaceae	*Searsia undulata* *Erica* spp.

1. Dodd and Bell 1993a; 2. Dodd and Bell 1993b; 3. Lambers et al. 2014; 4. Groom and Lamont 2015; 5. Jacobsen et al. 2007b; 6. Bhaskar et al. 2007; 7. Poole et al. 1981; 8. Rhizopoulou and Mitrakos 1990; 9. Martinez-Vilalta et al. 2002; 10. West et al. 2012; 11. Jacobsen et al. 2007a.

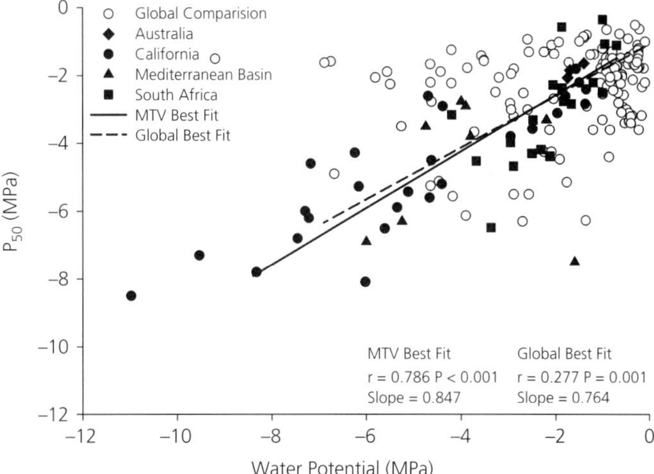

Figure 6.12   **Cavitation resistance and water status.** Cavitation resistance to dehydration (estimated as the negative pressure where 50% of vascular function is lost; P_{50}) plotted in relation to the maximum level of dehydration a plant experiences in the field (estimated as water potential during the driest part of the year) for mostly evergreen mediterranean-type vegetation (MTV) species and a global dataset that includes a range of species from deserts, seasonal tropical forests, and temperate forests. (Modified from Choat et al. 2012.) More negative pressure values in the xylem develop during dehydration and species with lower values of water potential are more resistant to dehydration-induced vascular dysfunction. The water potential values are mostly measured at predawn (except Australia) when the plant is in equilibrium with the soil and the leaves are at or close to equilibrium with the xylem pressure. Each data point is a mean for a species. Data for MTV are from Canham et al. (2009), Jacobsen et al. (2007a, b), Jacobsen et al. (2012), and Martinez-Vilalta et al. (2002). The best fit line and statistics are from a standardized major axis regression and correlation analysis, and, for MTV (solid line), includes the outlier in the lower right corner, which is *Ilex aquifolium*.

Dehydration-tolerant species experience tissue dehydration during drought. Such plants have adaptive traits that enable this dehydration to occur while avoiding permanent damage. One such trait is resistance to dehydration-induced cavitation. As already discussed, cavitation can occur following freezing and thawing; however, it can also occur during dehydration by a different mechanism. In this case, the tension on the water in the vascular system increases to extract water from increasingly dry soils that tenaciously cling to the moisture. A critical point is reached where the water can no longer support the tension and gas enters the vessels, whereupon the water is displaced by gas. The gas expands to fill vessels, forming emboli, and this leads to reduced conductive capacity to support transpiring leaves. Species widely differ in dehydration-induced cavitation resistance. The level of dehydration (minimum water potential) correlates with cavitation resistance such that greater dehydration is exhibited by species with greater resistance to cavitation (Figure 6.12). This indicates that this trait is a key component of the dehydration tolerator strategy. This makes sense for evergreen species that have to keep evergreen leaves supplied with water.

The relationship in Figure 6.12 is also important for showing that MTV, insofar as data are available (no data were available for Chile and data for Australia are limited), follows a similar trend. This trend (slope of the best fit line) is also similar to that found globally among woody angiosperm species (Figure 6.12). However, MTV differs in the range of dehydration experienced by species and in the high levels of cavitation resistance displayed by some species. Some desert species may get as dehydrated as chaparral shrubs, but they do not have high resistance to cavitation (Jacobsen et al. 2008). One study found that chaparral and fynbos vegetation more closely resembled one another than other more arid communities, including the Succulent Karoo of South Africa and the Mojave Desert (Jacobsen et al. 2009). The extreme levels of dehydration experienced by MTV species and their high cavitation resistance are associated with the MTC and long hot rainless summers. Moreover, the high resistance to cavitation arises because evergreen MTV species need to supply their leaves with water during the dry season; otherwise, they experience shoot dieback (Davis et al. 2002) or even mortality during intense droughts (Venturas et al. 2016).

There are other traits of adaptive importance to dehydration tolerators. These include the ability to maintain turgor pressure and relative hydration of cells in response to drying soils as well as dense tissues with thick cell walls (Bartlett et al. 2012). This latter trait is indicative of the sclerophyllous leaves of MTV shrubs and is discussed above in the section on leaves. Tissue density extends to the vascular tissue as well and dehydration-tolerant species generally have denser xylem than dehydration-avoiding ones. This is hypothesized to be important in resisting implosion of water-filled vessels under highly negative pressures as occurs when tissues are dehydrated (Jacobsen et al. 2005).

6.4 Demography and population dynamics: the key role of fire

6.4.1 Fire regime

Populations of the plants and animals that inhabit MTC regions are strongly influenced by disturbance in the form of fire. Fire will occur given sufficient continuous fuel (plant biomass) that allows for it to spread, dry conditions that reduce moisture content of fuel so that it is combustible, and an ignition source (Case Study 14). Oxygen is also required and must be considered over geologic and evolutionary timescales (Pausas and Keeley 2009). The concept of **fire regime** describes key aspects of fires that characterize a particular region, such as the type of fire (ground, surface, or crown), extent of the burned area, return interval, fire intensity, and seasonal timing of fire. The key factors that determine a particular fire regime include fuel structure, ignition source, seasonality, and productivity. The fuel structure in MTV shrublands is generally continuous, which leads to a crown fire where the aboveground portions of plants are burned. Depending on the productivity of the site and time since the last fire, they may be more or less intense fires with greater fuel build-up being associated with more intense fires. The return interval of the fires is variable within and among MTV in the different MTC regions, but generally ranges from 10 to 100 years. The seasonality of fires is during the predictable MTC dry season during the late summer and fall periods when potential fuel is converted into available fuel as plant biomass drops in water content. Wind patterns are also an important factor that lead to extensive burning in the MTC regions (Keeley et al. 2012).

The fuel structure is an important driver of these fire regimes and has recently been examined by Keeley et al. (2012). There are many different types of communities within MTC regions, including shrublands, woodlands, grasslands, and forest, and these differ in fire regime (see Chapter 4). The shrublands in these regions will generally have continuous and abundant enough fuel to carry a crown fire provided they have had sufficient time to accumulate fuel following the preceding fire. One study found that MTV in South Africa had low flammability from years 1 to 6 following fire (Seydack et al. 2007). In areas with abundant herbaceous growth, the time to flammability may be shorter. This is a common feature of degraded shrublands that have been invaded by alien grasses or that are degraded and invaded by native grasses in the case of the Mediterranean region (see Chapter 8). Areas at the more arid end of the MTV spectrum that are degraded, such as areas in the eastern Mediterranean (Batha-type vegetation), may not have the fuel load to sustain a predictable fire regime and more such area may occur under a warming and drying environment (Case Study 14).

The fire regimes of MTV in the different MTC regions are generally similar, but there are some important differences. The most fundamental difference among the regions is found in Chile. Chile has limited natural ignition sources owing to the Andes Mountains blocking summer convective storms moving west from Argentina; nevertheless, present day Chile has ample fires that are ignited by human activities. The Andes uplift occurred during the Miocene, beginning some 20 million years ago (Ma) and ending about 2.5 Ma and there was likely fire before this time. This history is an important context for interpreting the differences in fire-related traits of Chilean vegetation compared to the other MTC regions (Montenegro et al. 2003; Chapter 2; Case Study 11).

The four other MTC regions have long histories of fire cycles that have shaped the evolution of their MTV. Evidence suggests that Australia and South Africa have shrubland fuel structure and fire activity going back to at least the early Tertiary (Case Study 11). By contrast, the northern hemisphere regions had more limited early Tertiary fire that lacked the same degree of predictability and intensity until climate changes in the late Tertiary and the

Case Study 11 Linking fire traits and historical fire regimes in the mediterranean-type environment

Juli G. Pausas, Centro de Investigaciones sobre Desertificacion, Consejo Superior de Investigaciones Científicas (CIDE-CSIC), Valencia, Spain

Many mediterranean shrublands and forests are subject to recurrent fires of relatively high intensity (crown fires), and plants have evolved traits to cope with them. These traits allow persistence at the level of an individual (traits related to resprouting) or population (traits related to post-fire seeding) (Pausas et al. 2004). To what extent are these strategies unique to mediterranean-type ecosystems (MTEs)?

Resprouting (R) is not unique to mediterranean-type climate (MTC) regions, and is instead a ubiquitous trait that occurs in all biomes and in many lineages, and it provides survival against many disturbance types (e.g. strong winds, fires, large animals). Resprouting is considered an ancestral trait that was already present in ancient flora (e.g. Mesozoic). However, one peculiarity of MTC regions is that resprouting in woody species is strongly binary; that is, there is a tendency for species to be either a good resprouter (most individuals resprout even after high-intensity fires) or a non-resprouter, while this is not necessarily true in other biomes. In other fire-prone biomes, many woody species may resprout or not depending on other factors (age, weather, fire characteristics, etc.); we call these weak resprouters. In MTC regions, weak resprouters may also occur but they are evolutionarily unstable; the reliable recurrence of intense fires has selected against weak resprouters and favoured traits for strong resprouting capacity (e.g. the acquisition of specialized resprouting structures like lignotubers or of different types of bud protection mechanisms). In other ecosystems, fires are typically less intense because they are more frequent (savannas), because the environment is wetter (tropical forests), or because fuels are low (arid ecosystems), and thus weak resprouters easily survive and persist throughout recurrent fires.

Post-fire seeding (S), that is the ability to accumulate a fire-resistant seed bank either in the canopy (serotiny) or in the soil, is quite unique to the mediterranean-type biome; most post-fire seeders are MTC region species. The reason, again, is the particular fire regime of the MTC regions. Ancestral environments were moister and were probably dominated by resprouters, and many of them were relatively weak resprouters as in many tropical forests today. With the increasing aridity during the Tertiary and the Quaternary, and the appearance of the MTC, fires become more frequent and intense. Such conditions not only selected against weak resprouters and favoured good resprouters; they also generated an alternative evolutionary pathway (Pausas and Verdú 2005; Pausas and Keeley 2014): some weak resprouters evolved the capacity to regenerate after fire from their seed bank; that is, in some species, seeds became fire-resistant or the fruits or cones became serotinous. With the acquisition of post-fire seeding, weak resprouters could cope with relatively intense fires by prolific post-fire recruiting. As the fire intensity increased, resprouting in these species became less relevant, and some lineages even lost their resprouting capacity altogether and became obligate post-fire seeders (Pausas and Keeley 2014).

The consequences of this evolutionary dynamic are that in MTC regions plants may or may not have traits for resprouting (R+, R–) as well as traits for post-fire seeding (S+, S–). Thus, four life histories are possible: obligate resprouters (R+S–), facultative species (R+S+), obligate post-fire seeders (R–S+), and species without any of these traits (R–S–).

Continued

Case Study 11 (*Continued*)

In the latter case, populations do not really persist after fire, but they may colonize from the surrounding (unburnt) vegetation (post-fire colonizers). These four broad functional types explain a large proportion of the variance of the vegetation dynamics in MTEs subject to crown fires (Pausas and Keeley 2014). However, the relative importance of each of these functional types varies among the different MTEs (Case Study 11 Figure). In the MTC regions of the northern hemisphere, obligate resprouters are the most common, while they are rare in Australia and South Africa. In these two southern MTC regions, post-fire seeders, including facultative post-fire seeders, are more prominent. In Chile, post-fire seeding is largely absent and obligate resprouters are abundant. These differences are tied to the different historical role of fire in the different regions. Australia and South Africa are believed to have had a high fire activity since at least the early Tertiary, which allowed fire-driven diversification over much of the Tertiary (He et al. 2011; Crisp et al. 2011). In the northern hemisphere, early Tertiary fire-prone landscapes were marginal, and lacked sufficient predictability and intensity to select for post-fire seeding until the climatic changes in the late Tertiary and the Quaternary (Keeley et al. 2012). Fire has not had an evolutionary role in central Chile since the uplift of the Andes (Miocene) that limited summer storms, and thus lightning and fire activity (Keeley et al. 2012). In conclusion, the relative abundance of the different fire-related traits is strongly tied to the historical fire regime in the mediterranean biome and at the global scale (Pausas 2015).

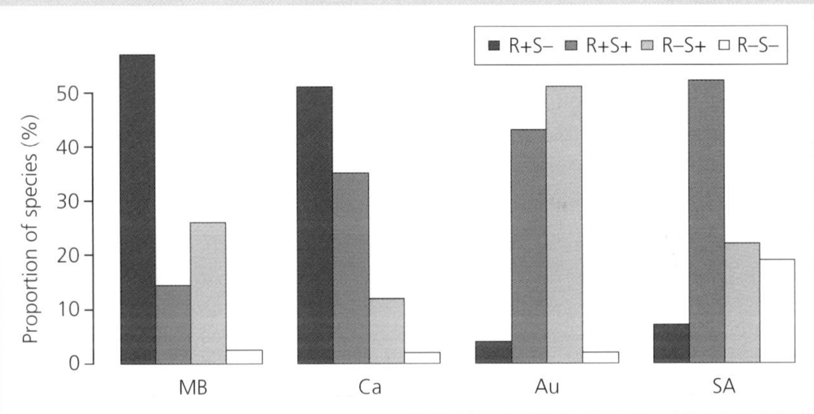

Case Study 11 Figure. Life histories of mediterranean-type climate region species. Proportion of woody species belonging to each of the four life histories (R+S–: obligate resprouters; R+S+: facultative species; R–S+: obligate seeders; and R–S–: others) in the Mediterranean Basin (MB), California (Ca), Australia (Au), and South Africa (SA); in Chile, post-fire seeders (R+S+, R–S+) are almost absent (not shown). (Based on Pausas et al. 2004.)

Quaternary created conditions for more extensive burning (Keeley et al. 2012). This difference in fire over evolutionary time has important implications for understanding the present day make-up of adaptive types found in these regions (Case Study 11).

6.4.2 Shrub response to fire: from individuals to populations

Organisms that inhabit MTC regions have specialized traits to persist in the context of a particular fire regime. It is important to note that the fire regime is what is important and not just fire per se. If the fire regime is altered it can have dramatic and degrading consequences for a particular system (Chapter 8). For example, fires that occur out of season (winter or spring) may hinder resprouting, owing to lack of stored carbohydrates (Moreira et al. 2012), and seedling recruitment and survival (Keeley et al. 2012). Fire opens up space, the ash layer is rich in nutrients, and seed predators are reduced, thus this creates opportunities for seedlings to establish and many species have specialized regeneration and reproduction to take advantage of this condition. There have been many excellent reviews of this topic (e.g. Bond and van Wilgen 1996; Groom and Lamont 2015; Keeley et al. 2012).

Plants in MTC regions have four different modes of recovering after a fire. The first is seedling recruitment from seed banks immediately following the fire (post-fire seeders). The second is when resprouts are stimulated to flower by fire and seedling recruitment occurs from seeds that are produced by these resprouts after a fire (fire-stimulated flowering). In this case, seedling recruitment occurs before the canopy closes, i.e. in the first few years following a fire, but not immediately following the fire. A third mode is resprouting followed by delayed seedling recruitment that occurs after the canopy has closed (typical in obligate resprouting species). A fourth mode is colonization where individuals, not present before fire, colonize post-fire via seed dispersing to the site from unburnt populations (opportunistic species). In some instances, fire refugia may occur at sites where species are protected by rocks or topography. Even rarer are species that only regenerate by resprouting and appear to be asexual or nearly so (Pausas and Keeley 2014).

Resprouting refers to the production of new shoots from buds that are protected from heat by bark, wood, or soil insulation and fuelled by stored carbohydrate reserves (Vesk and Westoby 2004). Resprouting can occur from aboveground stems, epicormic resprouting, or from basal structures that include woody swellings at the soil surface (lignotubers or burls), rhizomes, bulbs, corms, or roots (Figure 2.7). This mode of regeneration, particularly basal resprouting, is common and is an ancient trait among perennial plants (Feild et al. 2004). Species that resprout after fire will often resprout after other disturbances, such as dieback caused by freezing temperatures or herbivory (Davis et al. 2007; Ramirez et al. 2012). Moreover, resprouting species may continuously sprout new shoots in the absence of disturbance, forming a canopy of shoots that differ in age (Keeley 1992). Although resprouting is an ancient trait, lignotubers are mostly found in MTV species (Keeley et al. 2012). Resprouting species will often have a population of different-aged individuals since resprouting allows individuals to survive multiple fires; however, it is often not possible to age long-lived resprouts so approximations are used (Stohlgren and Rundel 1986).

Many of the characteristic plant lineages of MTC regions do not resprout after fire (Table 6.2) and instead they persist at sites through the germination of fire-cued seeds. The species that time their recruitment to occur following fire commonly form an even age population comprised of a cohort that dates to the last fire. Among seed banks there are two types. This first is where seeds are stored in the canopy of a plant where seeds are contained in woody cones (dried woody fruit in the case of angiosperms), which is called serotiny (Figure 2.8). The second type is when species form a soil seed bank, which is common among woody perennials and annuals. Dispersal for these seeds are ballistic or they simply fall from the plant and the seeds are incorporated into the soil over time. By contrast, some seeds are dispersed by ants (myrmecochory) and such seeds contain a fleshy nutrient-rich attractant for the ants called an elaiosome (Groom and Lamont 2015). Myrmecochory

Table 6.2 Common genera of post-fire seeders. Species numbers for each genera within each region are included (modified from Keeley et al. 2012)

Region and genus	Family	Number of species
Australia		
Adenanthos	Proteaceae	15
Banksia	Proteaceae	38
Grevillea	Proteaceae	138
Hakea	Proteaceae	77
California		
Arctostaphylos	Ericaceae	48
Ceanothus	Rhamnaceae	23
Chile		
None		
Mediterranean		
Cistus	Cistaceae	18
Erica	Ericaceae	1
Genista	Fabaceae	1
Ulex	Fabaceae	1
South Africa		
Aspalathus	Fabaceae	81
Cliffortia	Rosaceae	60
Erica	Ericaceae	390
Leucadendron	Proteaceae	27
Leucospermum	Proteaceae	23
Podalyria	Fabaceae	4
Paranomus	Proteaceae	9
Protea	Proteaceae	28

is common in Australia and Africa, and less common in the other MTC regions (Keeley et al. 2012; Chapter 3). Serotinous species are common among South African (e.g. *Protea* and *Leucadendron*) and Australian (e.g. *Banksia, Eucalyptus*, and *Hakea*) MTV species, but are rare in Chile, California (e.g. *Hesperocyparis* spp. and *Pinus* spp.), and the Mediterranean (e.g. *Pinus* spp.). Serotiny is more common in the vegetation of Australia and South Africa than any other system on Earth (Keeley et al. 2012). The low nutrient soils and reliable rainfall in South Africa and Australia have been important factors leading to the abundance of serotinous species in those regions relative to other MTC regions (Lamont and Enright 2000). The evolution of serotiny is likely related to herbivory pressures exacerbated by nutrient-poor soils (Keeley et al. 2012).

Seed banks are cued to germinate in response to a particular aspect of fire (see also Chapter 9). Seed dormancy refers to a state where a seed will not germinate even though the conditions for germination are suitable. Dormancy can serve to delay germination to a time when survival of the plant is favourable. In the case of serotinous plants, the fire triggers cones to open and deposit the seeds on the soil following a fire and germination ensues with sufficient moisture. Seeds generally lack dormancy and will readily germinate (Lamont et al. 1991). In other species, with soil-stored seed banks, the heat of the fire cracks a thick seed coat that allows for water uptake (imbibition) and germination. Other seeds require exposure to chemicals associated with fire to break dormancy (Keeley and Bond 1997). One source of chemical cues is found in the smoke from burned plant material (Wicklow 1977). Another source is from chemicals that are leached from charred plant remains (charate; Keeley and Pizzorno 1986).

Another group of species rapidly resprouts after fire and their reproduction is stimulated by fire; therefore, these species produce seeds in the months or years following fire. These species are commonly herbaceous perennials (Figure 2.7h), small shrubs that are woody at the base (suffrutescents), and many are monocots like Australian species in the grass tree genus *Xanthorrhoea*. Such taxa will often produce abundant seed in the first year or two after fire.

Species that either recruit from stored seeds or exhibit fire-stimulated flowering and seed release contrast with many obligate resprouting species. Some resprouting species recruit new seedlings in between fires in the understorey of mature stands. For these resprouting species, flowering and seed germination are not fire-cued and they are thus independent of fire for their sexual reproduction.

6.4.3 Evolution of traits in response to fire

Fire has shaped the traits of species inhabiting MTC regions in numerous ways (Case Study 11). Some excellent studies have taken advantage of newer data describing the evolutionary relationships (phylogenies) among important groups of MTC region plant species that are dated and thus can be examined in the context of palaeoenvironmental conditions (Bond and Midgley 2003; Crisp et al. 2011; He et al. 2011). The role of fire in shaping plant traits has been questioned (Bradshaw et al. 2011), but there is a strong case for fire as a shaper of plant traits (Keeley et al. 2012).

One fascinating hypothesis is that some plants have evolved to be more flammable (Mutch 1970). One version of this idea, the 'kill thy neighbour' hypothesis, holds that post-fire recruiting species benefit from being more flammable because by doing so they harm less flammable resprouters that would compete against their offspring after a fire (Bond and Midgley 1995; Schwilk and Kerr 2002). Another idea is that retaining fuel in the canopy so that it readily burns protects seeds and basal organs from a long hot ground fire that could damage tissues (Gagnon et al. 2010). There are a number of ways a plant can become more flammable, such as by producing flammable chemicals, retaining dead branches, leaves, or florets in the canopy, and producing thin branches with greater surface to volume ratios (Cowan and Ackerly 2010; Pausas et al. 2017); however, one of the controversies is that such traits may evolve for other reasons (Keeley et al. 2012). The evolution of flammability hypothesis continues to engender critical inquiry and a refinement of ideas (He et al. 2011; Midgley 2013; Pausas et al. 2017).

Different species have different ways in which they persist after fire (Case Study 11). Fire creates seedling recruitment opportunities in its aftermath and post-fire recruiting species take advantage of this. Some species both resprout (R+) and recruit seedlings (S+) from a dormant fire-cued seed bank after fire and they are often called facultative seeders (R+S+). In this terminology, facultative means optional since recruiting seedlings is not essential because individuals can persist by resprouting alone. Some species have lost the ability to resprout (R−) and only regenerate after fire from seedling recruitment from a fire-cued seed bank, and these are termed obligate seeders (R−S+). It should be emphasized that seeder is shorthand in this terminology and that it is not enough simply to produce seeds that are present in a post-fire environment. The critical factor is that those seeds produce seedlings that ultimately survive to adulthood, which requires further specialization in many cases. Another category of post-fire persistence involves resprouting post-fire with no post-fire seedling recruitment (R+S−), and these are referred to as obligate resprouters. For obligate resprouters, seedling recruitment is fire-independent and occurs in between fires where seedlings germinate and recruit in mature stands of vegetation. Obligate resprouters commonly have fruit that is widely dispersed by vertebrates.

Resprouting and fire-independent recruitment are ancient traits that are ancestral in many taxa (Wells 1969). This means that the evolution of post-fire obligate seeders has required evolution of traits to form fire-cued seed banks and the loss of resprouting (Keeley et al. 2012). This latter evolutionary step is intriguing because the loss of resprouting is risky as it concentrates the entire reproductive output of an individual into one point in time after a fire, which makes an individual vulnerable to recruitment failure. Further stoking interest in the evolution of obligate seeding is the observation that many obligate seeding taxa are endemic to MTC areas (Table 6.2), have their centre of distribution in an MTC region, contain many species, and many of these species have narrowly restricted distributions. These factors point to these species as having their origins in an MTC (neoendemics), thus they have long been considered emblematic of MTC regions and indeed they have been used to define MTC regions on a floristic basis.

Many models have been offered to explain the evolution of obligate seeders (reviewed in Keeley et al. 2012). At the core of most models is life history theory, which examines key stages that characterize a species' life, such as age of maturation, number of offspring produced, longevity, and resource investment in reproduction and persistence (Bell 2001). According to this theory, natural selection works to fine tune these events to maximize fitness. Broadly, there is a tradeoff between persistence (resprouting) and recruitment (seeding and survival of seedlings), and this is an important context for understanding the evolution of different post-fire regeneration types. There are clear costs associated with resprouting that obligate seeders do not incur, such as allocation of carbohydrates to storage to fuel regrowth of shoots after fire and production and maintenance of a bank of buds (Pate et al. 1990). Obligate seeders are often single-stemmed and some may attain taller stature than multi-stemmed resprouters; however, resprouters can attain greater age and stature. Although some of the costs are clear, they manifest differently in different species and at different stages in plant development and evolutionary history (phylogeny) is important too (Bell 2001; Keeley et al. 2012; Pratt et al. 2012a). Because facultative seeders, obligate resprouters, and obligate seeders differ in many key life history parameters they have commonly been referred to as different life history types.

Also key in the models to explain the evolution of these different life history types is the fire regime through time (Pausas and Keeley 2014). A recent eco-evolutionary model posits that obligate seeding can evolve in fire regimes that limit resprouting and thereby create more seedling recruitment opportunities. Such a fire regime is one that is predictable and of high intensity, whereby resprouters are harmed and space is cleared for seeders to establish. The fire regime would also need to be long enough so that it did not commonly kill plants prior to reaching reproductive maturity and short enough so that plants and seed banks do not die off.

Other environmental and evolutionary factors are hypothesized to be important in the evolution of obligate seeders. In more arid areas, water limitations may limit resprouting more than obligate seeding, which may favour post-fire seeding species over time (Pratt et al. 2014). Successful recruitment of seedlings after fire has been linked to a suite of adaptive traits related to water limitations (Pratt et al. 2007). Obligate seeders in California and the Mediterranean are highly tolerant of dehydration, which has been hypothesized to be associated with selection at the seedling stage. Shallow rooted seedlings in an open post-fire environment must be under strong selection to survive these hot and dry conditions that accompany the summer rainless period and they have evolved the traits that allow them to thrive in this regeneration niche (Pratt et al. 2008; Vilagrosa et al. 2014). Interestingly, this pattern does not seem to hold generally in South African and Australian MTC regions (Canham et al. 2009; Pratt et al. 2012a); however, obligate seeders increase in dominance in drier areas in Australia (Keeley et al. 2012). There are some reasons why this might be the case. First, there is limited data from these regions and more studies may find evidence for this pattern, at least among some lineages. Second, the different history of these regions with respect to fire regime has led to differences in life history type evolution (Case Study 11). Finally, the combination of more reliable rainfall in these regions (Cowling et al. 2005), coupled with nutrient-impoverished soils, offers an explanation. Resprouters are more competitive at sites with more resources that can support greater productivity (Clarke et al. 2005; Knox and Clarke 2005; Keeley et al. 2012; Case Study 11). The nutrient-impoverished soils mean that MTV in these regions will still have low productivity (Chapter 7) and will favour seeders under the right fire regime. The more reliable rainfall may mitigate the level of dehydration that seedlings experience and select for high dehydration tolerance. A final point is that South Africa and southwest Australia may have had a longer history of fire regime and soil conditions that limit productivity and resprouting, thus explaining their greater numbers of obligate seeding species relative to other MTC regions (Chapter 2; Keeley et al. 2012).

It is important to consider reproduction and heredity of these different life history types (Wells 1969). Long-lived resprouters may accumulate somatic mutations that ultimately lead to their sterility (Wiens et al. 2012). Thus, at dry sites where resprouter recruitment is poor, the resprouters may slowly drop out of the community over time. Obligate seeders are killed by fire and recruit a new generation of sexually produced offspring after each fire. This means that obligate seeders have shorter generation times, non-overlapping generations, and a reshuffling of the gene pool after every fire. These factors should accelerate evolutionary change and may be associated with diversification of obligate seeders. This hypothesis predicts greater genetic diversity in obligate seeders, which has received mixed support from genetics studies, and there are alternative hypotheses that may contribute to genetic variation in resprouters, such as the accumulation of somatic mutations mentioned above (Segarra-Moragues and Ojeda 2010; Verdú et al. 2007).

Fire regime changes over evolutionary timescales have been used to explain differences in life history types across MTC regions (Case Study 11). Chile, which lacks a long-term predictable fire regime, has virtually no obligate seeding species (Table 6.2). Chile and the northern hemisphere MTC regions also contain more obligate resprouters than South Africa and Australia, which has been linked to less predictable high-intensity fires in the early Tertiary and the existence of patches of shrublands in marginal habitats in a matrix of woodland and forest (Keeley et al. 2012). This did not change until the late Tertiary and the Quaternary, when the climate would have favoured such a fire regime and an expansion of MTV. By contrast, a fire regime favourable for obligate seeding was in place for longer in South Africa and Australia (Case Study 11; Table 6.2).

Resprouting, while clearly adaptive for many species in MTC environments, has been shaped by fire as a selective force (Figure 2.7); however, demonstrating that resprouting is an adaptation to fire is not simple. Resprouting is an ancient trait that is well represented among vascular plants and inferring the selective forces early in land plant evolution is challenging. Resprouting is well represented among species inhabiting environments where fire is not a naturally occurring disturbance, such as tropical rainforests. In such cases there are other disturbances, such as herbivory and wind-throw, that trigger resprouting. Nevertheless, basal lignotubers appear to be specialized structures that develop from a very early age and thus are not a wound response and may be linked to fire (Keeley et al. 2012; Verdaguer and Ojeda 2005). The role of the lignotuber is to store buds that can regenerate after fire and it often contains abundant carbohydrates that can, along with the roots, supply part of the fuel that is used to produce new shoots. Lignotubers are unusually well represented among species in MTC regions, and were likely shaped and reinforced by fire regime factors. In Chilean MTV, many species have lignotubers that likely evolved prior to the maximum Andean uplift in the late Miocene.

Literature cited

Ackerly DD. 2004a. Adaptation, niche conservatism, and convergence: comparative studies of leaf evolution in the California chaparral. The American Naturalist 163: 654–71.

Ackerly DD. 2004b. Functional strategies of chaparral shrubs in relation to seasonal water deficit and disturbance. Ecological Monographs 74: 25–44.

Bannister P. 2007. Godley review: a touch of frost? Cold hardiness of plants in the southern hemisphere. New Zealand Journal of Botany 45: 1–33.

Bartlett MK, Scoffoni C, Sack L. 2012. The determinants of leaf turgor loss point and prediction of drought tolerance of species and biomes: a global meta-analysis. Ecology Letters 15: 393–405.

Bauerle TL, Richards JH, Smart DR, Eissenstat DM. 2008. Importance of internal hydraulic redistribution for prolonging the lifespan of roots in dry soil. Plant, Cell & Environment 31: 177–86.

Bell DT. 2001. Ecological response syndromes in the flora of southwestern Western Australia: fire resprouters versus reseeders. The Botanical Review 67: 417–40.

Bhaskar R, Valiente-Banuet A, Ackerly DD. 2007. Evolution of hydraulic traits in closely related species pairs from mediterranean and nonmediterranean environments of North America. New Phytologist 176: 718–26.

Bjorkman O, Mooney HA, Ehleringer J. 1975. Photosynthetic responses of plants from habitats with contrasting thermal environments. Carnegie Institution of Washington Yearbook 74: 743–8.

Bond WJ, Midgley JJ. 1995. Kill thy neighbour: an individualistic argument for the evolution of flammability. Oikos 1: 79–85.

Bond WJ, Midgley JJ. 2003. The evolutionary ecology of sprouting in woody plants. International Journal of Plant Sciences 164: S103–S114.

Bond WJ, Van Wilgen BW. 1996. Fire and plants. Chapman and Hall, London.

Boorse GC, Ewers FW, Davis SD. 1998. Response of chaparral shrubs to below-freezing temperatures: acclimation, ecotypes, seedlings vs. adults. American Journal of Botany 85: 1224–30.

Bradshaw SD, Dixon KW, Hopper SD, Lambers H, Turner SR. 2011. Little evidence for fire-adapted plant traits in Mediterranean climate regions. Trends in Plant Science 16: 69–76.

Brodribb TJ, Holbrook NM. 2005. Water stress deforms tracheids peripheral to the leaf vein of a tropical conifer. Plant Physiology 137: 1139–46.

Canham CA, Froend RH, Stock WD. 2009. Water stress vulnerability of four *Banksia* species in contrasting ecohydrological habitats on the Gnangara Mound, Western Australia. Plant, Cell & Environment 32: 64–72.

Choat B, Jansen S, Brodribb TJ, Cochard H, Delzon S, Bhaskar R, Bucci SJ, Feild TS, Gleason, SM, Hacke UG, Jacobsen AL, Lens F, Maherali H, Martínez-Vilalta J, Mayr S, Mencuccini M, Mitchell PJ, Nardini A, Pittermann J, Pratt RB, Sperry JS, Westoby M, Wright IJ and Zanne AE 2012. Global convergence in the vulnerability of forests to drought. Nature 491: 752–5.

Clarke PJ, Knox KJ, Wills KE, Campbell M. 2005. Landscape patterns of woody plant response to crown fire: disturbance and productivity influence sprouting ability. Journal of Ecology 93: 544–55.

Coley PD, Barone JA. 1996. Herbivory and plant defenses in tropical forests. Annual Review of Ecology and Systematics 27: 305–35.

Comstock JP, Mahall BE. 1985. Drought and changes in leaf orientation for two California chaparral shrubs: *Ceanothus megacarpus* and *Ceanothus crassifolius*. Oecologia 65: 531–5.

Cowan PD, Ackerly DD. 2010. Post-fire regeneration strategies and flammability traits of California chaparral shrubs. International Journal of Wildland Fire 19: 984–9.

Cowling RM, Campbell BM. 1980. Convergence in vegetation structure in the mediterranean communities of California, Chile and South Africa. Vegetatio 43: 191–7.

Cowling RM, Ojeda F, Lamont BB, Rundel PW, Lechmere-Oertel R. 2005. Rainfall reliability, a neglected factor in explaining convergence and divergence of plant traits in fire-prone mediterranean-climate ecosystems. Global Ecology and Biogeography 14: 509–19.

Crisp MD, Burrows GE, Cook LG, Thornhill AH, Bowman DMJS. 2011. Flammable biomes dominated by eucalypts originated at the Cretaceous-Palaeogene boundary. Nature Communications 2: Article 193.

Davis SD, Sperry JS, Hacke UG. 1999. The relationship between xylem conduit diameter and cavitation caused by freezing. American Journal of Botany 86: 1367–72.

Davis SD, Ewers FW, Sperry JS, Portwood KA, Crocker MC, Adams GC. 2002. Shoot dieback during prolonged drought in *Ceanothus* (Rhamnaceae) chaparral of California: a possible case of hydraulic failure. American Journal of Botany 89: 820–8.

Davis SD, Ewers FW, Pratt RB, Brown PL, Bowen TJ. 2005. Interactive effects of freezing and drought on long distance transport: a case study of chaparral shrubs of California. Pages 425–35 in Holbrook NM, Zwieniecki M, eds. Vascular transport in plants. Academic Press, San Diego.

Davis SD, Helms AM, Heffner MS, Shaver AR, Deroulet AC, Stasiak NL, Vaughn SM, Leake CB, Lee HD and Sayegh ET. 2007. Chaparral zonation in the Santa Monica Mountains: the influence of freezing temperatures. Fremontia 35: 12–15.

Diffenbaugh NS, Swain DL, Touma D. 2015. Anthropogenic warming has increased drought risk in California. Proceedings of the National Academy of Sciences 31: 3931–6.

Dodd J, Bell DT. 1993a. Water relations of the canopy species in a *Banksia* woodland, Swan Coastal Plain, Western Australia. Austral Ecology 18: 281–93.

Dodd J, Bell DT. 1993b. Water relations of understorey shrubs in a *Banksia* woodland, Swan Coastal Plain, Western Australia. Austral Ecology 18: 295–305.

Duker R, Cowling RM, Preez DR, Vyver ML, Weatherall-Thomas CR, Potts AJ. 2015. Community-level assessment of freezing tolerance: frost dictates the biome boundary between Albany subtropical thicket and Nama-Karoo in South Africa. Journal of Biogeography 42: 167–78.

Ewers FW, Lawson MC, Bowen TJ, Davis SD. 2003. Freeze/thaw stress in *Ceanothus* of southern California chaparral. Oecologia 136: 213–19.

Feild TS, Arens NC, Doyle JA, Dawson TE, Donoghue MJ. 2004. Dark and disturbed: a new image of early angiosperm ecology. Paleobiology 30: 82–107.

Field CB. 1991. Ecological scaling of carbon gain to stress and resource. Pages 35–65 in Mooney HA, Winner WE, Pell EJ, eds. Response of plants to multiple stresses. Academic Press, San Diego.

Field C, Merino J, Mooney H. 1983. Compromises between water-use efficiency and nitrogen-use efficiency in five species of California evergreens. Oecologia 60: 384–9.

Field CH, Mooney HA. 1986. Photosynthesis—nitrogen relationship in wild plants. Pages 25–55 in Givnish T, ed. On the economy of plant form and function: Proceedings of the Sixth Maria Moors Cabot Symposium, evolutionary constraints on primary productivity, adaptive patterns of energy capture in plants, Harvard Forest, August 1983. Cambridge University Press, Cambridge.

Flexas J, Diaz-Espejo A, Gago J, Gallé A, Galmés J, Gulías J and Medrano H. 2014. Photosynthetic limitations in Mediterranean plants: a review. Environmental and Experimental Botany 103: 12–23.

Gagnon PR, Passmore HA, Platt WJ, Myers JA, Paine CE, Harms KE. 2010. Does pyrogenicity protect burning plants? Ecology 91: 3481–6.

Gould SJ, Lewontin RC. 1979. The spandrels of San Marco and the Panglossian paradigm: a critique of the adaptationist programme. Proceedings of the Royal Society of London B: Biological Sciences 205: 581–98.

Gould SJ, Vrba ES. 1982. Exaptation—a missing term in the science of form. Paleobiology 8: 4–15.

Griffin JR. 1978. Maritime chaparral and endemic shrubs of the Monterey Bay region, California. Madroño 25: 65–81.

Groom PK, Lamont BB. 1999. Which common indices of sclerophylly best reflect differences in leaf structure? Écoscience 6: 471–4.

Groom PK, Lamont B. 2015. Plant life of southwestern Australia: adaptations for survival. de Gruyter, Open.

Hassiotou F, Evans JR, Ludwig M, Veneklaas EJ. 2009. Stomatal crypts may facilitate diffusion of CO_2 to adaxial mesophyll cells in thick sclerophylls. Plant, Cell & Environment 32: 1596–611.

He T, Lamont BB, Downes KS. 2011. Banksia born to burn. New Phytologist 191: 184–96.

Higgins KB, Lamb AJ, van Wilgen BW. 1987. Root systems of selected plant species in mesic mountain fynbos in the Jonkershoek Valley, south-western Cape Province. South African Journal of Botany 53: 249–57.

Hoffmann AJ, Hoffmann AE. 1976. Growth pattern and seasonal behavior of buds of *Colliguaya odorifera*, a shrub from the Chilean mediterranean vegetation. Canadian Journal of Botany 54: 1767–74.

Jacobsen AL, Ewers FW, Pratt RB, Paddock WA, Davis SD. 2005. Do xylem fibers affect vessel cavitation resistance? Plant Physiology 139: 546–56.

Jacobsen AL, Agenbag L, Esler KJ, Pratt RB, Ewers FW, Davis SD. 2007a. Xylem density, biomechanics and anatomical traits correlate with water stress in 17 evergreen shrub species of the Mediterranean-type climate region of South Africa. Journal of Ecology 95: 171–83.

Jacobsen AL, Pratt RB, Ewers FW, Davis SD. 2007b. Cavitation resistance among 26 chaparral species of southern California. Ecological Monographs 77: 99–115.

Jacobsen AL, Pratt RB, Davis SD, Ewers FW. 2008. Comparative community physiology: nonconvergence in water relations among three semi-arid shrub communities. New Phytologist 180: 100–13.

Jacobsen AL, Esler KJ, Pratt RB, Ewers FW. 2009. Water stress tolerance of shrubs in Mediterranean-type climate regions: convergence of fynbos and succulent karoo communities with California shrub communities. American Journal of Botany 96: 1445–53.

Jacobsen AL, Roets F, Jacobs SM, Esler KJ, Pratt RB. 2012. Dieback and mortality of South African fynbos shrubs is likely driven by a novel pathogen and pathogen-induced hydraulic failure. Austral Ecology 37: 227–35.

Jordan GJ, Weston PH, Carpenter RJ, Dillon RA, Brodribb TJ. 2008. The evolutionary relations of sunken, covered, and encrypted stomata to dry habitats in Proteaceae. American Journal of Botany 95: 521–30.

Keeley JE. 1987. Fruit production patterns in the chaparral shrub *Ceanothus crassifolius*. Madroño 34: 273–82.

Keeley JE. 1992. Recruitment of seedlings and vegetative sprouts in unburned chaparral. Ecology 73: 1194–208.

Keeley JE, Bond WJ. 1997. Convergent seed germination in South African fynbos and Californian chaparral. Plant Ecology 133: 153–67.

Keeley JE, Bond WJ, Bradstock RA, Pausas JG, Rundel PW. 2012. Fire in Mediterranean ecosystems: ecology, evolution and management. Cambridge University Press, Cambridge.

Keeley SC, Pizzorno M. 1986. Charred wood stimulated germination of two fire-following herbs of the California chaparral and the role of hemicellulose. American Journal of Botany 73: 1289–97.

Knox KJ, Clarke PJ. 2005. Nutrient availability induces contrasting allocation and starch formation in resprouting and obligate seeding shrubs. Functional Ecology 19: 690–8.

Kummerow J. 1983. Comparative phenology of Mediterranean-type plant communities. Pages 300–17 in Kruger FJ, Mitchell DT, Jarvis JUM, eds. Mediterranean-type ecosystems: the role of nutrients. Springer, Heidelberg.

Kummerow J, Krause D, Jow W. 1978. Seasonal changes of fine root density in the southern Californian chaparral. Oecologia 37: 201–12.

Kummerow J, Montenegro G, Krause D. 1981. Biomass, phenology, and growth. Pages 69–96 in Miller PC, ed. Resource use by chaparral and matorral: a comparison of vegetation function in two mediterranean type ecosystems. Springer, Heidelberg.

Lambers H, Colmer TD, Hassiotou F, Mitchell PM, Poot P, Shane MW and Veneklaas EJ. 2014. Carbon and water relations. Pages 129–46 in Lambers H, ed. Plant life on the sandplains in Southwest Australia: a global biodiversity hotspot. UWA Publishing, Crawley.

Lamont BB, Enright NJ. 2000. Adaptive advantages of aerial seed banks. Plant Species Biology 15: 157–66.

Lamont BB, Le Maitre DC, Cowling RM, Enright NJ. 1991. Canopy seed storage in woody plants. The Botanical Review 57: 277–317.

Lamont BB, Groom PK, Cowling RM. 2002. High leaf mass per area of related species assemblages may reflect low rainfall and carbon isotope discrimination rather than low phosphorus and nitrogen concentrations. Functional Ecology 16: 403–12.

Lawrence WT. 1987. Gas exchange characteristics of representative species from the scrub vegetation of central Chile. Pages 279–304 in Tenhunen JD, Catarina FM, Lange OL, Oechel WC, eds. Plant response to stress: functional analysis in mediterranean ecosystems. Springer, Heidelberg.

Limousin JM, Misson L, Lavoir AV, Martin NK, Rambal S. 2010. Do photosynthetic limitations of evergreen *Quercus ilex* leaves change with long-term increased drought severity? Plant, Cell & Environment 33: 863–75.

Loveless AR. 1961. A nutritional interpretation of sclerophylly based on differences in the chemical composition of sclerophyllous and mesophytic leaves. Annals of Botany 25: 168–84.

Manion PD. 1981. Tree disease concepts. Prentice Hall, New York.

Martínez-Vilalta J, Prat E, Oliveras I, Piñol J. 2002. Xylem hydraulic properties of roots and stems of nine Mediterranean woody species. Oecologia 133: 19–29.

McDonald PG, Fonseca CR, Overton J, Westoby M. 2003. Leaf-size divergence along rainfall and soil-nutrient gradients: is the method of size reduction common among clades? Functional Ecology 17: 50–7.

McDowell N, Pockman WT, Allen CD. 2008. Mechanisms of plant survival and mortality during drought: why do some plants survive while others succumb to drought? New Phytologist 178: 719–39.

Mediavilla S, Garcia-Ciudad A, Garcia-Criado B, Escudero A. 2008. Testing the correlations between leaf life span and leaf structural reinforcement in 13 species of European Mediterranean woody plants. Functional Ecology 22: 787–93.

Midgley JJ. 2013. Flammability is not selected for, it emerges. Australian Journal of Botany 61: 102–6.

Midgley JJ, Enright NJ. 2000. Serotinous species show correlation between retention time for leaves and cones. Journal of Ecology 88: 348–51.

Mitchell PJ, Veneklaas EJ, Lambers H, Burgess SS. 2008. Leaf water relations during summer water deficit: differential responses in turgor maintenance and variation in leaf structure among different plant communities in south-western Australia. Plant, Cell & Environment 31: 1791–802.

Montenegro G, Aljaro ME, Kummerow J. 1979. Growth dynamics of Chilean matorral shrubs. Botanical Gazette 140: 114–19.

Montenegro G, Araya S, Aljaro ME, Avila G. 1982. Seasonal fluctuations of vegetative growth in roots and shoots of Central Chilean shrubs. Oecologia 53: 235–7.

Montenegro G, Gómez M, Díaz F, Ginocchio R. 2003. Regeneration potential of Chilean matorral after fire: an updated view. Pages 381–409, Veblen TT, Baker WL, Montenegro G, Swetnam TW, eds. Fire and climatic change in temperate ecosystems of the western Americas. Springer, Heidelberg.

Mooney HA. 1977. Convergent evolution in Chile and California: Mediterranean climate ecosystems. Dowden, Hutchinson and Ross, Inc.

Mooney HA. 1983. Carbon-gaining capacity and allocation patterns of mediterranean-climate plants. Pages 103–19 in Kruger FJ, Mitchell DT, Jarvis JUM, eds. Mediterranean-type ecosystems: the role of nutrients. Springer, Heidelberg.

Mooney HA, Dunn EL. 1970. Convergent evolution of Mediterranean-climate evergreen sclerophyll shrubs. Evolution 24: 292–303.

Mooney HA, Parsons DJ, Kummerow J. 1974. Plant development in Mediterranean climates. Pages 255–67 in Leith H, ed. Phenology and seasonality modeling. Springer, Heidelberg.

Mooney HA, Harrison AT, Morrow PA. 1975. Environmental limitations of photosynthesis on a California evergreen shrub. Oecologia 19: 293–301.

Mooney HA, Field C, Gulmon SL, Rundel P, Kruger FJ. 1983. Photosynthetic characteristics of South African sclerophylls. Oecologia 58: 398–401.

Moreira B, Tormo J, Pausas JG. 2012. To resprout or not to resprout: factors driving intraspecific variability in resprouting. Oikos 121: 1577–84.

Mutch RW. 1970. Wildland fires and ecosystems—a hypothesis. Ecology 51: 1046–51.

Oechel WC, Lawrence W, Mustafa J, Martínez J. 1981. Energy and carbon acquisition. Pages 151–83 in Miller PC, ed. Resource use by chaparral and matorral: a comparison of vegetation function in two mediterranean type ecosystems. Springer, Heidelberg.

Oertli JJ, Lips SH, Agami M. 1990. The strength of sclerophyllous cells to resist collapse due to negative turgor pressure. Acta Oecologica 11: 281–9.

Orshan G, Le Roux A, Montenegro G. 1984. Distribution of monocharacter growth form types in mediterranean plant communities of Chile, South Africa and Israel. Bulletin de la Société Botanique de France. Actualités Botaniques 131: 427–39.

Parker VT, Pratt RB, Keeley JE. 2016. Chaparral. Pages 479–501 in Mooney HA, Zavaleta E, eds. Ecosystems of California. University of California Press, California.

Pausas JG. 2015. Bark thickness and fire regime. Functional Ecology 29: 315–27.

Pausas JG, Keeley JE. 2009. A burning story: the role of fire in the history of life. BioScience 59: 593–601.

Pausas JG, Keeley JE. 2014. Evolutionary ecology of resprouting and seeding in fire-prone ecosystems. New Phytologist 204: 55–65.

Pausas JG, Verdú M. 2005. Plant persistence traits in fire-prone ecosystems of the Mediterranean basin: a phylogenetic approach. Oikos 109: 196–202.

Pausas JG, Bradstock RA, Keith DA, Keeley JE. 2004. Plant functional traits in relation to fire in crown-fire ecosystems. Ecology 85: 1085–100.

Pausas JG, Pratt RB, Keeley JE. 2016. Towards understanding resprouting at the global scale. New Phytologist 209: 945–54.

Pausas JG, Keeley JE, Schwilk DW. 2017. Flammability as an ecological and evolutionary driver. Journal of Ecology 105: 289–97.

Pate JS, Froend RH, Bower BJ, Hansen A, Kuo J. 1990. Seedling growth and storage characteristics of seeder and resprouter species of Mediterranean-type ecosystems of SW Australia. Annals of Botany 65: 585–601.

Poole DK, Roberts SW, Miller PC. 1981. Water utilization. Pages 123–49 in Miller PC, ed. Resource use by chaparral and matorral: a comparison of vegetation function in two mediterranean type ecosystems. Springer, Heidelberg.

Pratt RB, Ewers FW, Lawson MC, Jacobsen AL, Brediger MM, Davis SD. 2005. Mechanisms for tolerating freeze–thaw stress of two evergreen chaparral species: *Rhus ovata* and *Malosma laurina* (Anacardiaceae). American Journal of Botany 92: 1102–13.

Pratt RB, Jacobsen AL, Golgotiu KA, Sperry JS, Ewers FW, Davis SD. 2007. Life history type and water stress tolerance in nine California chaparral species (Rhamnaceae). Ecological Monographs 77: 239–53.

Pratt RB, Jacobsen AL, Mohla R, Ewers FW, Davis SD. 2008. Linkage between water stress tolerance and life history type in seedlings of nine chaparral species (Rhamnaceae). Journal of Ecology 96: 1252–65.

Pratt RB, Jacobsen AL, Jacobs SM, Esler KJ. 2012a. Xylem transport safety and efficiency differ among fynbos shrub life history types and between two sites differing in mean rainfall. International Journal of Plant Sciences 173: 474–83.

Pratt RB, Jacobsen AL, Hernandez J, Ewers FW, North GB, Davis SD. 2012b. Allocation tradeoffs among chaparral shrub seedlings with different life history types (Rhamnaceae). American Journal of Botany 99: 1464–76.

Pratt RB, Jacobsen AL, Ramirez AL, Helms AM, Traugh CA, Tobinet MF, Heffner MS, Davis SD. 2014. Mortality of resprouting chaparral shrubs after a fire and during a record drought: physiological mechanisms and demographic consequences. Global Change Biology 20: 893–907.

Ramirez AR, Pratt RB, Jacobsen AL, Davis SD. 2012. Exotic deer diminish post-fire resilience of native shrub communities on Santa Catalina Island, southern California. Plant Ecology 213: 1037–47.

Read J, Sanson GD. 2003. Characterizing sclerophylly: the mechanical properties of a diverse range of leaf types. New Phytologist 160: 81–99.

Rhizopoulou S, Mitrakos K. 1990. Water relations of evergreen sclerophylls. I. Seasonal changes in the water relations of eleven species from the same environment. Annals of Botany 65: 171–8.

Rotondi A, Rossi F, Asunis C, Cesaraccio C. 2003. Leaf xeromorphic adaptations of some plants of a coastal Mediterranean macchia ecosystem. Journal of Mediterranean Ecology 4: 25–36.

Schwilk DW, Kerr B. 2002. Genetic niche-hiking: an alternative explanation for the evolution of flammability. Oikos 99: 431–42.

Segarra-Moragues JG, Ojeda F. 2010. Postfire response and genetic diversity in *Erica coccinea*: connecting population dynamics and diversification in a biodiversity hotspot. Evolution 64: 3511–24.

Seydack AH, Bekker SJ, Marshall AH. 2007. Shrubland fire regime scenarios in the Swartberg Mountain Range, South Africa: implications for fire management. International Journal of Wildland Fire 16: 81–95.

Sniderman JK, Jordan GJ, Cowling RM. 2013. Fossil evidence for a hyperdiverse sclerophyll flora under a non–Mediterranean-type climate. Proceedings of the National Academy of Sciences 110: 3423–8.

Specht RL. 1988. Foliar analyses. Pages 63–80 in Specht RL (ed.) Mediterranean-type ecosystems. Springer, Netherlands.

Stohlgren TJ, Rundel PW. 1986. A population model for a long-lived, resprouting chaparral shrub: *Adenostoma fasciculatum*. Ecological Modelling 34: 245–57.

Turner IM. 1994. A quantitative analysis of leaf form in woody plants from the world's major broadleaved forest types. Journal of Biogeography 21: 413–19.

Venturas MD, MacKinnon ED, Dario HL, Jacobsen AL, Pratt RB, Davis SD. 2016. Chaparral shrub hydraulic traits, size, and life history types relate to species mortality during California's historic drought of 2014. PloS One: 11: p.e0159145.

Verdaguer D, Ojeda F. 2005. Evolutionary transition from resprouter to seeder life history in two Erica (Ericaceae) species: insights from seedling axillary buds. Annals of Botany 95: 593–9.

Verdú, M, Dávila P, García-Fayos P, Flores-Hernández N, Valiente-Baunet A. 2003. 'Convergent' traits of mediterranean woody plants belong to pre-mediterranean lineages. Biological Journal of the Linnean Society 78: 415–27.

Verdú M, Pausas JG, Segarra-Moragues JG, Ojeda F. 2007. Burning phylogenies: fire, molecular evolutionary rates, and diversification. Evolution 61: 2195–204.

Vesk PA, Westoby M. 2004. Sprouting ability across diverse disturbances and vegetation types worldwide. Journal of Ecology 92: 310–20.

Vilagrosa A, Hernández EI, Luis VC, Cochard H, Pausas JG. 2014. Physiological differences explain the co-existence of different regeneration strategies in Mediterranean ecosystems. New Phytologist 201: 1277–88.

Von Willert DJ, Herppich M, Miller JM. 1989. Photosynthetic characteristics and leaf water relations of mountain fynbos vegetation in the Cedarberg area (South Africa). South African Journal of Botany 55: 288–98.

Wells PV. 1969. The relation between mode of reproduction and extent of speciation in woody genera of the California chaparral. Evolution 23: 264–7.

West AG, Dawson TE, February EC, Midgley GF, Bond WJ, Aston TL. 2012. Diverse functional responses to drought in a Mediterranean-type shrubland in South Africa. New Phytologist 195: 396–407.

Wicklow DT. 1977. Germination response in *Emmenanthe penduliflora* (Hydrophyllaceae). Ecology 58: 201–5.

Wiens D, Allphin L, Wall M, Slaton MR, Davis SD. 2012. Population decline in *Adenostoma sparsifolium* (Rosaceae): an ecogenetic hypothesis for background extinction. Biological Journal of the Linnean Society 105: 269–92.

Wright IJ, Cannon K. 2001. Relationships between leaf lifespan and structural defences in a low-nutrient, sclerophyll flora. Functional Ecology 15: 351–9.

Wright IJ, Reich PB, Westoby M, Ackerly DD, Baruch Z, Bongers F, Cavender-Bares J, Chapin T, Cornelissen JH, Diemer M, Flexas J, Garnier E, Groom PK, Gulias J,

Hikosaka K, BB Lamont, T Lee, W Lee, C Lusk, JJ Midgley, M-L Navas, Ü Niinemets, J Oleksyn, N Osada, H Poorter, P Poot, L Prior, V I Pyankov, C Roumet, S C Thomas, MG Tjoelker, EJ Veneklaas and Villar R. 2004. The worldwide leaf economics spectrum. Nature 428: 821–7.

Yates MJ, Anthony Verboom G, Rebelo AG, Cramer MD. 2010. Ecophysiological significance of leaf size variation in Proteaceae from the Cape Floristic Region. Functional Ecology 24: 485–92.

7 Ecosystems processes

Abstract

Ecosystems are assemblages of organisms interacting with one another and their environment (Chapter 1). Key to the functioning of ecosystems is the flow of energy, carbon, mineral nutrients, and water in these systems. The numerous processes involved are chiefly driven by climate, soil, and fire (Chapter 2). In cases where the key drivers are the same in different areas, then ecosystems should converge in their structure and function, which has been a motivation for comparing across mediterranean-type climate (MTC) regions. Convergence of MTC regions has been evaluated, but such comparisons at the ecosystem level are challenging because ecosystems are complex and dynamic entities. Here we review carbon, nutrient, and water dynamics of mediterranean-type ecosystems in the context of ecosystem function. As nutrients in soils are low in some MTC regions, we review how this has led to unique adaptations to meet this challenge.

7.1 Ecosystem structure and primary productivity

Plants are commonly the focus of ecosystem studies because they have a dominant influence on all the major exchanges in the ecosystems they inhabit. Green plants capture energy from sunlight and use it to fix carbon from carbon dioxide gas in the atmosphere in the process called **photosynthesis** (Chapter 6). This carbon is fixed into new **biomass**, which has a similar energy content regardless of what it is (leaves, stems, flowers, etc.); therefore, biomass is commonly used as a measure of energy. In this way, plants produce the carbon and energy that flows into ecosystems and they are referred to as **primary producers**. The biomass produced forms the base of the food web and organisms that are herbivores are primary consumers. Secondary consumers are carnivores and feed on the primary consumers and so on. Biomass that is not consumed eventually falls to the ground and

The Biology of Mediterranean-Type Ecosystems. Karen J. Esler, Anna L. Jacobsen, and R. Brandon Pratt,
Oxford University Press (2018). © Karen J. Esler, Anna L. Jacobsen, and R. Brandon Pratt 2018.
DOI 10.1093/oso/9780198739135.001.0001

is decomposed. Plants give the ecosystem its characteristic structure and are the defining feature of the ecosystems. This structure has a major influence on virtually every aspect of ecosystem function. This biomass also forms the fuel that affects the fire regime (Chapter 6). Mediterranean-type climate (MTC) regions contain an abundance of evergreen shrubs that form a closed canopy when provided with enough moisture and low to moderate disturbance. The biomass of shrubs per unit area of ground has been compared across MTC regions, as have some other aspects of stand structure (Table 7.1). Numerous studies have been conducted to assess how biomass is affected by fire, while others have examined standing biomass or production over short timescales. Aspects of this topic have been reviewed elsewhere (Hilbert and Canadell 1995; Rambal 2001).

A clear pattern in all MTC regions is that a period of rapid biomass accumulation occurs in the years following fire, and that, after this initial rise, a maximum is reached after 5–50 years (Table 7.1; Black 1987; Burrows and McCaw 1990; Rambal 2001). Woodlands, such as those dominated by *Quercus* spp. in the Mediterranean, may continue to accrue biomass after 50 years (Rambal 2001). Some vegetation undergoes a decline after this maximum is reached and many years may pass without another fire. This is evidenced by a decline in new biomass produced (Rundel and Parsons 1979) and a steady accumulation of dead biomass and litter. In one South African study most of the mountain fynbos biomass on a site was dead after 37 years (van Wilgen 1982). One mechanism that could lead to such a decline in productivity is a failure of nutrient cycling to sustain continued production of biomass (Marion and Black 1988). Thus, on nutrient-poor sites and among species that are not nitrogen-fixers, this decline may be more likely or occur earlier after fire (see Section 7.2). Another factor is that, as biomass accumulates, the proportion of carbon obtained via photosynthesis has to go towards sustaining the living biomass in the process of **respiration** (the liberation of CO_2 in the metabolic conversion of glucose), which would compete with, and limit, the amount available for new growth. Other factors related to a build-up of dead biomass are the mortality of short-lived species (van Wilgen and Forsyth 1992), periodic droughts (Davis et al. 2002), and density-dependent thinning processes (Schlesinger and Gill 1980).

Another way to understand energy and carbon flow through ecosystems is to measure how they exchange CO_2 with the atmosphere. The atmospheric CO_2 that is taken up by the plants (**gross primary production**) is balanced against that lost via respiration. **Net primary production** is the CO_2 fixed in photosynthesis minus the CO_2 respired back to the atmosphere by plants. Other respiring species, chiefly soil organisms, will also have a large effect on the amount of CO_2 given off by an ecosystem. The exchanged CO_2 of an entire ecosystem is referred to as **net ecosystem exchange** (NEE). This can be determined using towers set up on a site that measure the changes in gas levels in the air above ecosystems. One such study found that a 100-year-old

Table 7.1 Stand structure and productivity. Aspects of aboveground stand structure and productivity for the evergreen sclerophyllous shrublands found in different mediterranean-type climate (MTC) regions

Vegetation	Dominant species or soil	Age (years)	Height (m)	Biomass[2] (g/m²)	Leaf mass (g/m²)	Leaf area index	Litter (g/m²)	Litterfall (g m⁻² yr⁻¹)	ANPP (g m⁻² yr⁻¹)	Biomass/time[3] (g m⁻² yr⁻¹)	Ref.[1]
Australia											
Kwongon	Sandplain Shallow soil[4,5]	15–16	0.70[6]	300			110			19	1
	Sandplain Deep soil[4,5]	15–16	1.70[6]	325			220			21	1
	Calcareous[4,5]	15–16	1.66[6]	250			350			16	1
Heath	*Banksia ornata, Allocasuarina pusilla*	0–5 5–10							64 & 160[7] 88[7]		2
Mallee	*Eucalyptus incrassata, Melaleuca uncinata*	0–5 5–10							114[7] 88[7]	90	2
Thicket	*Acacia, Melaleuca, Allocasuarina*	5 >38	>2	450 4090					110 380	108	3
California											
Chaparral	*Ceanothus megacarpus*	5		2055	228					411	4
		12		3708	336			585	175	309	4
		21		6337	465					302	4
	C. megacarpus	22	5.30	7624	457			801	1056	347	5
	Adenostoma fasciculatum	2	0.30	593	232	1.8				297	
		6	0.45	793	200	1.5				132	6
		16	0.75	1486	286	2.2				93	6
		37	0.75	1400	159	1.2				38	6
		>60	0.90	1370	156	1.2				23	6
	A. fasciculatum	23	<2.0	2127	438			83	362	92	7
	A. fasciculatum C. gregii	22	1.67	2308	503	2.7	355	273	671	104	8

Continued

Table 7.1 Continued

Vegetation	Dominant species or soil	Age (years)	Height (m)	Biomass[2] (g/m²)	Leaf mass (g/m²)	Leaf area index	Litter (g/m²)	Litterfall (g m^{-2} yr^{-1})	ANPP (g m^{-2} yr^{-1})	Biomass/time[3] (g m^{-2} yr^{-1})	Ref.[1]
Chile											
Matorral	*Cryptocarya alba*	15	1.21	738	348	2.0	245	163	253	49	8
Mediterranean											
Phrygana	*Genista acanthoclada, Thymus capitatus*	40	<1	790		0.50		155		20	9 in 10
Phrygana	*Phlomis fruticosa, Euphorbia acanthothamnos*	10	<1	1111	209	1.7		210	202	111	11
Garrigue	*Quercus coccifera*	17	>1	2350	400			228	140	138	12 in 10
	Q. coccifera	30	>1	3150	560			261	110	105	12 in 10
South Africa											
Fynbos	Proteoid, Restioid, Ericoid	4	~1	914			117			228	13
Fynbos	*Phylica cephalantha*	11	~1	1731			76			157	14
	Graminoid, Restioid	4	0.4	710			53			178	15
	Protea repens	4	0.4	620			33			155	
	Widdringtonia nodiflora	21	1.0	5100			1430			242	
		37		7600			5300			205	

ANPP, Aboveground net primary productivity.

1. 1. Westcott et al. 2014, 2. Specht 1969, 3. Mappin et al. 2003, 4. Schlesinger and Gill 1980, 5. Gray 1982, 6. Rundel and Parsons 1979, 7. Mooney and Rundel 1979, 8. Mooney et al. 1977, 9. Tsiourlis 1990, 10. Rambal 2001, 11. Margaris 1976, 12. Rapp and Lossaint 1981, 13. Mitchell et al. 1986, 14. Low 1983, 15. van Wilgen 1982.

2. Aboveground of both live and dead biomass.

3. This is the biomass column divided by time since fire.

4. Proteaceae, Myrtaceae, Fabaceae, Ericaceae, and Cyperaceae and Restionaceae.

5. The sandplain sites (n=18) are nutrient-poor and the calcareous sites (n=17) have higher nutrients.

6. Values are approximate and are maximum heights in plots.

7. Values do not appear to include reproductive output, thus will be an underestimate.

California chaparral stand dominated by chamise (*Adenostoma fascicula-tum*) could maintain significant positive CO_2 uptake during normal rainfall years, whereas during droughts the system was a source of CO_2, giving off more of the gas than it took up on an annual basis (Luo et al. 2007). Two important conclusions from this are that old stands of shrubs within MTC regions can continue to be productive carbon sinks in the long absence of fire, and that the MTC and water limitations have a strong effect on CO_2 exchange in these ecosystems.

Productivity is commonly determined as the amount of biomass produced over time. This parameter is referred to as net primary productivity (NPP) and has the units of mass of plants produced per ground area per year ($g\,m^{-2}\,y^{-1}$). NPP values can vary broadly from year to year depending on rainfall totals and timing, and they will also be higher in younger stands (Table 7.1); moreover, soil nutrients affect NPP (Mooney et al. 1977), as do landscape factors such as aspect and slope (Marion and Black 1988). This variation complicates a simple comparison across the MTC regions (Rambal 2001). To make such comparisons one would need to match sites in the context of key drivers (Mooney et al. 1977). Moreover, NPP includes organic secretions and belowground growth, which is rarely measured and, when it is, it contains considerable uncertainty (Rambal 2001). **Aboveground NPP** (ANPP) is a more common measurement. Roots are important because they comprise a large portion of biomass and they cannot produce their own carbohydrates via photosynthesis; therefore, roots are an important sink for carbon. Because of these factors, the data available are at best useful for general comparisons of the range of ANPP across different systems. Table 7.1 provides a non-exhaustive list of studies examining ANPP. The values reported for ANPP here do not include litter in the ANPP since shoots were measured directly in biomass estimates, thus they will differ from some of the values reported in Rambal (2001). Data are only included for iconic evergreen MTC region shrublands. As a general statement, more open and arid shrublands (e.g. sage scrub or succulent karoo) will have lower productivity (Gray 1982; Rutherford 1978), while low to mid-elevation habitats with substantial numbers of trees will have higher productivity (Rambal 2001).

The range of ANPP for MTC region evergreen shrublands is broad (64–1056), with a mean value of 265 $g\,m^{-2}\,year^{-1}$ (Table 7.1). This value is less than more mesic forested systems and greater than more arid systems (Chapin et al. 2011). The highest values come from chaparral, with Gray (1982) reporting an ANPP of 1056 $g\,m^{-2}\,year^{-1}$ from a coastal stand of a nitrogen-fixing plants, which is comparable to temperate forest (950 $g\,m^{-2}\,year^{-1}$; Chapin et al. 2011). The lowest values come from Australia; this low value is similar to two old stands, one a California chamise-chaparral stand and another stand of short Mediterranean phrygana vegetation from Greece (Table 7.1).

One would predict that lower values would occur in Australia and South Africa in areas of low nutrients. The available ANPP data are too limited to assess this so we divided the total standing biomass per area by the time since fire, which gives a coarse estimate of ANPP for further comparisons (biomass/time in Table 7.1). These results partially support this prediction, with the highest in California and lowest in Australian kwongon vegetation. The fynbos productivities are not low in this analysis in spite of generally nutrient-poor soils (Low 1983). The fynbos sites in the studies varied in rainfall from high to very high in the case of the mountain fynbos sites (mean = 1744 mm; van Wilgen 1982). These high rainfall values should promote productivity because water would not be as limiting. By contrast, sites in Australia that are presented in Table 7.1 may be both nutrient-limited and relatively arid, leading to both strong nutrient and moisture limitations of productivity (Westcott et al. 2014).

Within a season, the MTC exerts a strong influence over production in the same manner as previously discussed for photosynthesis and growth (Chapter 6). In cases where there is a longer growing season, production may be higher. For example, a protracted growing season may be a partial explanation for the extremely high ANPP value reported for the maritime-influenced *Ceanothus* chaparral by Gray (1982). In another comparison, Chile had a longer growing season than California owing to a broader range of growth forms, a more equable climate, and a more open stand structure. In both Chile and California growth was concentrated in the later winter and spring but was more spread throughout the year in Chile than California (Mooney et al. 1977), and this peak growth pattern is common to the MTC (Margaris 1976). The reliable and evenly distributed rainfall common over much of the MTC regions of South Africa and Australia may lead to a generally longer productive season (Cowling et al. 2005).

Since leaves are the main photosynthetic organs of plants, they have a large impact on productivity. Being evergreen, mediterranean-type vegetation (MTV) can maintain some level of photosynthesis throughout most of the year, albeit at a low rate during the dry season and the cool winter (Figure 6.5). Plants make a wide range of adjustments to modulate photosynthesis when moving from an individual leaf (Chapter 6) to a whole canopy (Rambal 2001). One important factor is the amount of leaf area a plant and a stand produces in its canopy that is measured as the **leaf area index** (LAI; leaf area per unit ground area). This parameter strongly correlates with productivity across biomes (Chapin et al. 2011). A closed canopy has an LAI of about 3 and MTV ranges from 0.49 in open phrygana in Greece to 2.65 in dense chaparral in California (Table 7.1). Values may be as high as 4.5 in MTC oak woodlands (Rambal 2001). Biomes such as tropical and temperate forests may have a LAI of 6, while less productive and more open deserts and tundra will have values of 1 or less (Chapin et al. 2011). Differences in leaf structure may affect LAI. Australian and South African MTV species have the greatest leaf mass area (LMA; Figure 6.2), which means that each leaf produced will cost more in terms of carbon per unit area (Mooney 1983), and greater cost may lower LAI with associated impacts on NPP. Mature Chilean matorral has been noted to be more open canopy (compared to chaparral) and contain more annual species in the openings, which leads to a lower LAI (Table 7.1; Mooney et al. 1977).

Reproductive structures are an important component of productivity. Reproduction is only possible if there is adequate moisture and carbon from current or stored photosynthate. If either are lacking, the shrubs will not flower or fruit will abort prior to going to seed. Some species produce flower buds in the season prior to flowering, which uncouples them from current conditions, while other species do not and rather are responsive to present conditions (Chapter 6). In mixed stands of shrubs a diversity of flowering times will support a broader diversity of animal pollinators, nectar consumers, and granivores, and sites degraded by factors such as high fire frequency will be impoverished in shrub diversity and some pollinators as well (Chapter 9). This highlights the connection between biodiversity and ecosystem function (see Hobbs et al. 1995 for other examples).

Belowground productivity is an important component to consider (Rambal 2001). Plants forage for limiting resources by allocating resources towards root growth to overcome water and nutrient limitations, and towards shoot growth to maximize light and CO_2 capture. In MTC regions, water and nutrients are limiting, thus allocating to roots is critical. A number of studies have examine belowground production in MTV (Mappin et al. 2003; references in Rambal 2001). One important finding is that MTV species broadly differ in how they partition biomass between root and shoots (Hellmers et al. 1955; Hilbert and Canadell 1995; Higgins et al. 1987; Mappin et al. 2003). Resprouting species commonly allocate more biomass to roots than shoots, and some resprouting species form large lignotubers that make up a large part of their belowground biomass (Chapter 6). Because of the large differences in root to shoot allocation and the difference between resprouting types there are likely to be large differences in belowground productivity depending on species composition within and across MTC regions.

7.2 Mineral nutrients

One of the dominant themes of MTC region comparisons has been differences in soil nutrients between the regions and the biological implications of this (e.g. Kruger et al. 1986). The MTC regions of South Africa and Australia have more extensive areas of nutrient-poor soils than the other regions, which should lead to functional differences when compared to more nutrient-rich sites (Barbour and Minnich 1990). The distribution of nutrient-poor soil is variable across the landscape and all five of the MTC regions will have a mosaic of soils of high and low fertility (Hopper 2009; Chapter 2).

Burning of fossil fuels and some land uses have resulted in pollution that has altered the nutrient status of large areas, including those in MTC regions (Chapter 8).

There are a number of nutrients that are essential for plant survival and production, but the two that are of chief importance to primary production are nitrogen and phosphorus (Rundel 1983). Both of these nutrients are generally taken up from the soil. To understand why there are differences in soil nutrients among regions we must briefly consider soil formation. Soil formation is a process controlled by climate, organisms, parent material, topography, and age. **Parent materials**, the rock types from which soils are derived, differ in mineral composition and they have varying amounts of phosphorus and very little or no nitrogen (Table 2.1). For example, the sandstone in the MTC region of South Africa is more phosphorus-poor than other parent materials in the region and in comparison to the mean of the Earth's crust (Cramer et al. 2014; Lambers et al. 2010). **Weathering** refers to the physical and chemical breakdown of soils into more stable products. It happens because of reactions between soil chemicals and the atmosphere or water. **Leaching** is another part of the process whereby soil chemicals are dissolved in water and transported downward and become unavailable or leave the system. Weathering is the chief source of inorganic phosphorus in young soils. As soils age, the phosphorus content becomes increasingly depleted, as losses are not replaced by inputs over time. Another factor is that the phosphorus becomes tightly bound to humus and minerals, rendering it unavailable to plants. In MTC regions of Australia and South Africa, for millions of years large areas have been exposed to the atmosphere and the leaching effects of rain, and this has led to highly weathered and infertile soils, particularly with respect to available phosphorus (Cramer et al. 2014; Table 7.2). By contrast, large areas of MTC regions in California, Chile, and the Mediterranean are more recently uplifted and younger, thus being

Table 7.2 Soil and foliar nutrient content. Mean and range (in parentheses) of total soil nitrogen (N) and phosphorus (P) from different mediterranean-type climate regions. Foliar values are geometric mean and range

Location	Soil content		Foliar content	
	Total N (mg/kg)	Total P (mg/kg)	N (mg/g)	P (mg/g)
Australia	894 (186–2700)	90 (13–540)	8.6 (2.2–27)	0.5 (0.1–3.8)
California	1523 (606–3000)	233 (34–520)	9.6 (6.4–21.8)	0.9 (0.5–2.8)
Chile	1225 (700–2400)	603 (172–1001)	13.6 (8.4–27.3)	1.7 (0.8–2.6)
Mediterranean	3086 (400–10,000)	276 (58–1000)	17.8 (6.6–38.8)	1.3 (0.4–3.4)
South Africa	973 (214–5973)	87 (18–310)	6.5 (3.0–17.1)	0.3 (0.1–1.1)

Data are from Stock and Verboom (2012) and references therein.

less weathered, less leached, and containing greater levels of phosphorus (Table 7.2). Since soil is heterogeneous, another way researchers have assessed the nutrient availability to plants is to measure foliar nutrient content directly. When this is done it shows the same general pattern as the soil (Table 7.2; see also Lambers et al. 2010).

Phosphorus is a non-renewable nutrient in natural environments where depositional inputs from rain and dust are lower than losses that occur over time. This contrasts with nitrogen, which receives fresh inputs, the largest source of which is from nitrogen-fixing bacteria that are both free-living and associated with some plant species. Thus, because nitrogen in a system can be added, it does not become depleted over long timescales if organisms are present to fix it. This does not mean that nitrogen is not limiting in MTC regions; in fact, it is in all MTC regions. One convincing way to demonstrate this is through fertilization studies that add nutrients and monitor how production responds. Such studies have found that adding nitrogen generally leads to a boost in production (Glyphis and Puttick 1989; Kummerow et al. 1982; McMaster et al. 1982; Witkowski et al. 1990). The same has been found in many studies for phosphorus (Kummerow et al. 1982; McMaster et al. 1982); however, species in South Africa and Australia show minimal phosphorus response and toxicity at some levels of addition, indicating that they are rigidly adapted to low levels (Hawkins et al. 2009; Shane et al. 2004; Witkowski et al. 1990).

7.2.1 Plant adaptations to low nutrients

Plants have evolved a host of traits that equip them to thrive in nutrient-poor soils. MTV species provide some of the most extreme and interesting adaptations to low-nutrient soils and have been a model system for study. This is most prominent in the nutrient-poor soils in Australian and South African MTC regions (Lamont 1982). Long-lived sclerophyllous evergreen leaves are important as an adaptive trait in low-nutrient environments, as already discussed in Chapter 6. A common feature of many MTV species is root associations with other organisms (Chapter 3), including the nearly ubiquitous mycorrhizal fungi (Brundrett 2009). Mycorrhizal fungi form mutualistic associations with roots where the fungi receive carbohydrates from plants and in return increase the capacity for water and nutrient uptake, including organic nitrogen (Cramer et al. 2014). Such associations are common worldwide. Curiously, there are some important groups that lack mycorrhizal associations in South African and Australian MTC regions (Allsopp and Stock 1993). The groups that lack the associations are those that typically have other specializations for nutrient uptake. One such group are species that form **cluster roots**, dense profusions of small roots that form along an elongated root (Chapter 3 and Case Study 5). These highly specialized roots are found in a number of taxa: Cyperaceae, Restionaceae, Fagales,

Cucurbitaceae, Rosales, and Fabaceae (Lambers et al. 2006). The cluster roots have a high surface area and secrete acids and phosphatases to solubilize mineral and organic phosphorus. The production of these roots depends on interactions with microbes and is affected by the nitrogen content of the soil (Case Study 5).

Another adaptive trait is the efficient use of phosphorus that has been described in some Australian MTV species. One means of efficiently using phosphorus is to mine it from senescing leaves before they are dropped. Species in MTC Australia, such as *Banksia* spp., are able to draw down phosphorus to very low levels before they drop their leaves (Lambers et al. 2015). Another way they achieve high efficiency is to replace phospholipids in developing leaves with galactolipids and sulfolipids (Lambers et al. 2015). Some Australian species can do this without compromising photosynthetic performance.

Another type of microbe and plant interaction is the formation of root structures that house nitrogen-fixing bacteria that are symbiotic with plants. Plants produce these structures in a complex and tightly regulated process that ultimately leads to a home and carbohydrate source for the microbes and a nitrogen source for the plants. These types of associations are found in a wide range of taxa, including ancient cycads (*Macrozamia* spp. in Australia) that associate with cyanobacteria (*Nostoc* spp.). By far the most common association is between bacteria in the genera *Rhizobium*, *Bradyrhizobium*, *Burkholderia*, and *Mesorhizobium*, called rhizobia, that reside in root nodules formed by plants in the legume family (Fabaceae), excluding the subfamily Caesalpinioideae (Table 7.3). The bacteria are housed in the centre of the nodule. Leghaemoglobin (a haem protein) is produced in the centre of the nodule to control the oxygen content of the bacterial environment since the nitrogen-fixing process is inhibited by abundant oxygen, yet the microbes require some oxygen for respiration. Another association forms between the filamentous Actinomycete bacteria in the genus *Frankia*, which are called Actinorhizal symbionts (Table 7.3; see also Figure 3.4b).

Table 7.3 Common nitrogen-fixing plant taxa

	Australia	California	Chile	Mediterranean	South Africa
Common nitrogen-fixing taxa	*Acacia* spp.[1] *Allocasuarina* spp.[2] *Casuarina* spp.[2] *Daviesia* spp.[1] *Gastrolobium* spp.[1] *Gompholobium* spp.[1]	*Astragalus* spp.[1] *Acmispon glaber*[1] *Ceanothus* spp.[2] *Cercocarpus betuloides*[2]	Rhamnaceae[2] *Sophora macrocarpa*	*Coriaria myrtifolia*[2] *Cystisus* spp.[1] *Genista* spp.[1] *Retama* spp.[1] *Ulex* spp.[1]	*Aspalathus* spp.[1] *Myrica* spp.[2] *Psoralea* spp.[1]

1. Legumes; 2. Actinorhizal.

Another adaptation common to South African and Australian MTC regions is carnivory. Carnivorous plants obtain a portion of their nutrients from eating animals and are typically found in areas with nutrient-poor soils. These plants trap insects on modified leaves with sticky secretions (sundews; Figure 7.1b, c, d; see Plate 12), pitfall traps into pitchers (pitcher plants; Figure 7.1e; see Plate 12), or by sucking them into bladders (bladderworts). After the insects are trapped, digestive secretions lead to the breakdown and release of nutrients (commonly nitrogen) that the plants absorb. Numerous sundews (*Drosera* spp.) are common in sunny bogs in the South African Cape region. The southwest Australian sandplains is a particularly diverse region for abundance and diversity of carnivorous plants (Lamont 1982). They are found in the genera *Byblis*, *Cephalotus*, *Drosera*, and *Utricularia*.

Parasitism is a mode of acquiring resources by tapping into the tissues of another plant and extracting the resources in the host's tissues. The common resources tapped are carbohydrates and water. The degree to which a plant relies on parasitism for carbohydrate acquisition can be partial (**hemiparasite**), where the parasite is green and does some photosynthesis, or total (**holoparasite**), where the plant does not photosynthesize and is not green. Parasites can tap into stems or roots using a modified root structure called a haustorium. Parasitic plants are common in ecosystems all over the globe and are found in all five MTC regions, with South African and Australian MTC regions standing out for the numbers of genera and endemic parasitic taxa (Lambers et al. 2014; Lamont 1982; Press and Graves 1995). A particularly spectacular species in southwest Australia is the tree parasite, *Nuytsia floribunda* (Christmas tree), which, at 10 m tall, is likely the largest parasite in the world (Figure 7.1a).

7.2.2 Nutrients, fire, and decomposition

Fire is a key factor in nutrient cycling in the MTC regions. In long unburned stands most of the nutrients in the system are in the standing biomass and the litter. After fire there is a pulse of nutrients as the biomass is consumed and the nutrients are released to the soil. Immediately after fire, nitrogen in the form of ammonium is high and in the months following fire pH changes and increased microbial activity converts ammonium to nitrate (Rundel 1983; Stock and Lewis 1986). Some of the nitrogen is lost depending on how intense the fire is, with hotter fires leading to greater losses via volatilization. Losses of other nutrients during burning are generally low. Other than volatilization, losses may also occur due to erosion and when rains fall, with both surface runoff and leaching occurring. Areas of steep slopes or on more gradual slopes when winter storms are especially heavy may lead to greater losses. Losses have been found to cease after the first winter rainy season following fire (van Wyk et al. 1992). Some loss of nutrients is offset by

Figure 7.1 **Carnivorous and parasitic plants.** Some of the more unusual modes of obtaining mineral nutrients include carnivory and parasitism, as illustrated with the following examples: the hemiparasitic tree *Nuytsia floribunda* and author Anna Jacobsen (a), a tiny sundew (*Drosera* sp.) from the Eneabba sandplains in southwest Australia (b), a sundew (*Drosera* sp.) from a bog in southwest Australia (c), a sundew (*Drosera* sp.) from a damp fynbos site in Betty's Bay in South Africa (d), and a pitcher plant (*Cephalotus follicularis*) in a bog in southwest Australia (e). The regions other than Australia and South Africa have very

rapid growth of annuals, geophytes, or short-lived legumes (e.g. *Acmispon glaber* in chaparral and *Aspalathus* spp. in fynbos) in the post-fire environment. These plants capture the nutrients in their tissues and keep them from being lost from the system. The loss of nutrients is chiefly nitrogen and without fresh inputs it would decline in the system over time (DeBano and Conrad 1978).

Another key way that nutrients cycle through biomass is via decomposition of litter. Litter inputs into MTV stands are similar to the amount of ANPP (Table 7.1). If decomposition was rapid then the litter would not accumulate; however, this is not the case and amounts of litter in these landscapes comprise a substantial amount of the total biomass (Table 7.1). Decomposition is limited by a number of factors. One is the MTC, which leaves only a narrow window where conditions are warm and moist enough to sustain high levels of metabolism (Figure 6.5). A second factor is the nutrient-poor status of the sclerophyllous MTV leaves and commonly dense woody tissues. Leaves with lower carbon-to-nitrogen ratios are decomposed faster and MTV species have generally high carbon-to-nitrogen ratios in the leaves they shed. Thus, without fire to mineralize litter it would build up and draw down the available nutrients in the system.

7.3 Hydrology

Hydrological studies have been undertaken in MTC regions, primarily from a practical perspective aimed at understanding how watersheds yield a given quantity and quality of fresh water for human uses. This is a high-priority political concern in all of these semi-arid regions (Le Maitre et al. 2000). The chief input in MTC regions is rainfall. The number of storms, seasonality, and quantity of rainfall does differ somewhat across the regions. Cowling et al. (2005) found that the Australian and South African MTC regions have less variable between-year winter rain and a higher frequency of small-to-moderate rainfall events in both winter and summer, and higher summer rain. California had the most variable regime and more intense summer drought, whereas the Andalusia region of Spain had the greatest number of large and intense downpours. These factors will affect how much runoff occurs and how much water will infiltrate into below-ground storage (Poole et al. 1981).

Figure 7.1   **(Opposite)** few carnivorous plants, but two interesting ones are *Drosophyllum lusitanicum*, found in the western Mediterranean, which can grow in dry alkaline soils (f). The California pitcher plant, *Darlingtonia californica*, grows in bogs and seeps in northern California and is unusual for attaining large sizes of up to 80 cm in height. (See Plate 13)

Source: Photos a, b, c, d, e, and g from R. Brandon Pratt and photo f from Karen J. Esler.

Vegetation structure and fire affect hydrology in a range of ways that have subtle to strong effects on water yields out of systems in streams and rivers. A key aspect of vegetation is how open the canopy is. In closed canopy conditions more water will be intercepted by the canopy and flow down stems, less will fall directly on the soil, and more will be lost from the leaves (both transpiration and canopy evaporation). The most extreme open canopy condition will be the first winter after fire, when water falls on bare ground and vegetation canopy area is minimal, leading to a drastic reduction in transpiration owing to low LAI. In the years after fire the stream flows from watersheds show a marked increase (Meixner and Wohlgemuth 2003; Scott and van Wyk 1992). If a watershed is cleared, such as through logging, then the water yield also increases. The loss of shrub cover also leads to increased surface runoff and sedimentation into streams that can degrade water quality and harm aquatic fauna (Horwitz et al. 2008). Long-term conversion of some shrublands to savannahs has led to sustained alterations of hydrology that will likely hinder the ability of the system to recover (Le Maitre et al. 2000; Wohlgemuth 2016). In South Africa, the clearing of woody alien species has been a priority to increase water yields (Dye and Jarmain 2004).

Another way that fire can affect hydrology is when burned soils become repellent to water (i.e. they become hydrophobic) owing to chemical changes occurring during the fire (DeBano 2000). When this happens it leads to a period after fire (months to years) when infiltration of water into soils is hindered because of this effect. This can lead to greater erosion and runoff following fires.

Vegetation can play a key role in hydrology by redistributing water in the soil through root systems. **Hydraulic redistribution** refers to the movement of water between soil regions through roots. Water moves down gradients of energy that are affected by pressure and solutes (**water potential**). The flow moves from higher water potential regions to lower ones and if the soil has lower water potential then it may suck water out of roots. This can happen when deep-rooted species tapping rich water sources move water along their root systems to areas of drier shallow soils and water leaks out of the roots and irrigates these soils (hydraulic lift; Burgess and Bleby 2006). The process can also work in reverse. The shallow soil layer, which dries out first during the summer dry season, is the most nutrient-rich region of the soil. It could be advantageous, from a mineral nutrition perspective, to keep these roots and nearby microorganisms hydrated and active to enhance nutrient availability. Hydraulic redistribution and its implications for ecosystem hydrological and nutrient cycling has not been well examined in MTC regions (however, see Hawkins et al. 2009 and Burgess and Bleby 2006). Given that these systems comprise a range of deep- and shallow-rooted species, have seasonal rainfall, and are nutrient-poor, it is likely that this process is important.

Literature cited

Allsopp N, Stock WD. 1993. Mycorrhizal status of plants in the Cape Floristic Region, South Africa. Bothalia 23: 91–104.

Barbour MG, Minnich RA. 1990. The myth of chaparral convergence. Israel Journal of Botany 39: 453–63.

Black CH. 1987. Biomass, nitrogen, and phosphorus accumulation over a southern California fire cycle chronosequence. Pages 445–58 in Tenhunen JD, Catarino FM, Lange OL, Oechel WC, eds. Plant response to stress. Springer, Heidelberg.

Brundrett MC. 2009. Mycorrhizal associations and other means of nutrition of vascular plants: understanding the global diversity of host plants by resolving conflicting information and developing reliable means of diagnosis. Plant and Soil 320: 37–77.

Burgess SS, Bleby TM. 2006. Redistribution of soil water by lateral roots mediated by stem tissues. Journal of Experimental Botany 57: 3283–91.

Burrows ND, McCaw WL. 1990. Fuel characteristics and bushfire control in banksia low woodlands in Western Australia. Journal of Environmental Management 31: 229–36.

Chapin III FS, Matson PA, Vitousek P. 2011. Principles of terrestrial ecosystem ecology, 2nd edn. Springer, Heidelberg.

Cowling RM, Ojeda F, Lamont BB, Rundel PW, Lechmere-Oertel R. 2005. Rainfall reliability, a neglected factor in explaining convergence and divergence of plant traits in fire-prone mediterranean-climate ecosystems. Global Ecology and Biogeography 14: 509–19.

Cramer MD, West AG, Power SC, Skelton R, Stock WD. 2014. Plant ecophysiological diversity. Pages 248–72 in Allsop N, Colville JF, Verboom GA, eds. Fynbos: ecology, evolution, and conservation of a megadiverse region. Oxford University Press, Oxford.

Davis SD, Ewers FW, Sperry JS, Portwood KA, Crocker MC, Adams GC. 2002. Shoot dieback during prolonged drought in *Ceanothus* (Rhamnaceae) chaparral of California: a possible case of hydraulic failure. American Journal of Botany 89: 820–8.

Davis GW, Richardson DM. 1995. Mediterranean-type ecosystems: the function of biodiversity. Springer, Heidelberg.

DeBano LF. 2000. The role of fire and soil heating on water repellency in wildland environments: a review. Journal of Hydrology 231: 195–206.

DeBano LF, Conrad CE. 1978. The effect of fire on nutrients in a chaparral ecosystem. Ecology 59: 489–97.

Dye P, Jarmain C. 2004. Water use by black wattle (*Acacia mearnsii*): implications for the link between removal of invading trees and catchment streamflow response: working for water. South African Journal of Science 100: 40–4.

Glyphis JP, Puttick MG. Phenolics, nutrition and insect herbivory in some garrigue and maquis plant species. 1989. Oecologia 78: 259–63.

Gray JT. 1982. Community structure and productivity in Ceanothus chaparral and coastal sage scrub of southern California. Ecological Monographs 52: 415–34.

Hawkins HJ, Hettasch H, West AG, Cramer MD. 2009. Hydraulic redistribution by *Protea* 'Sylvia' (Proteaceae) facilitates soil water replenishment and water acquisition by an understorey grass and shrub. Functional Plant Biology 36: 752–60.

Hellmers H, Horton JS, Juhren G, O'Keefe J. 1955. Root systems of some chaparral plants in southern California. Ecology 36: 667–78.

Higgins KB, Lamb AJ, Van Wilgen BW. 1987. Root systems of selected plant species in mesic mountain fynbos in the Jonkershoek Valley, south-western Cape Province. South African Journal of Botany 53: 249–57.

Hilbert DW, Canadell J. 1995. Biomass partitioning and resource allocation of plants from Mediterranean-type ecosystems: possible responses to elevated atmospheric CO_2. Pages 76–101 in Moreno JM, Oechel WC, eds. Global change and mediterranean-type ecosystems. Springer, New York.

Hobbs RJ, Richardson DM, Davis GW. 1995. Mediterranean-type ecosystems: opportunities and constraints for studying the function of biodiversity. Pages 1–42 in Davis GW, Richardson DM, eds. Mediterranean-type ecosystems: the function of biodiversity. Springer, Heidelberg.

Hopper SD. 2009. OCBIL theory: towards an integrated understanding of the evolution, ecology and conservation of biodiversity on old, climatically buffered, infertile landscapes. Plant and Soil 322: 49–86.

Horwitz P, Bradshaw D, Hopper S, Davies P, Froend R, Bradshaw F. 2008. Hydrological change escalates risk of ecosystem stress in Australia's threatened biodiversity hotspot. Journal of the Royal Society of Western Australia 91: 1–11.

Kruger FJ, Mitchell DT, Jarvis JU. 1986. Mediterranean-type ecosystems: the role of nutrients. Springer, Heidelberg.

Kummerow J, Avila G, Aljaro ME, Araya S, Montenegro G. 1982. Effect of fertilizer on fine root density and shoot growth in Chilean matorral. Botanical Gazette 143: 498–504.

Lambers H, Shane MW, Cramer MD, Pearse SJ, Veneklaas EJ. 2006. Root structure and functioning for efficient acquisition of phosphorus: matching morphological and physiological traits. Annals of Botany 98: 693–713.

Lambers H, Brundrett MC, Raven JA, Hopper SD. 2010. Plant mineral nutrition in ancient landscapes: high plant species diversity on infertile soils is linked to functional diversity for nutritional strategies. Plant and Soil 334: 11–31.

Lambers H, Shane MW, Laliberté E, Swarts ND, Teste FP, Zemunik G. 2014. Plant mineral nutrition. Pages 101–27 in Lambers H, ed. Plant life on the sandplains in Southwest Australia, a global biodiversity hotspot. University of Western Australia Publishing, Crawley.

Lambers H, Finnegan PM, Jost R, Plaxton WC, Shane MW, Stitt M. 2015. Phosphorus nutrition in Proteaceae and beyond. Nature Plants 4: 15109.

Lamont B. 1982. Mechanisms for enhancing nutrient uptake in plants, with particular reference to mediterranean South Africa and Western Australia. The Botanical Review 48: 597–689.

Le Maitre DC, DB Versfeld, Chapman RA. 2000. The impact of invading alien plants on surface water resources in South Africa: a preliminary assessment. Water SA 26: 397–408.

Low AB. 1983. Phytomass and major nutrient pools in an 11-year post-fire coastal fynbos community. South African Journal of Botany 2: 98–104.

Luo H, Oechel WC, Hastings SJ, Zulueta R, Qian Y, Kwon H. 2007. Mature semiarid chaparral ecosystems can be a significant sink for atmospheric carbon dioxide. Global Change Biology 13: 386–96.

Mappin KA, Pate JS, Bell TL. 2003. Productivity and water relations of burnt and long-unburnt semi-arid shrubland in Western Australia. Plant and Soil 257: 321–40.

Margaris NS. 1976. Structure and dynamics in a phryganic (East Mediterranean) ecosystem. Journal of Biogeography 3: 249–59.

Marion GM, Black CH. 1988. Potentially available nitrogen and phosphorus along a chaparral fire cycle chronosequence. Soil Science Society of America Journal 52: 1155–62.

McMaster GS, Jow WM, Kummerow J. 1982. Response of *Adenostoma fasciculatum* and *Ceanothus greggii* chaparral to nutrient additions. Journal of Ecology 70: 745–56.

Meixner T, Wohlgemuth PM. 2003. Climate variability, fire, vegetation recovery, and watershed hydrology. Pages 651–6 in Proceedings of the First Interagency Conference on Research in the Watersheds, Benson, Arizona.

Mitchell DT, Coley PG, Webb S, Allsopp N. 1986. Litterfall and decomposition processes in the coastal fynbos vegetation, south-western Cape, South Africa. The Journal of Ecology 74: 977–93.

Mooney HA. 1983. Carbon-gaining capacity and allocation patterns of mediterranean-climate plants. Pages 103–19 in Kruger FJ, Mitchell DT, Jarvis JUM, eds. Mediterranean-type ecosystems: the role of nutrients. Springer, Heidelberg.

Mooney HA, Rundel PW. 1979. Nutrient relations of the evergreen shrub, *Adenostoma fasciculatum*, in the California chaparral. Botanical Gazette 140: 109–13.

Mooney HA, Kummerow J, Johnson AW, Parsons DJ, Keeley S, Hoffmann A, Hays RI, Giliberto J and Chu C. 1977. The producers—their resources and adaptive responses. Pages 85–143 in Mooney H, ed. Convergent evolution in Chile and California. Dowden, Hutchinson and Ross, Pennsylvania, Inc., USA.

Poole DK, Roberts SW, Miller PC. 1981. Water utilization. Pages 123–49 in Miller PC, ed. Resource use by chaparral and matorral. Springer, Heidelberg.

Press M, Graves J. 1995. Parasitic plants. Springer, Heidelberg.

Rambal S. 2001. Hierarchy and productivity of mediterranean-type ecosystems. Pages 315–44 in Roy J, Saugier B, Mooney HA, eds. Terrestrial global productivity. Academic Press, San Diego, USA.

Rapp M, Lossaint P. 1981. Some aspects of mineral cycling in the garrigue of southern France. Pages 289–301 in di Castri F, Goodall DW, Specht RL, eds. Ecosystems of the world. Elsevier Scientific Publishing, New York, USA.

Rundel PW. 1983. Impact of fire on nutrient cycles in mediterranean-type ecosystems with reference to chaparral. Pages 192–207 in Kruger FJ, Mitchell DT, Jarvis JUM, eds. Mediterranean-type ecosystems: the role of nutrients. Springer-Verlag, Heidelberg.

Rundel PW, Parsons DJ. 1979. Structural changes in chamise (*Adenostoma fasciculatum*) along a fire-induced age gradient. Journal of Range Management 32: 462–6.

Rutherford MC. 1978. Karoo-fynbos biomass along an elevational gradient in the western Cape. Bothalia 12: 555–60.

Schlesinger WH, Gill DS. 1980. Biomass, production, and changes in the availability of light, water, and nutrients during the development of pure stands of the chaparral shrub, *Ceanothus megacarpus*, after fire. Ecology 61: 781–9.

Scott DF, Van Wyk DB. 1992. The effects of fire on soil water repellency, catchment sediment yields and streamflow. Pages 216–39 in Van Wilgen BW, Richardson DM, Kruger FJ, van Hensbergen HJ, eds. Fire in South African mountain fynbos. Springer, Heidelberg.

Shane MW, Szota C, Lambers H. 2004. A root trait accounting for the extreme phosphorus sensitivity of *Hakea prostrata* (Proteaceae). Plant, Cell & Environment 27: 991–1004.

Specht RL. 1969. A comparison of the sclerophyllous vegetation characteristic of mediterranean type climates in France, California, and southern Australia. II. Dry matter, energy, and nutrient accumulation. Australian Journal of Botany 17: 293–308.

Stock WD, Lewis OA. 1986. Soil nitrogen and the role of fire as a mineralizing agent in a South African coastal fynbos ecosystem. The Journal of Ecology 74: 317–28.

Stock WD, Verboom GA. 2012. Phylogenetic ecology of foliar N and P concentrations and N: P ratios across mediterranean-type ecosystems. Global Ecology and Biogeography 21: 1147–56.

Tsiourlis GM. 1990. Phytomasse, productivité primaire et biogéochimie des écosytèmes mediterranées phrygana et maquis (Ile de Naxos, Grèce). Doctoral dissertation, Thesis, Free University of Brussels, Brussels.

Van Wilgen BW. 1982. Some effects of post-fire age on the above-ground plant biomass of fynbos (macchia) vegetation in South Africa. Journal of Ecology 1: 217–25.

Van Wilgen BW, Forsyth GG. 1992. Regeneration strategies in fynbos plants and their influence on the stability of community boundaries after fire. Pages 54–80 in van Wilgen BW, Richardson DM, Kruger FJ, van Hensbergen HJ, eds. Fire in South African mountain fynbos. Springer, Heidelberg.

Van Wyk DB, Lesch W, Stock WD. 1992. Fire and catchment chemical budgets. Pages 240–57 in van Wilgen BW, Richardson DM, Kruger FJ, van Hensbergen HJ, eds. Fire in South African mountain fynbos. Springer, Heidelberg.

Westcott VC, Enright NJ, Miller BP, Fontaine JB, Lade JC, Lamont BB. 2014. Biomass and litter accumulation patterns in species-rich shrublands for fire hazard assessment. International Journal of Wildland Fire 23: 860–71.

Witkowski ET, Mitchell DT, Stock WD. 1990. Response of a Cape fynbos ecosystem. Acta Oecologica 11: 311–26.

Wohlgemuth P. 2016. Long-term hydrologic research on the San Dimas Experimental Forest, southern California: lessons learned and future directions. Pages 227–32 in Stringer CE, Krauss KW, Latimer JS, eds. 2016. Headwaters to estuaries: advances in watershed science and management—Proceedings of the Fifth Interagency Conference on Research in the Watersheds, North Charleston, South Carolina. e-General Technical Report SRS-211: 211.

The Modern Stage: Transformation

8 Transformation

Abstract

Extensive habitat loss and habitat conversion has occurred across all mediterranean-type climate (MTC) regions, driven by increasing human populations who have converted large tracts of land to production, transport, and residential use (land-use, land-cover change) while simultaneously introducing novel forms of disturbance to natural landscapes. Remaining habitat, often fragmented and in isolated or remote (mountainous) areas, is threatened and degraded by altered fire regimes, introduction of invasive species, nutrient enrichment, and climate change. The types and impacts of these threats vary across MTC regions, but overall these drivers of change show little signs of abatement and many have the potential to interact with MTC region natural systems in complex ways.

8.1 Threats to mediterranean-type climate regions

Large-scale human transformation of all mediterranean-type climate (MTC) regions has led to them prominently featuring in the global categorization of biodiversity hotspots (Myers et al. 2000; Marchese 2015), i.e. thirty-five areas on the Earth's land surface where high levels of endemism correspond with high levels of threat (Chapter 1). The types and impacts of these threats vary across MTC regions (Underwood et al. 2009a), and include habitat loss through land-use change, natural disturbance regime changes (particularly fire), invasions, and anthropogenic climate change. These threats are interactive and their magnitude fluctuates across and within MTC regions, but underlying them is a positive association with human population density increases, estimated to be 1.8 per cent growth per year across global biodiversity hotspots, higher than the 1.3 per cent global average (Cincotta et al. 2000).

Because of their mild climates, MTC regions continue to be popular places to live and work. All MTC regions contain major metropolitan centres, and these

The Biology of Mediterranean-Type Ecosystems. Karen J. Esler, Anna L. Jacobsen, and R. Brandon Pratt,
Oxford University Press (2018). © Karen J. Esler, Anna L. Jacobsen, and R. Brandon Pratt 2018.
DOI 10.1093/oso/9780198739135.001.0001

Table 8.1 Mediterranean-type climate (MTC) region cities and population centres. Urban localities of regional or administrative significance (>1 million people) in the five MTC regions of the world, ordered according to population size

City	Country	Population (2016)	Population density (km²)	Annual rate of change (%)
Los Angeles	USA	15,135,000[a]	2400[a]	0.3[b]
Santiago	Chile	6,265,000[a]	6400[a]	0.9[b]
Madrid	Spain	6,240,000[a]	4700[a]	1.4[b]
San Francisco	USA	5,955,000[a]	2100[a]	0.1[b]
Barcelona	Spain	4,740,000[a]	4400[a]	1.2[b]
Rome	Italy	3,930,000[a]	3500[a]	0.6[b]
Cape Town	South Africa	3,865,000[a]	4700[a]	1.9[b]
Naples	Italy	3,700,000[a]	3600[a]	−0.1[b]
Algiers	Algeria	3,675,000[a]	8300[a]	1.3[b]
Casablanca	Morocco	3,240,000[a]	11,900[a]	0.8[b]
Izmir	Turkey	3,170,000[a]	10,600[a]	2.1[b]
San Diego	USA	3,110,000[a]	1600[a]	1.0[b]
Aleppo	Syria	3,665,000[a]	14,200[a]	3.1[b]
Athens	Greece	3,480,000[a]	6000[a]	−0.3[b]
Lisbon	Portugal	2,685,000[a]	2800[a]	0.5[b]
Tunis	Tunisia	2,240,000[a]	5100[a]	0.8[b]
Beirut	Lebanon	2,225,000[a]	3300[a]	2.6[b]
Damascus	Syria	2,605,000[a]	14,400[a]	1.6[a]
Perth	Australia	1,785,000[a]	1000[a]	1.8[b]
Marseille	France	1,610,000[a]	2300[a]	0.6[b]
Porto	Portugal	1,480,000[a]	1900[a]	0.2[b]
Adelaide	Australia	1,150,000[a]	1300[a]	0.6[b]
Rabat	Morocco	1,880,000[a]	10,200[a]	1.8[b]
Hamah	Syria	1,300,000[a]	25,100[a]	6.0[a]
Marrakech	Morocco	1,210,000[a]	15,100[a]	2.8[b]
Adana	Turkey	1,125,000[a]	6900[a]	3.2[b]

[a] Obtained from demographia.com, accessed 16/02/2017; [b] Obtained from un.org, accessed 16/02/2017, based on 2000–2016 average annual rate of change.

are projected to expand (Table 8.1). Only two of twenty-six cities with populations >1 million people show a declining rate of change, with California and the Mediterranean topping MTC regions with regard to urban growth. Indeed, between 1990 and 2000, population density in MTC regions increased by 13 per cent (Underwood et al. 2009a). The threats to MTC region biodiversity are therefore likely to continue. Sala et al. (2000) predicted that, of all terrestrial biomes, MTC regions would experience the greatest proportional loss in biodiversity by 2100, driven primarily by changes in land use and climate.

8.2 Human interactions with mediterranean-type climate region landscapes

To understand current and future threat patterns to MTC regions, it is necessary briefly to consider the timing and duration of hominid occupation in these regions (Figure 8.1). These patterns of occupation have had variable influences on the degree and rates of transformation in each region.

Some of the earliest evidence for the presence of modern humans, *Homo sapiens*, and therefore our interaction with an MTC landscape, is found in the Greater Cape Floristic Region of southern Africa (Deacon and Deacon 1999; Klein 2001; Henn et al. 2011; Marean et al. 2014). At this time, approximately 195–123 thousand years ago (ka), westerly-driven frontal systems brought winter rain to the west and, to some extent, the south coast. It has been hypothesized that the interplay between intertropical convergence zone-driven summer rains to the north and east may have provided a template for a migratory system for grazing ungulates along the (now submerged) coastline (Marean 2010a; Copeland et al. 2016). The juxtaposition of these ungulates available for hunting (Klein 1983; Rector and Verrelli 2010; Faith 2011), together with an abundance of edible geophytes (plants with underground storage organs) (Singels et al. 2016), a predictable, rewarding resource-rich (i.e. copious marine invertebrates) coastline (De Vynck et al. 2016), and an abundance of conveniently placed freshwater springs has been hypothesized to have provided an ideal resource base to facilitate the persistence of modern humans through a particularly cold glacial phase of the Pleistocene (Marean 2010b). Furthermore, that MTC region biodiversity supported our own species though a population bottleneck (Ambrose 1998; Blum and Jakobsson 2011) provides a compelling narrative for modern conservation efforts in these systems.

Further evolution of social complexity (e.g. ship-based hunting, permanent settlements, warfare, slavery, and money) did not occur in the early Cape coastal hunter-gatherer communities (Marean et al. 2014). Around 115–106 ka, small populations of *Homo sapiens* had reached the coastline of northeast Africa, and low population densities are likely to have occurred in southern Europe from 95 to 72 ka (Timmermann and Friedrich 2016). It was the Mediterranean that hosted the transition from forager to food-producing economies. This occurred by approximately 10 ka in Israel and 8 ka in the western Mediterranean (Blondel et al. 2010). Cultivation of wine grapes spread throughout the Mediterranean around 7 ka (Viers et al. 2013). In the warmer climate of the Holocene, the Mediterranean Basin and its islands witnessed the rise and fall of powerful civilizations (e.g. the Greek, Roman, and Ottoman empires). Kingdoms and city-states, supported by agricultural development, were established by 4 ka in the Eastern Mediterranean and Middle East. Partially hastened by drought, flourishing eastern Mediterranean

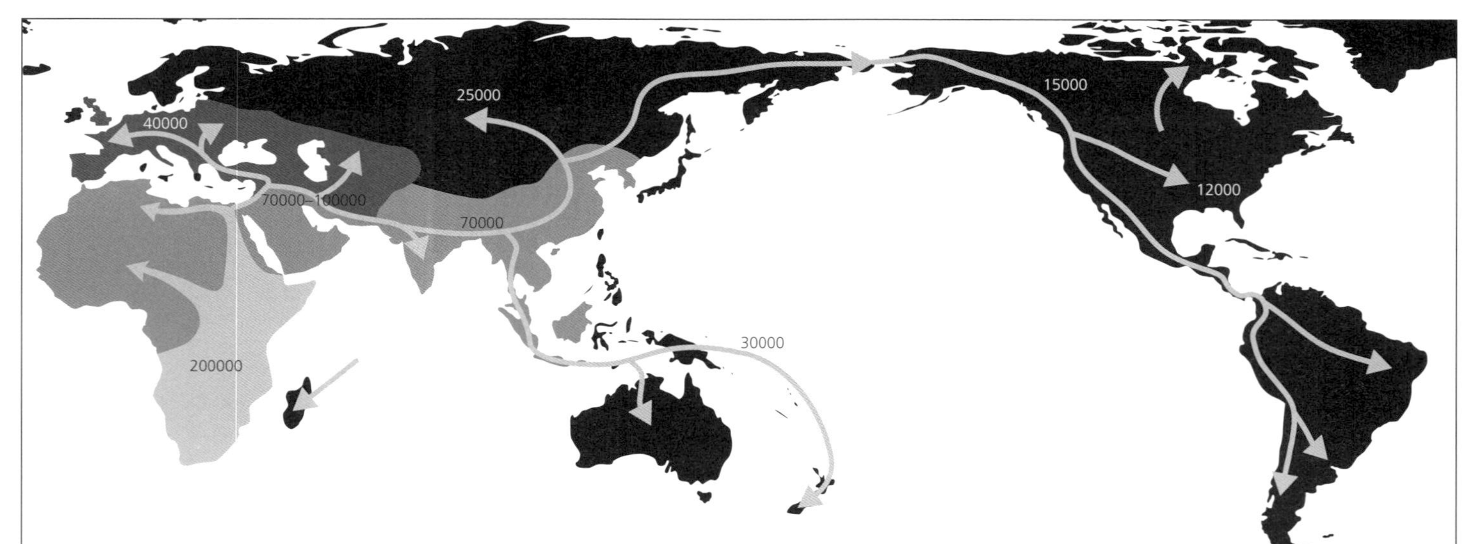

Figure 8.1   **Early human migrations.** A likely route of early *Homo sapiens* migration out of Africa is shown in the lightest shade of grey, with the approximate years before present of their successful establishment in a new area included as the numbers indicated on the map. The darker areas of grey indicate regions that were occupied by other human lineages during the arrival of *H. sapiens*, including *H. erectus* (intermediate grey) and *H. neanderthalensis* (dark grey). Regions shown in black were not populated by human lineages prior to the arrival of *H. sapiens*. (Modified from https://commons.wikimedia.org/w/index.php?curid=34697001, as posted by NordNordWest.)

civilizations collapsed 3.2 ka, but emerging rural settlements persisted through adapted agro-pastoral activities (Kaniewski et al. 2015). Such civilizations and cultures shaped land-use practices and resource-management approaches to the extent that the landscape of this MTC region is today regarded as a function of its long-term interaction with people (Butzer 2005). The high plant diversity and currently low extinction rates (Greuter 1994) of this region are suggested to be intricately tied to its long exposure to large-scale western colonization (di Castri 1981). Today, population density and percentage urban area of this region is high, second only to California, and is mostly concentrated along coastal and low–mid elevation areas (Underwood et al. 2009a).

Around 60 ka, modern humans crossed the Indonesian archipelago (accessible owing to lower sea levels) into Papua New Guinea and the north of Australia (Timmermann and Friedrich 2016). By between 50 and 48 ka, indigenous people occupied the Kwongan. Regional shifts in fire frequency, indicating impact on the landscape, are noted then (Kershaw et al. 1997; 2006) and again during the period of European colonization in the nineteenth and early twentieth centuries (Lynch et al. 2007). The impact of European colonization in MTC regions of Australia is largely associated with large-scale conversion of natural areas to production landscapes. More land is converted to agriculture than in any other MTC region, despite low population densities and proportionally low levels of urbanization (Underwood et al. 2009a).

Although evidence has emerged for potential earlier timing of human occupation in North America (24 ka; Bourgeon et al. 2017), the prevailing view is that dispersing groups moved into the Americas across the Bering land bridge during a short window period between 14 and 10 ka (Timmermann and Friedrich 2016). Little is known about the early stages of New World colonization, but indications are that early Holocene Native American populations in California impacted the landscape by providing a source of ignitions just over 10,000 years ago (Keeley et al. 2012). For the past 5000 years these people likely had significant influence on fire regime, particularly at low elevations. Large-scale European colonization of this 'new world' MTC region occurred in the nineteenth century, followed by rapid transformation of the landscape, particularly the coastal lowlands. In 2000, population density in California-Baja California was the highest of the five MTC regions, with 246 people km^{-2} (Underwood et al. 2009a).

Modern humans moved rapidly from North America to South America along the Pacific coastline around 14–12 ka (Dillehay 1999). In central Chile, pre-Inca and pre-Columbian people (Araucanians) kept llamas, used fire to clear sites, and even cultivated crops in the river valleys, but it was the arrival of Europeans in the mid-sixteenth century (1536–1541) that significantly changed the Chilean MTC region landscape (Aronson et al. 1998). While Chile has experienced the greatest recent (1990–2000) change in population density compared to the other MTC regions, its population density is low

Table 8.2 Human impact comparisons between the five mediterranean-type climate regions of the world

Region	Area[a] (km²)	Modern human colonization (ka)	Human population[a] (km²)	% Urban area[a]	% Agriculture[a]
Australia	785,274	50–48[b]	5	0.3	37
California	118,340	14–10[b]	246	9.2	6
Chile	146,839	14–12[c]	46	0.7	24
Mediterranean	2,022,672	115–72[b]	123	1.5	29
South Africa	95,554	195–123[d]	57	1.0	24

[a] Underwood et al. 2009a; [b] Timmermann and Friedrich 2016; [c] Dillehay 1999; [d] Marean 2010a.

(forty-six people km^{-2} in 2000), second only to Australia at five people km^{-2} (Underwood et al. 2009a).

In summary, although modern human origins are linked back to the Cape Region of South Africa, it is the Mediterranean that boasts the longest history of agricultural activity and human settlement. People colonized Australia, California, and Chile long before the wave of European colonization started in the mid-1600s (Table 8.2). Early people impacted natural disturbance regimes, notably fire regimes, but it is the colonial history that signals the start of major habitat transformation in most MTC regions.

8.3 Land-use, land-cover change

Human-controlled ecosystems now predominate globally, resulting in calls to recognize a new geological epoch, termed the Anthropocene (Crutzen 2002; Smith and Zeder 2013). Resulting land-use and land-cover changes, largely because of conversion to production landscapes, are the primary drivers of terrestrial biodiversity loss, and are considered to be the greatest threat to biodiversity in MTC regions (Underwood et al. 2009a), causing habitat loss, fragmentation, and degradation. It has been estimated that **biodiversity intactness**, the average proportion of natural diversity remaining in a local ecosystem, has transgressed safe limits (loss greater than 10 per cent of abundance or 20 per cent of species) in 58.1 per cent of the world's land surface, occupied by 71.4 per cent of the human population (Newbold et al. 2016). This transgression includes nine of fourteen terrestrial biomes, with MTC regions third on the list of most affected biomes (Newbold et al. 2016). Indeed, an analysis of key threats to biodiversity in the global mediterranean-type biome noted a negative correlation between remaining natural area and numbers of threatened mammals and plants—the less natural area remaining, the more species threatened (Underwood et al. 2009a).

Complex interactions and feedbacks drive land-use and land-cover change in MTC regions. There is some understanding at local and regional levels of direct ecological drivers of these changes, such as the role of fire and drought in driving **land-use and land-cover change (LULC)**. Less well understood is how social drivers, such as population growth trends, land management policies, and demand for resources (food, fuel), influence these ecological changes, and how these ecological changes, in turn, feed back to influence social processes that drive LULC. The literature can be confusing as these drivers may interact at different scales. Given regional variation in social and cultural history (Section 8.1), contexts for land-use management vary, as do natural land cover patterns and fire regimes. Considering advances in technology (e.g. remote sensing) and our ability to generate and analyse large, spatially explicit, datasets (e.g. using high-performance computers), the opportunity is ripe for interdisciplinary teams to generate a more holistic and synthetic understanding of the social-ecological drivers of LULC change. A good example of how ecological and socioeconomic drivers can be integrated using a modelling framework is provided by Fischer et al. (2005), who focused on socioeconomic and climate change impacts on agriculture in southern Africa. This kind of understanding is essential for the development of management responses and conservation strategies (Chapter 9), but current literature is patchy and lacks integration. Given the complexity and constraints described above, only broad LULC trends are described below for each region, with a sample of the literature referred to in each case.

With their long history of human occupation, the 'old world' landscapes of the Mediterranean have been significantly influenced to the extent that traditional land-use practices have moulded these landscapes for centuries (Figure 8.2; see Plate 13; Antrop 2004; Agnoletti 2007; Salvati et al. 2016). Early forms of agro-pastoral use of Mediterranean woodlands and shrublands remained relatively unchanged from Greek and Roman times up to the middle of the twentieth century (Caravello and Giacomin 1993). Where traditional practices have remained intact, associated areas are now considered to be cultural landscapes worthy of conservation (Rossler 2006; UNESCO 2012), a feature unique to this region. More recently, urbanization and agricultural intensification, particularly at low elevations and along the coastline, has altered these landscapes (Antrop 2004; Salvati et al. 2017), impacting biodiversity and ecosystem services (Metzger et al. 2006). Juxtaposed to the urbanization and intensification trends in the lowlands of the Mediterranean, depopulation and land abandonment has occurred in the more mountainous areas away from cities (e.g. Falcucci et al. 2007; Serra et al. 2008; Marull et al. 2015; Salvati et al. 2017). Natural reforestation after abandonment of vineyards, pastures, and rain-fed arable land is common in the Mediterranean (Poyatos et al. 2003; Salvati and Sabbi 2011; Millington et al. 2007), but these increases in forest areas have not necessarily led to improved ecological connectivity or increases in diversity (Marull et al. 2015). In central Italy the

Figure 8.2 **Mediterranean region traditional land use.** Maritime pine (*Pinus pinaster*) forests have been managed for the production of resin (a). Wounds are opened for resin to flow from trees to be collected in pots. Dehesas are managed as multifunctional landscapes for agriculture, sylviculture, and grazing that represent a traditional form of landscape management that has been utilized for centuries in Mediterranean oak forest communities (b). This image shows a dehesa with cork oak (*Quercus suber*) that has been harvested recently and with pasture for cattle grazing. (See Plate 14)

Source: Photos from Martin D. Venturas.

opposite has occurred, a reduction in forested area has been attributed to an increase in forest fires and overgrazing (Salvati et al. 2017). Overgrazing leading to declining soil productivity and desertification has affected the more arid parts of the Mediterranean (Vallejo et al. 2012).

In contrast to the Mediterranean, the 'new world' MTC regions have undergone rapid transformation since their colonization by European settlers, and over a much shorter timeframe (Underwood et al. 2009a; Table 8.2). In California, urban expansion, forest cutting, and fire dominate land-cover change (Sleeter et al. 2011). To support growing regional populations, large parts of the Central Valley and chaparral and oak woodland habitats were converted to suburban housing between 1973 and 2000 (Figure 8.3; Sleeter et al. 2011). The original range of sage scrub has been reduced by 80–85 per cent through urbanization and anthropogenic nitrogen deposition (Reid and Murphy 1995; Allen et al. 1998). Agriculture has remained relatively stable in some areas, while sharp declines have occurred in other area as farmers sold off valuable land to urban developers as urbanization has sprawled into once remote areas (Underwood et al. 2009a; Case Study 13). At the urban–wildland interface and lowlands where housing densities are intermediate, anthropogenic ignitions and fire extent peak (Syphard et al. 2007), and habitat type conversion from shrubland to herbaceous, alien-dominated communities has occurred (see Figure 2.9, Figure 8.6 below, and Case Study 12). Conversely, higher elevation coniferous forest has experienced increased fire-return intervals (FRIs), associated with fire suppression policies (Section 8.5). Forest composition in California has shifted towards increased dominance

Figure 8.3   **Urban and suburban development in southern California.** Urban expansion has resulted in the development of housing communities within previously intact chaparral shrub communities and the clearing of large areas of shrubland from around homes as part of fire protection ordinance (a). The Los Angeles Basin, which encompasses the greater Los Angeles area, has seen rapid urban development and expansion (b).

Source: Photos from Anna L. Jacobsen.

Case Study 12 Australian acacias—super invaders of mediterranean-type ecosystems

David M. Richardson, Centre for Invasion Biology, Department of Botany and Zoology, Stellenbosch University, Stellenbosch, South Africa

Ecosystems in all five regions of the world with mediterranean-type climates (MTCs) have been heavily invaded by alien plants; however, different life forms and species dominate the invasive floras of the different regions. The particular assemblages of invasive species that occur and flourish in each region are the result of the different introduction histories as well as ecological reasons that have allowed particular life forms and species to establish and proliferate in particular settings. In four of the five MTC regions, i.e. all but South Africa, herbaceous species are the most widespread and damaging invasive plants. Invasive alien trees are becoming more prominent in all regions, but the Cape Floristic Region (CFR) of South Africa easily claims the prize for 'Tree Invasions World Capital'. The invasive flora in this region is strikingly different from those of the other four regions in that trees and shrubs are prominent among the most widespread and influential invaders. Species of *Acacia*, *Eucalyptus*, and *Hakea* from climatically analogous regions in Australia, and *Pinus* from California and the Mediterranean have transformed vast tracts of fynbos vegetation in the CFR.

Australian acacias (hereafter 'acacias') comprise 1012 recognized species that were previously grouped in *Acacia* subgenus *Phyllodineae*. Acacias are present in most major biogeographical regions in Australia, but hotspots of diversity occur in the MTC regions

Continued

Case Study 12 (*Continued*)

of the southwestern and southeastern parts of the continent. Many acacias have been moved around the world by humans and widely planted over the past 250 years (Richardson et al. 2011). They are especially problematic invaders in the CFR, but their importance as invasive species is increasing in all the other regions and also within Australia, where many species have become invasive outside their native ranges.

More than one-third of all Australian acacia species are known to have been planted for various reasons outside their native ranges, and 23 species are known to be invasive (spreading over considerable distances from planting sites)—at least 15 species are invasive in MTC regions (Case Study 12 Table). About 70 species are known to have been introduced to South Africa alone, most of them more than 150 years ago. Fewer species were introduced to other MTC regions, and widespread planting of these species is also more recent in these regions. The CFR has the largest number of invasive Australian acacia species (13), followed by Australia (10), the Mediterranean (8), and California and Chile (both 5) (Rejmánek and Richardson 2013). *Acacia dealbata, A. longifolia*, and *A. melanoxylon* are known to be invasive in all five MTC regions, and *A. cyclops* and *A. mearnsii* are both invasive in four regions. Unlike in some other genera, no traits clearly separate invasive from non-invasive species in Australian acacias (Gibson et al. 2011). The current invasion status in this group is strongly influenced by the history of

Case Study 12 Table. Invasive Australian acacia species in mediterranean-type climate regions of the world (from Rejmanek and Richardson 2013)

	California	Mediterranean Basin	Australia	South Africa	Chile
Acacia baileyana F. Muel.			*	*	
Acacia cyclops A Cunn. Ex G. Don	*	*	*	*	
Acacia dealbata Link	*	*	*	*	*
Acacia decurrens Willd.			*	*	
Acacia elata A Cunn. Ex Benth.				*	
Acacia implexa Benth.				*	
Acacia longifolia (Andrews) Willd.	*	*	*	*	*
Acacia mearnsii De Wild.		*	*	*	*
Acacia melanoxylon R. Br.	*	*	*	*	*
Acacia paradoxa DC.	*		*	*	
Acacia podalyrifolia A. Cunn. Ex G. Don				*	*
Acacia pycnantha Benth.		*	*	*	
Acacia retinoides Schltr.		*			
Acacia saligna (Labil.) H. L. Wendl.		*	*	*	
Acacia stricta (Andrews) Willd.				*	

introduction and plantings—the earliest introductions and those that are most widely planted are the most widespread invaders (Castro-Díez et al. 2011).

There is clearly a substantial invasion debt for acacias in all MTC regions (many more species are likely to be introduced and become invasive, and many species that have already been introduced in particular regions are likely to occupy greater areas and cause more impacts). Large parts of these regions are climatically suitable for many acacia species and new introductions and plantings are continuing. Even if no further introductions took place, many species already introduced will become naturalized and invasive, spread further, and cause greater impacts over time (Rouget et al. 2016).

Invasion dynamics vary among species and ecosystems. Some species have the capacity to resprout vigorously following fire or physical damage, and many have other adaptations for dealing with key features of MTC regions such as fires, low-nutrient soils, and summer drought. Adaptations that allow acacias to establish and proliferate across a wide range of habitats in MTC regions include the capacity to form arbuscular mycorrhizas and to fix nitrogen by forming associations with rhizobia (co-introduced species of bacteria from their Australian range; Ndlovu et al. 2013), having generalist pollination systems, prolific seed production, efficient seed dispersal strategies, and accumulating large persistent seed banks that have fire-, heat- or disturbance-triggered germination cues. The processes shown in the Case Study 12 Figure for *A. saligna* in fynbos show how these invaders create a reinforcing feedback loop that promotes their own abundance, creating a regime shift that transforms invaded ecosystems from species-rich fynbos shrublands to species-poor acacia woodlands (Gaertner et al. 2014). Invaded ecosystems have radically increased biomass, altered fuel properties, and radically changed nutrient cycling and litter production. A meta-analysis of the impact of plant invasions on native species richness in MTC regions showed two acacias (*A. melanoxylon* and *A. saligna*) to have the highest negative effect size (Gaertner et al. 2009). Changes to biotic and abiotic features associated with these invasions have profound implications for the biota of these ecosystems and change their potential to deliver a range of goods and services, thereby altering human perceptions relating to value (Le Maitre et al. 2011).

Invasions by acacias pose huge threats to biodiversity and ecosystem functioning in the South African CFR and parts of the Mediterranean, and their impacts are increasing in other regions. Despite differences in the suite of species, details of invasion dynamics, impacts, human perceptions, and management options, there is much to be learned from experiences of dealing with acacias in one region that can be applied in other MTC regions. For example, seed-attacking biological control agents are well established in the CFR and lessons from this experience can be applied in other regions. Successes and failures in resolving conflicts of interest regarding acacias need to be shared to improve interventions.

The global experiment in biogeography involving the exchange of species between isolated parts of the world with similar environments (like MTC regions) has great potential for shedding light on aspects of the biodiversity of these regions. For example, the spectacular success of Australian acacias as invaders of MTC regions of the world offers great opportunities to improve our understanding of how the Australian environment and its evolutionary history have served as a factory for a highly diverse flora that is extremely well adapted for survival, growth, and proliferation in many parts of the world (Richardson et al. 2011).

Continued

Case Study 12 (*Continued*)

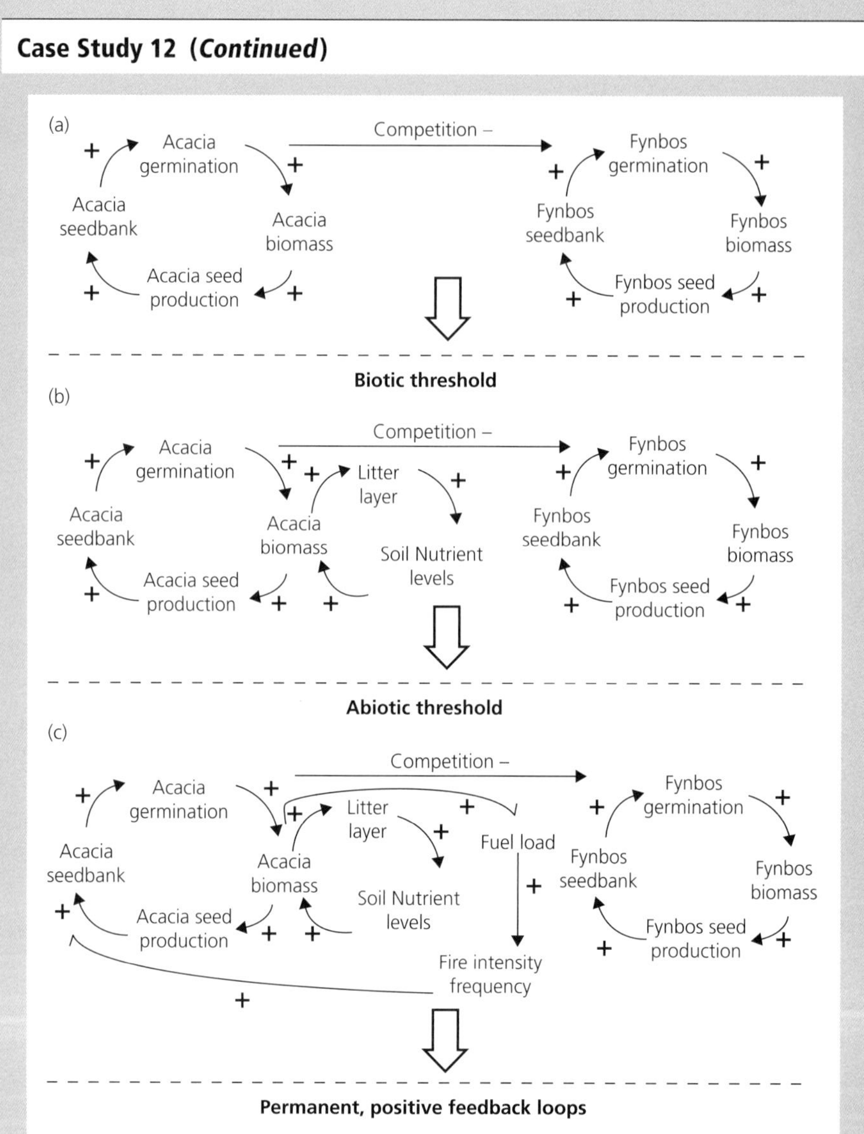

Case Study 12 Figure. Feedback loops in *Acacia* invasion dynamics. Acacias can create a reinforcing feedback loop that promotes their own abundance, creating a regime shift that transforms invaded ecosystems from species-rich fynbos shrublands to species-poor acacia woodlands. The system shifts from (a) disturbed fynbos under recent acacia invasion with biotic changes to (b) a hybrid ecosystem where biotic changes (both structural and functional) are accompanied by abiotic changes to (c) a **novel ecosystem** where permanent positive feedback loops are established. (Redrawn from Richardson and Gaertner 2013.)

Case Study 13 Land-use changes in an urbanizing world: a comparison between the city of Cape Town, South Africa and Los Angeles County, USA

Patricia M. Holmes, Biodiversity Management Branch, Environmental Management Department, City of Cape Town, South Africa

Alexandra D. Syphard, Conservation Biology Institute, Corvallis, Oregon, USA

Introduction

Land-use changes in mediterranean-type climate (MTC) regions are associated with a range of environmental impacts such as habitat loss and fragmentation, altered fire regimes, exotic species invasion, pollution, and altered hydrology. Here we compare two MTC region coastal cities that have experienced rapid urbanization while also being recognized as urban biodiversity hotspots: City of Cape Town (CCT) and Los Angeles County (LAC) (Case Study 13 Table). We explore similarities and differences in the impacts of land-use change, particularly urbanization, on biodiversity between the two areas.

Biophysical environment

The CCT is a highly diverse corner of the Cape Floristic Region (CFR) that supports a disproportionate number of threatened plant species: one-third (3300) of the CFR's species occur here in <3% of the CFR area; and 18% of South Africa's threatened plants occur here in 0.1% of the country's area, with thirteen species globally extinct in the wild (Raimondo et al. 2009; Rebelo et al. 2011). This high floral diversity is mirrored by the

Case Study 13 Table. A comparison of two urban mediterranean-type climate regions: City of Cape Town, South Africa and Los Angeles County, USA based on information available for 2015

	City of Cape Town	Los Angeles County
Area (km²)	2462	12,308
Floral region	Cape	Californian
Mountain ranges (highest altitude)	Cape Peninsula, Kogelberg (1200 m)	Santa Monica, San Gabriel (3100 m)
Natural vegetation remaining (%)*	38.5	58.0
Natural fire-return interval (years)	15	55
Native plant species	3300	2400
IUCN threatened species (% total)	9.7	4.2
Extinct species	13	23
Current population (millions)	3.86 (in 2013)	10.02 (in 2013)
Unemployment (%)	27.0	7.2
Income inequality (Gini Index)	67.0	55.0

*Includes some areas dominated by invasive exotic species. IUCN, International Union for Conservation of Nature.

Continued

Case Study 13 (*Continued*)

fauna and is partly explained by the high habitat diversity resulting from steep topographical, rainfall, and edaphic gradients between the mountain ranges and the lowlands (high beta and gamma diversity; Simmons and Cowling [1996]). The predominant vegetation type is fynbos. The richer, shale-derived soils in lowland areas support renosterveld, most of which were converted to agriculture by the mid-1900s. Both fynbos and renosterveld shrublands are fire-maintained, with average natural cycles of 15 and 7 years, respectively. However, now only the mountains burn regularly, with the majority of the lowlands too fragmented to carry natural fires. Invasion by exotic plant species and accelerating urban sprawl since the 1970s has increased the threatened status of many lowland ecosystems and species (Holmes et al. 2012).

Although LAC is the most populous county in the United States, it also hosts exceptional biodiversity, with large expanses of native vegetation protected through public agencies and land conservation organizations. The high levels of species richness and endemism in the region stem partly from its topographic complexity, with an elevational gradient that ranges from sea level to more than 3100 m and a strong coastal to inland temperature and precipitation gradient (Tamrazian et al. 2008). The primary vegetation type is chaparral shrubland, which forms a mosaic across the landscape with other vegetation communities, including California sage shrubland, exotic and native grassland, oak woodland, and mixed conifer forest at the highest elevations. The native vegetation consists of approximately 2400 plant species that support hundreds of bird, reptile, and mammal populations. This region has one of the highest concentrations of threatened and endangered species in the United States (Dobson et al. 1997). The landscape is highly fire-prone, with a natural fire regime of periodic large crown fires driven by extremely hot and dry Santa Ana winds that occur annually (Keeley et al. 2012).

History and socioeconomic drivers of land-use change

Although the CCT environment has been used by human populations for millennia, human impact increased following European settlement in 1654 (Allsopp et al. 2014). The city footprint and population remained small until the 1920s, increasing sharply after World War II and again from the late 1980s following the demise of apartheid influx control (Case Study 13 Figure). High fertility rate and rural-to-urban migration is driving high population increase and urban sprawl, resulting in the conversion of natural ecosystems; this trend is set to continue.

Los Angeles was founded in 1781, shortly following European settlement, and the county was formed in 1850 when California became a state. The region has experienced several population booms, owing to several historical catalysts. The first boom occurred after the linking of the city with the rail line in 1876, and subsequent periods of growth have occurred in response to factors such as oil discovery and the gold rush, the automobile, and the development of the film industry in the early twentieth century. The explosive population growth in Los Angeles has resulted in massive land-use change, primarily through housing development and urbanization.

In the early twentieth century, conversion to agriculture had the largest spatial impact on natural vegetation in CCT: by the 1940s most fertile areas (36% of land area) had been

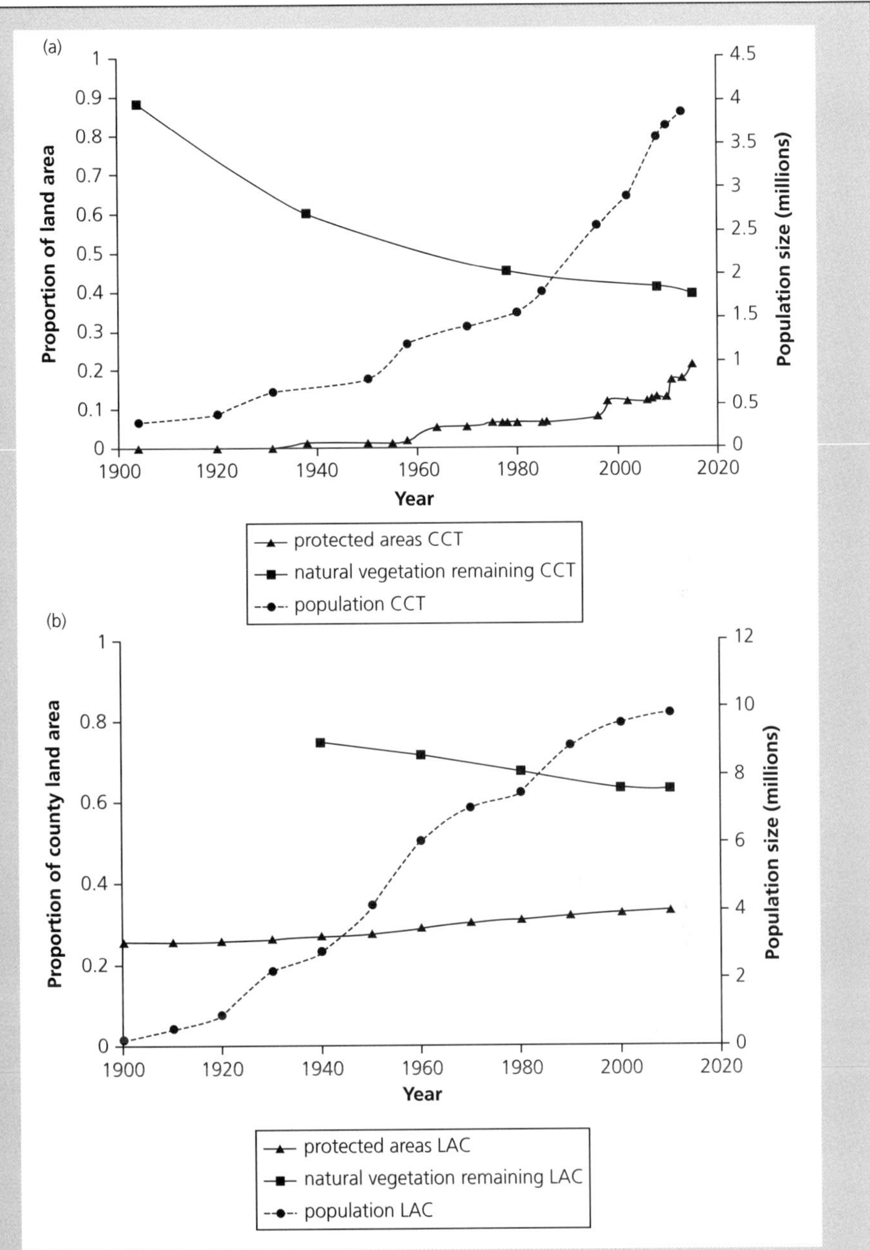

Case Study 13 Figure. Demographic and land-use changes. In the City of Cape Town (CCT, a) and Los Angeles County (LAC, b) over the past century, natural vegetation has declined (solid line and squares, left axis), the increase in protected area network (solid line and triangles, left axis) and human populations have grown (dashed line and circles, right axis).

Continued

Case Study 13 (*Continued*)

converted. This century, crops suited to nutrient-impoverished acid sands or hydroponics (e.g. potatoes, carrots, proteas, and blueberries) are converting some lowland fynbos vegetation areas. Urbanization has had the second largest impact, to date covering one-third of the total land area. Both agriculture and urban development largely are confined to lowland areas, thus lowland biodiversity is highly threatened and under-conserved compared to the mountains (Rebelo et al. 2011).

For LAC in the early twentieth century, conversion of natural habitat into agricultural lands also was the most dominant form of land-use change, and by the 1950s, approximately 21% of the landscape had become croplands, with citrus and other fruit trees becoming especially extensive. At the time, Los Angeles was considered the top agricultural county in the USA. As population growth started to explode in the middle of the century, however, farming was largely wiped out in favour of commercial and residential development. In fact, more than 80% of the croplands were converted to urban land over a relatively short timeframe (Syphard et al. unpublished data). Urban and residential development now represents the top contributor of direct habitat loss.

The CCT accounts for 70% of the Western Cape Province's gross domestic product, with its largest economic sector being the services (particularly finance) sector. However, there is a high unemployment level of 27%, with limited growth in the manufacturing sector to employ unskilled labour. The high unemployment rate, particularly among the youth, coupled with low educational levels, is a major challenge. A large informal economy has developed, with an increasing trend for illicit harvesting of natural products: more than 400 flora and fauna species are gathered in CCT for medicinal and other uses (Petersen et al. 2012). In addition, overstocking with cattle and goats is degrading natural vegetation in areas adjacent to informal settlements.

Invasive exotic species often enter a new country through introductions into cities, and especially port cities such as CCT (Gaertner et al. 2016). Invasive exotic trees and shrubs, mostly introduced for forestry or horticulture, invade and suppress fynbos and are the second largest threat to biodiversity following direct habitat loss.

Human settlements in CCT have altered the natural fire regime in two ways that lead to vegetation change (Allsopp et al. 2014): firstly, by interrupting the natural spread of fires driven by strong summer south-easterly winds from lowland to upland areas on the eastern slopes of the Cape Peninsula, resulting in fynbos conversion to thicket and forest; and secondly, by increasing ignition frequency, particularly around informal settlements, leading to too-frequent fires that promote grasses over shrubs.

As in the CCT, urban expansion in LAC has indirectly contributed to the conversion of woody shrublands to exotic grasslands, first by providing conduits for invasion. Also, urban expansion contributes to increased human fire ignitions and fire frequency (Syphard et al. 2007), which may further be facilitated by grasslands. Weedy annual grasses are highly flammable and tolerate high fire frequency and extend the fire season. Thus, a grass-fire cycle may be establishing that could further drive vegetation type-conversion, with the potential for large-scale vegetation change (Keeley et al. 2012; Lippitt et al. 2013; Keeley and Brennan 2012). As humans cause more than 95% of the fires in the region (Syphard et al. 2007), expansion of development into wildlands has also changed the pattern and impact of wildfires, resulting in substantial social and economic losses in recent decades (Syphard et al. 2012). Ironically, fuel management aimed to protect communities by reducing fire spread may exacerbate potential conversion of shrubs to grasses by providing a conduit for invasive spread. Thus, while humans have altered fire in two ways in the CCT, increased fire frequency in LAC has been the primary change and threat. Nevertheless, the impact of short-interval fires is becoming a convergent issue among all mediterranean-type ecosystems (Syphard et al. 2009).

Concluding comparison

Despite very different current socioeconomic profiles, there have been some similarities in land-use changes between CCT and LAC. For example, in the first half of the twentieth century agriculture was the main driver of vegetation conversion in both areas. The extent of conversion to agriculture was limited by mountain ranges in both areas, and in CCT by soil fertility limitations (fynbos areas) unsuited to crop production.

Population growth and urbanization started to increase in both areas from the 1920s and again after World War II. This was more rapid in LAC compared to CCT until the 1980s, after which CCT's growth rate increased rapidly. Until recent decades, LAC urbanization resulted in the conversion of both agricultural lands and native shrublands, whereas in CCT, urbanization consistently impacted natural vegetation on infertile soils. In both areas, lowland vegetation types are under-conserved compared to mountain vegetation types, and these tend to be locations most favourable for development.

The protected area network is increasing slowly in both cities and there is scope to increase it further. However, other drivers of change, including invasion by exotic species, altered fire regimes, and intensified harvesting of flora and fauna are negatively impacting biodiversity even within protected areas. Climate change is an additional and increasing driver of change that threatens natural communities.

Although land-use and biodiversity planning are important municipal tools to prioritize areas for urban development and conservation, respectively, on the ground operational management is also essential to ensure persistence of biodiversity; for example, by reducing negative impacts of invasive exotic species and promoting a natural fire regime.

Rural people continue to move to the cities and for CCT the challenge relates to minimizing the impacts of new formal and informal settlements while trying to conserve key lowland biodiversity sites for future generations. The human population trajectory for CCT is likely to continue rising steeply over the next few decades. Despite extensive areas of existing protected lands, sprawling patterns of residential development continue in LAC, although recent demographic trends suggest that growth may be slowing.

As we move into the future, efforts to conserve increasingly restricted, threatened, and fragmented natural communities will become both more important and more difficult. Indeed, these challenges are faced across all mediterranean-type ecosystems where urban centres are located within and near biodiversity hotspots. Solutions to this conundrum will be multifaceted, and innovations from many disciplines will be required to promote a sustainable development path that includes dignified, but densified, living

Continued

Case Study 13 (*Continued*)

spaces and access to local quality open spaces that conserve fragments of highly threatened biodiversity. Climate change may also necessitate the preservation of habitat linkages that facilitate anticipated species' range shifts. For CCT, time is running out for conservation efforts as the increasing requirement for human settlements directly impacts on the few remaining and highly threatened lowland natural areas. By contrast, managing the ongoing, indirect impacts of development on existing conservation lands may present the greatest challenge in LAC. Here, biodiversity is likely to be most profoundly threatened by human-caused increases in fire frequency and exotic species.

of oaks relative to pines, while large trees have declined by up to 50 per cent (1930s–2000s) (McIntyre et al. 2015). These patterns are consistent with increases in water deficit associated with climate change, but are likely exacerbated by changes in fire regimes.

The arrival of Europeans in Chile saw a dramatic alteration of natural disturbance regimes as a result of firewood harvest, cattle grazing, and the introduction of anthropogenic fire, leading to changes in vegetation composition and structure. More recent impacts to vegetation are as a result of urban and agricultural expansion. The major vegetation type of central Chile prior to the mid-sixteenth century was a dense and diverse woodland dominated by sclerophyllous trees and shrubs (a combination of matorral and *Prosopis chilensis* woodlands). Today it is predominantly alien-dominated annual grasslands (Holmgren et al. 2000) and a uniform *Acacia caven*-dominated anthropogenic savanna called espinal (Aronson et al. 1998) (Figure 8.4; see Plate 13). *Acacia caven* is widespread in South America and has invaded owing to a combination of land-clearing practices and the introduction of domestic livestock (Ovalle et al. 1990).

Since European colonization, widespread clearing of land for agriculture (mostly cereals) and grazing has also resulted in significant loss of habitat in the Australian MTC region (Yates and Hobbs 1997). Of the five regions, Australia has seen the highest percentage of conversion to agriculture (Table 8.2; Underwood et al. 2009a). Before large-scale agricultural development in southwestern Western Australia, livestock grazing (sheep) occurred extensively, changing habitat conditions and driving biotic invasions in the remnant natural fragments of the wheat belt (Hobbs 2001). In the late 1960s, forest clear-felling increased substantially in Western Australia, leading to a rise in groundwater levels and the mobilization of salt in the soil profile (Borg et al. 1988). The conversion of native forest and its replacement with annual crops and pastures has resulted in persistent increases in saline groundwater discharge, while flooding has increased (Borg et al. 1988). Dryland salinity affected about 10 per cent of agricultural land in 2000 and

Figure 8.4 **A degraded Chilean landscape.** *Acacia caven* is visible in the foreground and as isolated individuals on the open slope (centre). These open communities are the product of increased fire on the landscape and grazing from introduced livestock. (See Plate 15)

Source: Photo from R. Brandon Pratt.

is predicted to increase in impact and extent, presenting challenges for ecological restoration (Taylor and Hoxley 2003) (see Chapter 9).

Low-nutrient, quartzitic sandstone soils of the South African Cape uplands are considered unsuitable for extensive agricultural production, so mountainous areas of this region have been spared from large-scale transformation by agriculture, apart from plantation forestry. In contrast, lowlands with their relatively richer shale-derived soils have a long history of cultivation, despite low yields (Hoffman 1997; van Wilgen et al. 2016). Cereal cropping and forage production (wheat, oats, and lucerne/alfalfa) are the major drivers of habitat conversion in these habitats and have transformed approximately a quarter of the Cape Floristic Region. Urbanization and invasion by woody alien plants have transformed less than 2 per cent of the MTC region, but are considered to be significant threats (Rouget et al. 2003; Underwood et al. 2009a). Infrastructural development, technological innovation, and demand have contributed to a steady increase in habitat transformation by agriculture, including expansion of area under vines, deciduous fruit, canola seed oil, potatoes, and rooibos tea production (van Wilgen et al. 2016). In an assessment of the conservation status of South Africa's plant species (the first assessment by a mega-diverse country), a 300 per cent increase in the number of species threatened with extinction by the rooibos industry alone was noted (from 37 threatened taxa listed in 1997 to 149 in 2009; Raimondo et al. 2009). Habitat degradation is the second largest threat to plant species after habitat loss, and includes overgrazing by livestock, altered fire regimes, and encroachment by invasive alien plants (Raimondo et al. 2009).

In conclusion, of the five MTC regions, the California region has standout levels of population density and urban area (Table 8.2), concentrated in the lowlands, and is followed by the Mediterranean, South Africa, Chile, and Australia. Agriculture has impacted large areas of MTC regions in Australia, the Mediterranean, and lowland regions of Chile and South Africa (Underwood et al. 2009a). These land-use and land-cover changes have resulted in significant transformation of the five MTC regions, but with regional differences in types and rates of change.

8.4 Habitat fragmentation

LULC drives habitat fragmentation, consequently affecting biodiversity though a range of interacting pathways. Direct effects include area reduction, which reduces population sizes and increases rates of stochastic extinction, and isolation, which restricts movement and edge effects (e.g. increased exposure to invasion and changed physical gradients). These, successively, may have cascading effects on interacting species.

Because of the LULC described in Section 8.3, all MTC regions have areas where natural habitat remains only as scattered, isolated fragments, mostly in a sea of agriculture, or urbanization in California, and often under private tenure. Certain vegetation types have been severely impacted by this transformation, particularly in lowland habitats (Underwood et al. 2009b). Less than 10 per cent of the Cape's renosterveld, a shrubland type that harbours a third of the Cape's endemic plants, remains after grain cropping (Newton and Knight 2005). In Australia, clearing of woodland, heath shrubland, and mallee has left as little as 2–3 per cent of these vegetation types in some areas (Hobbs 2001). Similarly, after agricultural expansion and urbanization in Spain's Valencia region, small habitat fragments that harbour a rich group of rare, endemic, and highly threatened plant species remain (Lumbreras 2001). The rich soils of the California Central Valley have largely been converted to agricultural uses, and remaining fragments have been heavily invaded by introduced annuals, resulting in a near complete loss of a number of shrubland, grassland, woodland, and vernal pool communities (Bartolome et al. 2007). Across the United States, California has the highest number of wildland–urban interface housing units (Radeloff et al. 2005). The physical and social barriers caused by such high levels of urbanization-led fragmentation threaten the survival of mammalian carnivores (Riley et al. 2003). These fragments of natural habitat present a special challenge to conservation planning and management (see Chapter 9). Specialist species remaining in such habitats may be subject to **extinction debt** (delayed extinctions);

that is, while they have initially survived habitat change they are still doomed to become locally or regionally extinct. In South African fynbos, a comparison between recently fragmented habitat patches associated with anthropogenically driven habitat loss <150 years ago, naturally isolated fynbos islands formed through climate-driven forest expansion in the Holocene, and extensive areas of relatively pristine 'mainland' habitat demonstrated the presence of extinction debt in fynbos plant communities, but not in bird communities (Sandberg 2013; Sandberg et al. 2016).

8.5 Fire regime changes and habitat type conversion

As noted in earlier chapters, wildfire is a key ecological and evolutionary driver in MTC regions, shaping community structure and species. However, increasing human pressures, alien plant invasion, and changing climate have altered fire regimes beyond historical ranges of variability, thereby changing community structure, threatening species, and altering ecosystem function (Figure 8.5). Too-frequent fires, owing to increased anthropogenic ignitions and incidence of weather conducive to fire, and too-infrequent fires, owing

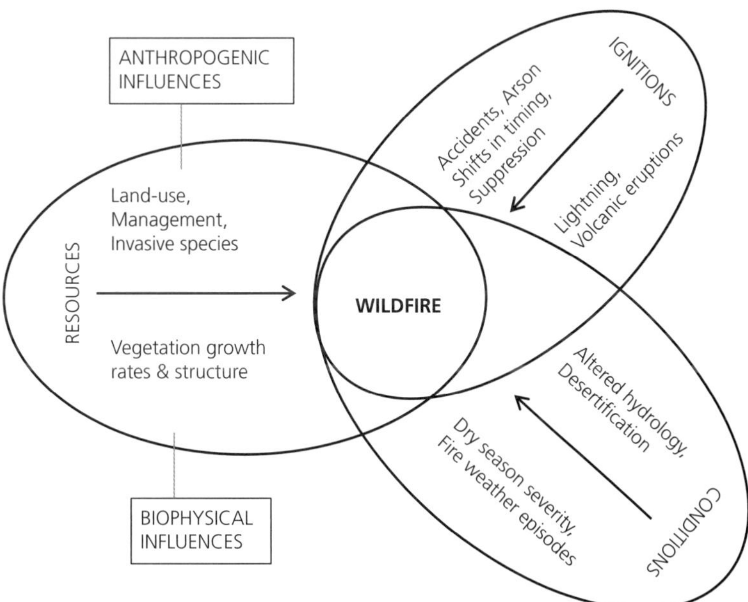

Figure 8.5   **Determinants of wildfire in mediterranean-type climate (MTC) regions.** Wildfire in MTC regions is determined by both biophysical and anthropogenic factors. Anthropogenic influences are those listed above the arrows in each ellipse. (Modified from Mann et al. 2016.)

to habitat fragmentation and policies of fire suppression, are key drivers of habitat transformation in these systems. A detailed treatment of fire ecology, evolution, and management can be found in the book by Keeley et al. (2012). Below is a brief summary of how fire regimes have changed in each MTC region and how these changes have impacted species and landscapes.

Although evidence is circumstantial, and impact is likely to have been localized owing to low population density, early South African Cape dwellers are believed to have used fire to facilitate hunting and foraging for geophytes since at least the start of the Holocene (Deacon 1983). Two thousand years ago, pastoralists may have manipulated grass dominance in the lowlands though the use of fire (renosterveld), but it is only when Europeans settled in the region in the mid-1600s that significant changes in fire regime occurred. Influenced by northern hemisphere thinking that fire would cause vegetation to deteriorate (Wicht 1945; van Wilgen et al. 2016), wildfire was initially suppressed by colonial authorities, a policy replaced in the 1970s with a prescribed burn regime. New information and political circumstance influenced fire management plans and practice (van Wilgen et al. 2016). Successive policies, namely burning to facilitate grazing, fire suppression, and promotion of natural fire regimes, appear to have had limited effect on fire patterns, which seem ultimately controlled by the rate of fuel accumulation, lightning occurrence, and weather (Seydack et al. 2007). Wilson et al. (2010), in a comparison of modelled fire periods for the Cape Floristic Region, demonstrate a 4-year decline in mean **fire-return interval** (FRI; the time between fires in a specific area) in recent decades (1951–1975, 22.75 years; 1976–2000, 18.75 years), linked to climate variability at local and global scales. Southey (2009) showed that in two regions of the Cape FRI had dropped by over half between 1970 and 2007.

Similar to Australian mediterranean-type vegetation (MTV) (Fitzpatrick et al. 2008; Hammill et al. 1998), the MTV flora of South Africa features species with small ranges, limited dispersal abilities, and sensitive life histories, including obligate reseeding shrubs that require sufficient time to mature and set seed between fires (Latimer et al. 2005; Williams et al. 2005; Schurr et al. 2007). A contraction of FRI therefore carries great significance as modern intervals are well below the time to flowering required by a swathe of long-lived obligate reseeding Proteaceae species, a functional group that acts as ecosystem engineers by providing spatially heterogeneous overstoreys (Cowling and Gxaba 1990). These species are more likely to be vulnerable under future climate change predictions of higher temperatures and lower rainfall. A shift in dominance towards species with life histories more tolerant of frequent fires, i.e. short-lived species and resprouting species, is likely.

When the structure, function, and composition of natural habitat undergoes long-term change due to changes in fire regime, land, or water use, it is termed **habitat type-conversion**. Through the use of fire, Native Americans may well have initiated habitat type conversion in coastal regions of California, but it is more recent changes associated with European/American colonization that dramatically expanded this conversion (Keeley 2002; Syphard et al. 2006; Keeley et al. 2012). Anthropogenic ignitions have increased in frequency in the twentieth century, linked to population growth (Keeley and Fotheringham 2003) and urban development. The consequences of such type-conversion are far-reaching and include impacts to native biodiversity (impacted sage scrub, for example, provides habitat for many endangered species; Davis et al. 1994), hydrological change (Williamson et al. 2004), changes in carbon storage capacity, nutrient cycling and leaking from ecosystems (Riggan et al. 1985), and changes in fire regime (Keeley et al. 2012).

Across fire-adapted MTC communities, there are common limits on how short the FRI may be before the ability of species to persist at the same level in a site is reduced. These limits include post-fire recruitment that was below replacement in coastal fynbos if FRI were less than 7 years (Kraaij et al. 2013), in Australian banksias at less than 10 years (Cowling et al. 1990), in California chaparral at less than 12 years for seeding species and with intervals of less than 6 years also threatening resprouting species persistence (Figure 8.6; Jacobsen et al. 2004; Lippitt et al. 2013), in Israel at intervals of less than 4–6 years (Tessler et al. 2015), and in Spain at FRI of less than 12 years (Malak et al. 2015). These minimum fire-return times suggest that historic fire regimes in all of these regions would have exceeded these FRIs and that, even for fire-adapted communities, fire regime is an important component in determining the resilience of communities to disturbance.

In California, 95 per cent of wildfires are currently initiated by humans (Syphard et al. 2007). Urban sprawl and incursion of housing developments into natural areas results in an uneasy relationship between people and wildfires. Damage to infrastructure is common and significant resources are directed towards fire suppression (Syphard et al. 2007), although contrasting views exist on the relative importance of fuel accumulation versus climate and weather in driving fire regimes in this region (Minnich 1983; Keeley et al. 1999). This has led to debates over the effectiveness of fuel abatement versus fire suppression as a management response. This debate may stem from the fact that two distinct fire regimes exist—those meteorologically driven by Santa Ana winds concentrated in high-wind corridors and coastal areas that do not depend on stand age, and those fires more sensitive to fuel accumulation occurring in higher elevation forests (Jin et al. 2015). Between 1900 and 2009, Santa Ana-driven fires accounted for 80 per cent of cumulative economic losses, amounting to $3.1 billion, whereas non-Santa Ana fires accounted for 70 per cent of total suppression costs. The outcome of too frequent fires in coastal areas is the widespread replacement of shrublands with herbaceous vegetation dominated by alien grasses and promoted in a grass fire cycle (Figure 8.6 and see Section 8.6.1 on invasive plants). With elevation, shrublands merge into mixed forests and then into coniferous

Figure 8.6　**Short fire-return interval (FRI) impacts.** The images are from a paired site in the Santa Monica Mountains, California from an area that burned only once in the past >80 years, in 1978 (a), and from an adjacent area that experienced a 4-year FRI, burning in 1978 and 1982 (b). Photos were taken in 2003, when both stands contained reproductively mature plants. The short FRI resulted in the loss of obligate seeding species and a decline in facultative resprouting species abundance. Gaps in the canopy formed in the site experiencing the short FRI were colonized by non-native annuals, predominantly *Bromus* and *Brassica*, and these gaps remained open 20 years after the fire. The site shown was one of the sites examined in Jacobsen et al. 2004.

Source: Photos from Anna L. Jacobsen.

forests, and the fire regime changes from one dominated by high-intensity crown fires to one featuring low-intensity surface fires.

8.6 Invasive species

Prior to European colonization, invasions were not a major threat to biodiversity in MTC regions, but in the last decades of the 1800s and early 1900s, escalating threats associated with invasions and their severity became apparent. Charles Elton (1958), widely regarded as an initiator of invasion-related research, initially proposed that invasions were directly associated with human disturbances (i.e. they were 'passengers' of disturbance; MacDougall and Turkington 2005). While this is true of many situations (e.g. Maskell et al. 2006), observations of widespread invasion of pristine ecosystems in the nutrient-poor, fire-prone Cape Region of South Africa modified the view of Elton (1958). Indeed, fire, in association with low-fertility Cape soils, enables aggressive shrubs and trees to invade otherwise pristine habitats. In these situations, invasive species can be seen as 'drivers' of disturbance (MacDougall and Turkington 2005). These early observations, originating from the third MEDECOS meeting in Stellenbosch, South Africa (1980), triggered an international programme on invasions funded by the Scientific Committee on Problems of the Environment (SCOPE) in the 1980s (Case Study 1), and led to decades of research into MTC region invasions (e.g. Groves and Di Castri 1991; Rejmánek and Randall 1994; Richardson and Pyšek 2008).

Today, invasive plant species are regarded as a major threat to MTC region biodiversity and ecosystem services (Figure 8.7; see Plate 14). A meta-analysis encompassing research in all five regions confirmed a significant decline in species richness caused by alien plant invasion, particularly at small spatial scales (Gaertner et al. 2009). Areas with a long invasion history were more strongly impacted, while, of growth forms, invasive trees caused the most decline in native species richness. Invasion is now understood to be because of **introduction dynamics** (how and when humans moved propagules around the planet), **invasibility** (the characteristics of the receiving environment), and **invasiveness** (traits of the invasive species).

8.6.1 Invasive plants

Pyšek et al. (2017), in a recent analysis of the Global Naturalized Alien Flora database (van Kleunen et al. 2015), noted that colder temperate and MTCs harbour nearly double as many naturalized aliens (19 per cent) compared to arid temperate, subtropical, and tropical climates (10 per cent). The five MTC regions differ considerably in their vulnerability to plant invasions (or community invasibility), the species that invade, and their contribution of

Figure 8.7   **Invasive alien plants.** Many non-native plants have been introduced into mediterranean-type climate (MTC) regions, have established themselves, and have subsequently spread into new habitats, displacing native species and causing harm in their introduced communities. Invasive species may be either woody or herbaceous and have impacted all five regions; most invasive species are from one of the other regions. Some examples of invasive species within MTC regions are shown, including red-eyed wattle (*Acacia cyclops*; a), which is native to Australia and invasive in California and South Africa; Monterey pine (*Pinus radiata*; b), which is native to California and invasive in South Africa; tree tobacco (*Nicotiana glauca*; c), which is native to Bolivia and Argentina and invasive in California and the Mediterranean; *Eucalyptus* spp. (d), which are native to Australia and invasive in some regions of California; brome (*Bromus* spp.) and oat (*Avena* spp.; e)—yellow grasses in both the foreground and background (Plate 14)—which are native to the Mediterranean and invasive in California; fountain grass (*Pennisetum setaceum*; f), which is native to northeastern Africa and invasive in California and South Africa; California poppy (*Eschscholzia californica*; g), which is native to California and invasive in Chile; and arum lily (*Zantedeschia aethiopica*; h), which is native to South Africa and invasive in Australia. (See Plate 16)

Source: Photo e from R. Brandon Pratt and all other photos from Anna L. Jacobsen.

non-native plant species to other MTC regions, with new world MTC regions richer in naturalized alien plants than old world ones (Pyšek et al. 2017).

The long history of agriculture, livestock grazing, and human settlement in the Mediterranean likely drove selection for invasive traits, e.g. rapid growth in open habitats and animal dispersal (Malo and Suarez 1997), among the species pool. These disturbance-adapted species provide the source for many species that invade other MTC regions, especially given the numerous trade links that Europe has with other parts of the world. Invading plant species are not considered a major challenge in most Mediterranean systems, except for wetlands and islands, an observation backed up by Pyšek et al. (2017). Between 1 per cent (Blondel et al. 2010) and 5 per cent (di Castri 1989; Le Floc'h et al. 1990) of the Mediterranean flora are non-native species, compared, for example, to approximately 20 per cent in California (Keeley et al. 2012). Because of the long association with people, Mediterranean invasive species are categorized into old (archaeophytes, introduced before 1500, typically weeds of arable land) and new (neophytes, typically escapees from crops and horticultural plants) groupings (Pyšek 1998). Most invasive species in the Mediterranean are weedy agricultural or roadside species that do not aggressively invade (Arianoutsou et al. 2010), and neophytes are frequently associated with the same habitats as archaeophytes (Chytrý et al. 2008). Shrublands in this region, dominated by rapidly resprouting shrubs, appear to have a higher level of resistance to disturbance and alien invasion (Trabaud 1990; Keeley et al. 2012), although, under increasing levels of disturbance and global change, it is predicted that invasive plants and their impacts will proliferate in the future (Lloret et al. 2004b; Gritti et al. 2006). Invasions in the other four MTC regions point towards their colonial history and, to some extent, trade links.

Because of its colonial history and trade links, California has more naturalized alien species from a wider variety of source countries compared with Chile (Arroyo et al. 2000; Jiménez et al. 2008). A large portion of the foothill and coastal ranges of California have undergone habitat-type conversion from chaparral and sage scrub to grasslands dominated by alien grasses such as wild oat, *Avena fatua*, several brome species, *Bromus* spp., and also forbs (e.g. thistles *Carduus pycnocephalus* and *Centaurea solstitialis*, fennel, *Foeniculum vulgare*, and mustard, *Brassica nigra*) from the Mediterranean (e.g. Freudenberger et al. 1987). Challenges with alien annual grasses have also been reported in Australia and Chile (Pignatti et al. 2002). These species are promoted in a grass fire cycle. As human-ignited fires become more frequent, seeder species that need time to build up their seed banks decline (Jacobsen et al. 2004), grasses invade, and their rapid fuel build-up and impact on soil water-holding capacity promotes further burning (Williamson et al. 2004). In California, the impact and extent of herbaceous invasive species, thought to have type-converted over 9 million hectares of native grassland and shrubland, is much greater than that of woody plants.

The Chilean MTC region, like California, has also experienced type-conversion to alien grasslands, and from the same set of Mediterranean grasses and forbs (Holmgren et al. 2000; Bustamante et al. 2005). Along with a novel set of herbivores (rabbits, cattle, horses, sheep, goats), these alien grasses were introduced by Spanish colonists who modified the landscape by clearing shrubland (Fuentes and Hajek 1979). Anthropogenic fires introduced to the area interact with these alien species to generate annual grass fire cycles, much like that described above for California (Keeley et al. 2012). The invasive woody legume species gorse, *Ulex europaeus*, and French broom, *Teline (Genista) monspessulana*, have also been favoured by fire in south-central Chile (Pauchard et al. 2008). Of the 690 alien plant species naturalized in continental Chile, 73 per cent are found in Central Chile; these are predominantly annuals with a Eurasian-North African origin (Arroyo et al. 2000).

In contrast to the three MTC regions discussed above, where herbaceous invasives predominate, South African fynbos has been extensively invaded by woody species such as Australian *Acacia*, *Eucalyptus*, and *Hakea* (shrubs in the family Proteaceae), and *Pinus* from California (Richardson et al. 1994). Insights into *Acacia* invasion can be found in a special issue on 'Human-mediated introductions of Australian acacias—a global experiment in biogeography' (Volume 17(5), Diversity and Distributions, 2011). *Pinus* and *Eucalyptus* were introduced by European colonizers to supply wood products (1652 onwards), while *Acacia* spp. were actively introduced to stabilize dunes and combat soil erosion (1830 onwards) (Richardson et al. 2003). The majority of invasive species in South Africa originate from southern temperate regions, notably Australia. Apart from experiencing similar ecological and cultural drivers (both were colonies of Britain), these MTC regions were connected via sea route stopover between Australia and Europe in the nineteenth century, prior to the opening of the Suez Canal (Wilson et al. 2014).

The woody invasive taxa in South Africa are considered to be **transformer species**, a subset of invasive plants that alter the composition, structure, and function of the ecosystem that they invade (Richardson et al. 2000). The life histories of these species, fast growing, prolific seed production, fire adaptation, are such that within a few fire cycles, closed stands can form, impacting biodiversity, hydrology, fire frequency, and biogeochemical cycles through the introduction of novel traits such as nitrogen fixation. The impacts of woody species have been extensively studied (e.g. Richardson and van Wilgen 2004; Gaertner et al. 2009; Richardson et al. 2014; Rundel et al. 2014). In particular, their significant impact on water supply (Le Maitre et al. 1996; de Lange and van Wilgen 2010) provided partial motivation for the establishment, in 1995, of the world's longest running invasive alien clearing programme, Working for Water (van Wilgen and Wannenburgh 2016) (see Chapter 9). While the focus of this programme is primarily on woody invasive species, the environmental problems associated with the invasion of grasses and other herbaceous species are gaining attention (e.g. Musil et al. 2005; Sharma et al. 2010), especially with concerns of increased invasion under future global change scenarios. After clearing *Acacia* spp., the legacy of elevated soil nitrates facilitates the dominance of herbaceous secondary invaders, hindering biodiversity recovery (see Chapter 9). Excellent overviews of the development of the scientific understanding and management of invasive alien plants in this region are provided by Wilson et al. (2014) and van Wilgen et al. (2016).

Colonial history is particularly relevant to invasion in Australia where local organizations deliberately introduced a wide range of species as ornamentals, for food and fodder, and medicinal use (Murray and Phillips 2012). Herbaceous species, including many geophytes from South Africa (e.g. *Romulea rosea, Morea flacida*, Arum lily *Zantedeschia aethiopica*), are particularly problematic, reducing species richness and altering successional pathways (Adair and Groves 1998). In policy and management spheres they are widely recognized as weeds threatening conservation, but few studies have measured environmental impact to biodiversity or ecological function.

8.6.2 Invasive animals

The primary focus on invasions in MTC regions has been on plants. Invasive animals have received less attention, but there are some notable species and impacts, primarily through predation on, and/or competition with, native species animal and plant communities.

The Argentine ant, *Linepithema humile*, has become established in all MTC regions (Ness and Bronstein 2004). The invasion by this human commensal species has been facilitated by habitat fragmentation and has potential to cause ecosystem-wide effects (Bolger et al. 2000). By displacing native ants through interference and exploitation competition it disrupts food webs and mutualisms. In California, horned lizard *Phrynosoma coronatum* populations have declined because Argentine ants have displaced their native ant food source (Suarez and Case 2002). In South Africa and California, seed dispersal of myrmecochorous (ant-dispersed) plants is disrupted by Argentine ants that, unlike native species, abandon seeds on the soil surface instead of buriying them, with compounding effects to plant community structure (Bond and Slingsby 1984; Christian 2001; Carney et al. 2003).

The most widely introduced vertebrates to MTC regions are fish—136 freshwater fish species have been introduced to MTC regions globally (Marr et al. 2013). Introduced predatory fish (e.g. rainbow trout *Oncorhynchus mykiss*, small mouth bass *Micropterus dolomieu*) eliminate native (often smaller bodied) endemic fish populations, shifting functional community composition. These shifts may have subtle down-stream impacts. For example, large-bodied, long-lived invasive fish may alter nutrient cycling in freshwater systems, reducing their transport to estuaries and inshore marine systems (Marr et al. 2013).

Effects of invasive animals have been particularly acute in Chile, which has seen large scale conversion of its landscape owing to the combined introduction of non-native herbivores (rabbits, hares, and domestic livestock) and herbaceous alien flora in the nineteenth century. Cattle likely facilitated the shift to *Acacia caven* savanna that is now dominant between the Coastal Range to the west and the Andes to the east, while rabbits prevent natural regeneration of matorral remnants on hillslopes (Holmgren 2002). In Australia, feral vertebrates such as cats *Felis catus* and foxes *Vulpes vulpes* have been actively controlled since the 1950s using predator baits (Calver and King 1986). Dramatic recovery of threatened native mammals has been recorded after such treatment (e.g. Friend 1990).

8.6.3 Pathogens

The study of pathogens across MTC regions is not well established, and much is to be learned in the era of DNA barcoding (a method that uses short genetic sequences from standard sections of the genome to identify species). The ecological role of pathogens in MTC regions is wide open for study and is relevant, since they may have significant negative effects, especially when interacting with other stresses. Native pathogens, for example, are thought to influence plant community structure and diversity; for example, the plant pathogen *Phellinus weirii* creates gaps in mixed conifer forests of the temperate biome, affecting the structure and ultimately increasing diversity (Hansen and Goheen 2000). In California, recent drought appears to have increased the susceptibility of a dominant coastal shrub species to a native pathogen, resulting in widespread mortality of *Malosma laurina* due to *Botryosphaeria dothidea* (Figure 8.8; see Plate 15; Aguirre et al. 2017). Most literature focuses on the transformation impacts of plant pathogens, which are believed to pose a major current and future threat to biodiversity in several MTC regions.

Characteristically spatially and temporally variable in their impacts, pathogens have the potential to drive rare and threatened plants to extinction (Fisher et al. 2012) and to cause long-term shifts in plant community composition, structure, and function (Castello et al. 1995; Shearer et al. 2007; 2013), with additive effects through the loss of flora-dependent fauna (Garkaklis et al. 2004; Bishop et al. 2010). In the southwestern Australian Floristic Region, approximately 40 per cent of plant species (2284 of 5710) are at risk from *Phytophthora cinnamomi* infection. The families Proteaceae, Ericaceae, and Papilionaceae are particularly susceptible to this fungal root disease, while the Myrtaceae, Mimosaceae, and Cyperaceae appear more resistant to infection (Shearer et al. 2004; 2013). *Phytophthora cinnamomi* is also widespread throughout the MTC region of South Africa (von Broembsen 1984), although its impacts on natural plant communities are poorly understood. *Phytophthora cinnamomi* has been linked to cork *Quercus suber* and Holm oak *Quercus*

Figure 8.8 **Pathogen impacts on mediterranean-type climate region plants.** Plants may exhibit impacts from either native or introduced pathogen species and these impacts may lead to changes in species dominance, abundance, or interactions that have large impacts on communities and ecosystems. In California, recent drought has altered the ability of plants to respond to a native pathogen, resulting in large-scale dieback and death of a local dominant shrub species. a) *Malosma laurina* dieback resulting from the pathogen *Botryosphaeria dothidea*. In Australia, introduction of the pathogen *Phytophthora cinnamomi* has resulted in widespread plant dieback, with a large percentage of the flora sensitive to this pathogen. b) A granite outcrop within Jarrah forest is shown; the Jarrah woodland has been heavily impacted by *Phytophthora* dieback. In South Africa, an introduced pathogen greatly impacted post-fire recovery in a species within an endemic plant family, Bruniceae (c). (See Plate 17)

Source: Photo in panel a from R. Brandon Pratt and photos in panels b and c from Anna L. Jacobsen.

ilex decline in southern Europe (Brasier 1996; Robin et al. 1998), while *P. ramorum* (sudden oak death) threatens mature tanoak *Notholithocarpus densiflorus* trees in California's oak woodlands (Cobb et al. 2012). Dry season dieback and mortality has been associated with a *Phytophthora* relative *Pythium* sp. in South African fynbos (Jacobsen et al. 2012). The latter species, previously undescribed, is recorded to have had local impacts on the widespread, emblematic, and sometimes dominant genera *Brunia* (Bruniaceae), *Leucadendron* (Proteaceae), and *Erica* (Ericaceae) (Figure 8.8; see Plate 15). Given concerns over the particular vulnerability of MTC regions to climate change, and the links between climate change and increased susceptibility to biotic stressors (Garrett et al. 2006; Cahill et al. 2008; Burgess et al. 2017), the threats associated with pathogens are of grave concern to MTC region conservation.

8.7 Nutrient enrichment

Anthropogenic activities have increased nutrient inputs into MTC systems in a variety of ways. Large-scale agriculture, reliant on fertilizers containing nitrogen (N) and phosphorus (P), elevates nutrients in adjacent remnant natural habitat fragments (Sharma et al. 2010), while extensive ploughing increases exposure of top soil to the atmosphere, exporting these nutrients as dust (Prospero et al. 1996). In Australia, widespread spraying of natural areas with phosphite to reduce impacts by the pathogen *Phytophthora cinnamomi* is a further source of elevated soil P (Lambers et al. 2013). Urban areas are also sources of pollutants. Anthropogenic nitrogen deposition can be significant downwind of major cities or in basins that develop temperature inversions that trap pollutants (Bobbink et al. 2010). These elevated nutrient inputs threaten sensitive species, change species composition, and increase invasion by herbaceous alien species.

Strongly leached soils, low in mineral nutrients, are characteristic of Australian and South African MTC regions (Chapter 7). In part, the high levels of plant diversity in these regions have been explained by adaptations to survive under these low-nutrient soils (Rundel et al. 2016). For example, cluster roots (Figure 3.4a), characteristic of Proteaceae and several other plant species, allow these plants to mobilize low levels of soil P through chemical means. Differential capacities to downregulate the P-uptake system means that certain species are highly sensitive to nutrient enrichment (Shane et al. 2004). These species are inferior competitors when soil P concentrations are elevated, leading to changes in species composition and plant diversity (Lambers et al. 2013).

The impact of elevated nitrogen on MTV has been little studied, but the levels of deposition have been examined. In the foothills and mountains of the Los Angeles Basin deposition levels are as high as 25–45 kg ha^{-1} year^{-1} (Bytnerowicz and Fenn 1996). Estimates put nitrogen deposition rates at 6–13 kg ha^{-1} year^{-1} in the Cape metropolitan area of South Africa (Wilson et al. 2009). Increases in herbaceous alien grasses, declines in native species, and impacts to mycorrhizal communities can occur at the relatively low loads of 10–15 kg N ha^{-1} (Bobbink et al. 2010). In MTC regions with large metropolitan areas or other forms of pollution generation, and topography that promotes temperature inversions, high levels of nitrogen deposition are likely. A risk assessment of G200 ecoregions (including the mediterranean-type biome) shows that MTC regions are already losing biodiversity due to N enrichment, and increased future deposition rates are likely to exacerbate this problem (Bobbink et al. 2010).

8.8 Climate change

Climate projections suggest that anthropogenic temperature increases, initiated and recorded in the twentieth century, will continue their upward trend (IPCC 2013). Despite some variability in predictions, a poleward shift in westerly rain-bearing systems may well result in a drier future for MTC regions, but the amplitudes in extremes will certainly increase. A variable, but overall drier future, with increasing temperatures, is therefore predicted for MTC regions (southwestern Australia: Bates et al. 2008; California: Seager et al. 2007; Chile: Magrin et al. 2014; East Mediterranean: Evans 2009; Kafle and Bruins 2009; South Africa: Hewitson and Crane 2006). Current regional climate data generally support these projections; for example, temperatures in the Mediterranean are currently ~1.3°C higher than during 1880–1920, compared to a global increase of ~0.85°C (Guiot and Cramer 2016), while variable, topographically nuanced trends in rainfall are noted (South Africa: Hoffman et al. 2011; California: McLaughlin et al. 2017).

Despite spatially heterogeneous projections of future water availability, global (Klausmeyer and Shaw 2009) and regional (Australia: Fitzpatrick et al. 2008; California: Loarie et al. 2008; Chile: Gutierrez et al. 2008; Mediterranean: Garzón et al. 2008; Kaniewski et al. 2014; South Africa: Midgley et al. 2002; Hannah et al. 2005) projections predict that changes in climate are likely to cause biodiversity loss in the decades to come. In a global comparison of projected spatial shifts in MTC extent, Klausmeyer and Shaw (2009) predicted a moderate expansion in global MTC, but uneven regional shifts explained by physical variation (Figure 8.9). Large projected contractions are predicted in South Africa and Australia where there is no contiguous land for southward expansion, whereas in California, Chile, Argentina, Greece, Turkey, Spain, and Portugal, topographic diversity and contiguous land potentially allows for future expansion. Since time to adapt and environmental

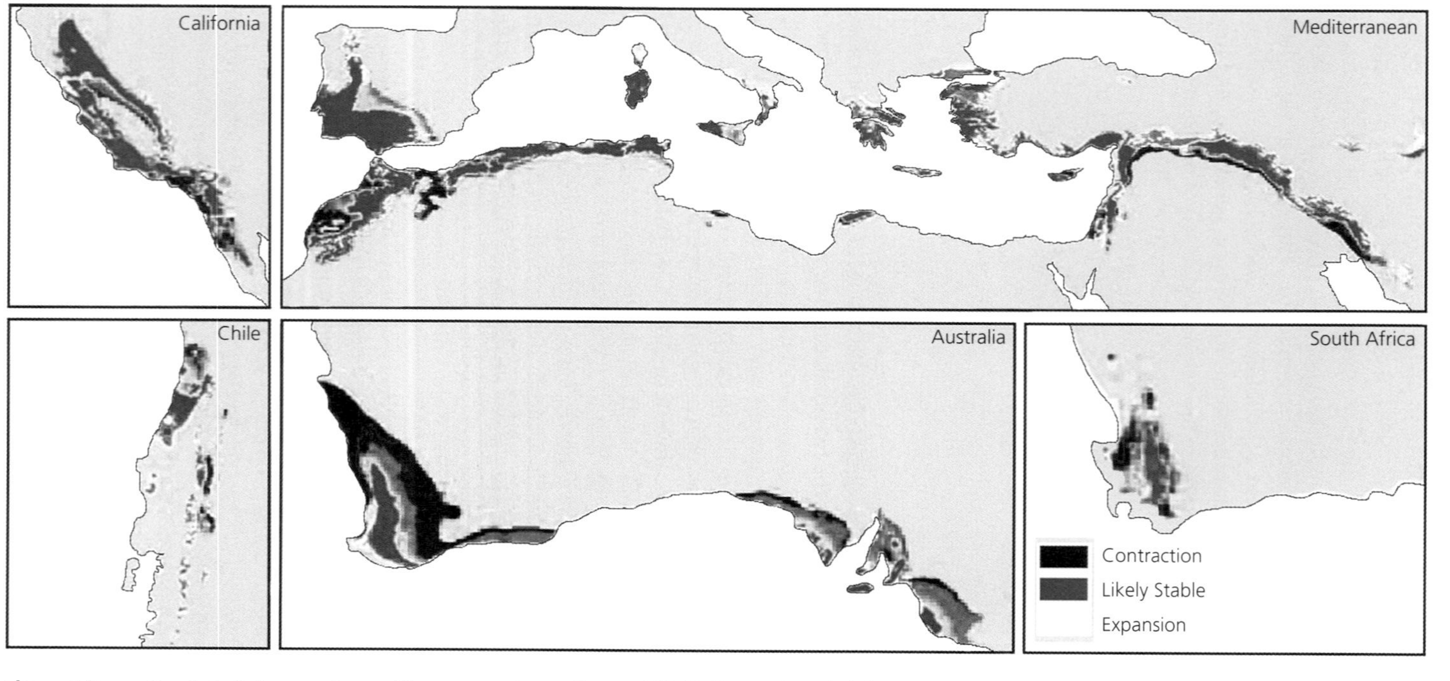

Figure 8.9 **Predicted changes in mediterranean-type climate (MTC).** The projected shifts in MTC extent over the next century (2070–2099 relative to 1960–1989) when modelled using a high-emissions scenario. Projected changes are for large amounts of climate loss (contraction), especially in the less topographically diverse MTC regions (shown as black) and for only minimal regions of MTC expansion (lightest shade of grey; very little of this shade is present). (Modified from Klausmeyer and Shaw 2009.)

(climatic and geologic) stability are significant factors in promoting relative plant species richness among MTC regions (Cowling et al. 2015; Rundel et al. 2016), these predicted changes are a major cause for concern. Predicted impacts may extend to other taxa. For example, low climatic variability and environmental stability are also linked to endemic bird species richness in southern Africa (Huntley et al. 2016). In a global review, Urban (2015) concluded that extinction risk will accelerate with increasing temperatures, and that this risk is greatest in regions that harbour diverse assemblages with high levels of endemism facing no-analogue climates.

Studies focusing on MTC regions have increasingly shown declines in ecosystem health in relation to drying and warming (Figure 8.10; see Plate 16). Drought-induced dieback and mortality has been recorded in Australia (Brouwers et al. 2013; Brouwers and Coops 2016; Matusick et al. 2012; 2013; Ruthrof et al. 2015b), California (Davis et al. 2002; Paddock et al. 2013; Venturas et al. 2016), South Africa (Hannah et al. 2007), and the Mediterranean (Peñuelas et al. 2001; Lloret et al. 2004b; del Cacho and Lloret 2011). Because of this drying and dieback, more frequent, more severe, and/or more intense fires result (Australia: Ruthrof et al. 2015a; California: Fried et al. 2004; Lenihan et al. 2003; Mediterranean: Symeonakis et al. 2015; Pausas 2004; South Africa: Wilson et al. 2010). The interaction between climate change and fire is not uniform across MTC region landscapes, as topography and vegetation type influence fire, and some fire regimes are more strongly controlled by other anthropogenic impacts such as management policy (Keeley and Syphard 2016). Lowlands are more likely to be impacted than uplands, where more refugia exist for species with contracting ranges (Klausmeyer and Shaw 2009; Beier and Brost 2010). Fragmented habitat, isolated by urbanization and agriculture, will be under greater stress as extremes are reached (Jump and Peñuelas 2005).

Some studies have found that the class of species most vulnerable to mortality during droughts are post-fire obligate seeding species (Causley et al. 2016; Paddock et al. 2013; Venturas et al. 2016; Figure 8.10a; see Plate 16; see Chapter 6). Among woody species in the fire-prone MTC regions, this group of species contains the greatest number of species and endemic taxa. If obligate seeders across MTC regions are generally the most vulnerable to more extreme low annual precipitation events and warming and drying then this could pose a major threat to biodiversity of the emblematic MTV. As already discussed, obligate seeders are also the most vulnerable to short FRIs. For these reasons, obligate seeders need to be a conservation and management priority.

While obligate seeders are commonly more vulnerable to drought there is the potential for fire and drought to interact, which may strain resprouters as well as seeders. Following a fire, resprouting is fuelled by stored carbohydrates in specialized woody structures (lignotubers; Chapter 3) and the root system. During drought, particularly long-term ones that last multiple years,

carbohydrate reserves may become diminished because stomatal closure during drought limits uptake of CO_2 and photosynthesis (Chapter 6; McDowell et al. 2008; Pausas et al. 2016). Because of this, if a fire occurs in the midst of a drought, resprouting may be limited owing to lack of stored reserves (Pausas et al. 2016). Also, if an intense drought occurs following a fire, resprouters may be affected by water limitations and may suffer high levels of mortality (Pratt et al. 2014). One of the reasons for this is that resprouters have tissues that are more sensitive to dehydration than comparable plants that are established and unburned. For example, resprouting tissues often shift to being more vulnerable to dehydration-induced cavitation of the vascular system (Chapter 6; Jacobsen et al. 2016; Pratt et al. 2014). Their leaf tissue is also more sensitive and loses turgor pressure (healthy levels of leaf hydration) more readily (Saruwatari and Davis 1989). One of the challenges of understanding drought effects on resprouting is that it can be difficult to separate drought effects from fire damage. The effect of drought can be apparent when plants survive to resprout after fire and subsequently suffer mortality during drought (Pratt et al. 2014). However, if individuals do not resprout after fire, it is commonly assumed that the fire was the agent of mortality. This assumption may not always be valid. Researchers found that, by using structures to capture rainfall to induce a drought, greater numbers of the chaparral resprouting species *Adenostoma fasciculatum* did not resprout after fire when compared to control plots (no rainfall manipulation) and plots that received double the normal rainfall (Pausas et al. 2016). The pattern of resprouting survival was greatest in the double-watered plots (~60 per cent), intermediate in the control plot (~40 per cent), and lowest in the drought treatment (~25 per cent). These results indicate that drought can be a key factor in resprout success and the interaction of drought and fire may lead to changing landscapes in MTC regions.

Increased frequency of fires under drying and warming conditions has led to habitat type conversion; i.e. permanent alterations in vegetation structure, function, and composition. In Australia, Brouwers and Coops (2016) recorded a shift from taller tree-dominated vegetation to lower shrub-dominated vegetation, whilst in southern California, coastal sage scrub has been extirpated from 80 to 85 per cent of its pre-European range, replaced by non-native grasslands (Reid and Murphy 1995). These climate-driven dieback and fire–climate interactions are likely to impact ecosystem processes such as nutrient cycling and carbon sequestration (Neary et al. 1999).

Figure 8.10   **(Opposite) Drought-induced mortality.** Recent droughts have led to dieback and mortality in mediterranean-type climate (MTC) regions. Panel (a) shows a die-off event for an obligate seeder *Banksia carlinoides* in the Eneabba sandplain regions of Australia; (b) shows mortality of chaparral shrubs during recent droughts (Venturas et al. 2016); and (c) shows the massive pine tree mortality that has occurred in California MTC areas during recent droughts. (See Plate 18)

Source: Photos from R. Brandon Pratt.

8.9 Conclusion

Extensive habitat loss and habitat conversion has occurred across all MTC regions, driven by increasing human populations, which have converted large tracts of land to production, transport, and residential use. Remaining habitat, often in isolated or remote (mountainous) areas, is threatened by altered fire regimes, introduction of invasive species, nutrient enrichment, and climate change. These drivers of change show few signs of abatement. Opportunities to mitigate future loss of biodiversity exist through conservation actions and ecological restoration, the topic of Chapter 9.

Literature cited

Adair RJ, Groves RH. 1998. Impact of environmental weeds on biodiversity: a review and development of a methodology. Environment Australia, Canberra.

Agnoletti M. 2007. The degradation of traditional landscape in a mountain area of Tuscany during the 19th and 20th centuries: implications for biodiversity and sustainable management. Forest Ecology and Management 249: 5–17.

Aguirre NM, Ochoa ME, Holmlund HI, Ewers FE, Davis SD. 2017. Hydraulic mechanisms of fungal-induced dieback in a keystone chaparral species during unprecedented drought in California. http://digitalcommons.pepperdine.edu/scursas/2017/posters/26/ (accessed April 16, 2017).

Allen EB, Padgett PE, Bytnerowicz A, Minnich R. 1998. Nitrogen deposition effects on coastal sage vegetation of southern California. Proceedings of the International Symposium on air pollution and climate change effects on forest ecosystems. USDA forest service, Pacific Southwest Research Station, Riversdale, California. https://www.fs.fed.us/psw/publications/documents/psw_gtr166/psw_gtr166_002_allen.pdf (accessed June 7, 2017).

Allsopp N, Anderson PML, Holmes PM, Melin A, O'Farrell PJ. 2014. People, the Cape Floristic Region, and sustainability. Pages 337–60 in Allsopp N, Colville JF, Verboom GA, eds. Fynbos: ecology, evolution, and conservation of a megadiverse region. Oxford University Press, Oxford.

Ambrose SH. 1998. Late Pleistocene human population bottlenecks, volcanic winter, and differentiation of modern humans. Journal of Human Evolution 34: 623–51.

Antrop M. 2004. Landscape change and the urbanization process in Europe. Landscape and Urban Planning 67: 9–26.

Arianoutsou M, Delipetrou P, Celesti-Grapow L, Basnou C, Bazos I, Kokkoris Y, Blasi C, Vilà M. 2010. Comparing naturalized alien plants and recipient habitats across an east–west gradient in the Mediterranean Basin. Journal of Biogeography 37: 1811–23.

Aronson J, Del Pozo A, Ovalle C, Avendaño J, Lavin A, Etienne M. 1998. Land use changes and conflicts in central Chile. Pages 155–68 in Rundel PW, Montenegro G, Jaksic F, eds. Landscape degradation and biodiversity in mediterranean-type ecosystems. Springer-Verlag, Berlin, Heidelberg.

Arroyo MTK, Marticorena C, Matthei O, Cavieres L. 2000. Plant invasions in Chile: present patterns and future predictions. Pages 385–424 in Mooney HA, Hobbs RJ, eds. Invasive species in a changing world. Island Press, Covelo.

Bartolome JW, Barry WJ, Griggs T, Hopkinson P. 2007. Valley grassland. Terrestrial Vegetation of California 3: 367–93.

Bates B, Hope P, Ryan B, Smith I, Charles S. 2008. Key findings from the Indian Ocean Climate Initiative and their impact on policy development in Australia. Climatic Change 89: 339–54.

Beier P, Brost B. 2010. Use of land facets to plan for climate change: conserving the arenas, not the actors. Conservation Biology 24: 701–10.

Bishop CL, Wardell-Johnson GW, Williams MR. 2010. Community-level changes in Banksia woodland following plant pathogen invasion in the Southwest Australian Floristic Region. Journal of Vegetation Science 21: 888–98.

Blondel J, Aronson J, Bodiou JY, Boeuf G. 2010. The Mediterranean region. Biological diversity though time and space. Oxford University Press, Oxford.

Blum MGB, Jakobsson M. 2011. Deep divergences of human gene trees and models of human origins. Molecular Biology and Evolution 28: 889–98.

Bobbink R, Hicks K, Galloway J, Spranger T, Alkemade R, Ashmore M, Bustamante M, Cinderby S, Davidson E, Dentener F, Emmett B. 2010. Global assessment of nitrogen deposition effects on terrestrial plant diversity: a synthesis. Ecological Applications 20: 30–59.

Bolger DT, Suarez AV, Crooks KR, Morrison SA, Case TJ. 2000. Arthropods in urban habitat fragments in southern California: area, age, and edge effects. Ecological Applications 10: 1230–48.

Bond W, Slingsby P. 1984. Collapse of an ant-plant mutalism: the argentine ant (*Iridomyrmex humilis*) and myrmecochorous Proteaceae. Ecology 65: 1031–7.

Borg H, Stoneman GL, Ward CG. 1988. The effect of logging and regeneration on groundwater, streamflow and stream salinity in the southern forest of Western Australia. Journal of Hydrology 99: 253–70.

Bourgeon L, Burke A, Higham T. 2017. Earliest human presence in North America dated to the Last Glacial Maximum: new radiocarbon dates from Bluefish Caves, Canada. PloS One 12.1: e0169486.

Brasier CM. 1996. *Phytophthora cinnamomi* and oak decline in Southern Europe: environmental constraints including climate change. Annales des Sciences Forestieres 53: 347–58.

Brouwers NC, Coops NC. 2016. Decreasing net primary production in forest and shrub vegetation across southwest Australia. Ecological Indicators 66: 10–19.

Brouwers NC, Mercer J, Lyons T, Poot P, Veneklaas E, Hardy G. 2013. Climate and landscape drivers of tree decline in a mediterranean ecoregion. Ecology and Evolution 3: 67–79.

Burgess TI, Scott JK, McDougall KL, Stukely MJ, Crane C, Dunstan WA, Brigg F, Andjic V, White D, Rudman T, Arentz F. 2017. Current and projected global distribution of *Phytophthora cinnamomi*, one of the world's worst plant pathogens. Global Change Biology 23: 1661–74.

Bustamante R, Pauchard A, Marticorena AJA, Cavieres L. 2005. Alien plants in Mediterranean-type ecosystems of the Americas: comparing floras at a regional and local scale. Pages 89–97 in Brunel S, ed. Invasive plants in Mediterranean-type regions of the world. Council of Europe Publ., Strasbourg, France.

Butzer KW. 2005. Environmental history in the Mediterranean world: cross-disciplinary investigation of cause and-effect for degradation and soil erosion. Journal of Archaeological Science 32: 1773–800.

Bytnerowicz A, Fenn ME. 1996. Nitrogen deposition in California forests: a review. Environmental Pollution 92: 127–46.

Cahill DM, Rookes JE, Wilson BA, Gibson L, McDougall KL, 2008. *Phytophthora cinnamomi* and Australia's biodiversity: impacts, predictions and progress towards control. Australian Journal of Botany 56: 279–310.

Calver MC, King DR. 1986. Controlling vertebrate pests with fluoroacetate: lessons in wildlife management, bio-ethics, and co-evolution. Journal of Biological Education 20: 257–62.

Caravello GU, Giacomin F. 1993. Landscape ecology aspects in a territory centuriated in Roman times. Landscape and Urban Planning 24: 77–85.

Carney SM, Brooke B, Holway D. 2003. Invasive Argentine ants (*Linepithema humile*) do not replace native ants as seed dispersers of *Dendromecon rigida* (Papaveraceae) in California, USA. Oecologia 135: 576–82.

Castello JD, Leopold DJ, Smallidge PJ. 1995. Pathogens, patterns, and processes in forest ecosystems. BioScience 45: 16–24.

Castro-Díez P, Godoy O, Saldaña A, Richardson DM. 2011. Predicting invasiveness of Australian acacias on the basis of their native climatic affinities, life history traits and human use. Diversity and Distributions 17: 934–45.

Causley CL, Fowler WM, Lamont BB, He T. 2016. Fitness benefits of serotiny in fire- and drought-prone environments. Plant Ecology 217: 773–9.

Christian CE. 2001. Consequences of a biological invasion reveal the importance of mutualism for plant communities. Nature 413: 635–9.

Chytrý M, Maskell LC, Pino J, Pyšek P, Vilà M, Font X, Smart SM. 2008. Habitat invasions by alien plants: a quantitative comparison among Mediterranean, subcontinental and oceanic regions of Europe. Journal of Applied Ecology 45: 448–58.

Cincotta RP, Wisnewski J, Engelman, R. 2000. Human population in the biodiversity hotspots. Nature 404: 990–2.

Cobb RC, Filipe JA, Meentemeyer RK, Gilligan CA, Rizzo DM. 2012. Ecosystem transformation by emerging infectious disease: loss of large tanoak from California forests. Journal of Ecology 100: 712–22.

Copeland SR, Cawthra HC, Fisher EC, Lee-Thorp JA, Cowling RM, le Roux PJ, Hodgkins J, Marean CW. 2016. Strontium isotope investigation of ungulate movement patterns on the Pleistocene Paleo-Agulhas Plain of the Greater Cape Floristic Region, South Africa. Quaternary Science Reviews 141: 65–84.

Cowling RM, Gxaba T. 1990. Effects of a fynbos overstorey shrub on understorey community structure: implications for the maintenance of community-wide species richness. South African Journal of Ecology 1: 1–7.

Cowling RM, Lamont BB, Enright NJ. 1990. Fire and management of south-western Australian banksias. Proceedings of the Ecological Society of Australia 16: 177–83.

Cowling RM, Potts AJ, Bradshaw PL, Colville J, Arianoutsou M, Ferrier S, Forest F, Fyllas NM, Hopper SD, Ojeda F, Procheş Ş. 2015. Variation in plant diversity in mediterranean-climate ecosystems: the role of climatic and topographical stability. Journal of Biogeography 42: 552–64.

Crutzen PJ. 2002. Geology of mankind. Nature 415: 23.

Davis F, Stine P, Stoms D. 1994. Distribution and conservation status of coastal sage scrub in southwestern California. Journal of Vegetation Science 5: 743–56.

Davis SD, Ewers FW, Sperry JS, Portwood KA, Crocker MC, Adams GC. 2002. Shoot dieback during prolonged drought in *Ceanothus* (Rhamnaceae) chaparral of California: a possible case of hydraulic failure. American Journal of Botany 89: 820–8.

Deacon HJ. 1983. The peopling of the fynbos region. Fynbos paleoecology: a preliminary synthesis. South African Scientific Programmes Report 75: 183–209.

Deacon HJ, Deacon J. 1999. Human beginnings in South Africa: uncovering the secrets of the Stone Age. David Philip Publications, Cape Town.

De Lange WJ, van Wilgen BW. 2010. An economic assessment of the contribution of biological control to the management of invasive alien plants and to the protection of ecosystem services in South Africa. Biological Invasions 12: 4113–24.

del Cacho M, Lloret F. 2011. Resilience of Mediterranean shrubland to a severe drought episode: the role of seed bank and seedling emergence. Plant Biology 14: 458–66.

De Vynck JC, Anderson R, Atwater C, et al. 2016. Return rates from intertidal foraging from Blombos Cave to Pinnacle Point: understanding early human economies. Journal of Human Evolution 92: 101–15.

Di Castri F. 1981. Mediterranean-type shrublands of the world. Pages 1–53 in Castri F, Goodall DW, Specht RL, eds. Mediterranean-type shrublands, collection: ecosystems of the world. Elsevier, Amsterdam.

Di Castri F. 1989. History of biological invasions with special emphasis on the Old World. Pages 1–30 in Drake JA, Mooney HA, di Castri F, eds. Biological invasions: a global perspective. John Wiley & Sons, New York.

Dillehay TD. 1999. The late Pleistocene cultures of South America. Evolutionary Anthropology 7: 206–16.

Dobson AP, Rodriguez JP, Roberts WM, Wilcove DS. 1997. Geographic distribution of endangered species in the United States. Science 275: 550–3.

Elton CS. 1958. The ecology of invasions by animals and plants. London, Metheun.

Evans JP. 2009. 21st century climate change in the Middle East. Climatic Change 92: 417–32.

Faith JT. 2011. Ungulate community richness, grazer extinctions, and human subsistence behavior in southern Africa's Cape Floral Region. Palaeogeography, Palaeoclimatology, Palaeoecology 306: 219–27.

Falcucci A, Maiorano L, Boitani L. 2007. Changes in land-use/land-cover patterns in Italy and their implications for biodiversity conservation. Landscape Ecology 22: 617–31.

Fischer G, Shah M, Tubiello FN, van Velhuizen H. 2005. Socio-economic and climate change impacts on agriculture: an integrated assessment, 1990–2080. Philosophical Transactions of the Royal Society B: Biological Sciences 360: 2067–83.

Fisher MC, Henk DA, Briggs CJ, Brownstein JS, Madoff LC, McCraw SL, Gurr SJ. 2012. Emerging fungal threats to animal, plant and ecosystem health. Nature 484: 186–94.

Fitzpatrick MC, Govem AD, Sanders NJ, Dunn RR. 2008. Climate change, plant migration, and range collapse in a global biodiversity hotspot: the Banksia (Proteaceae) of Western Australia. Global Change Biology 14: 1337–52.

Freudenberger DO, Fish BE, Keeley JE. 1987. Distribution and stability of grasslands in the Los Angeles basin. Bulletin of the Southern California Academy of Sciences 86: 13–26.

Fried JS, Torn MS, Mills E. 2004. The impact of climate change on wildfire severity: a regional forecast for northern California. Climatic Change 64: 169–91.

Friend JA. 1990. The numbat *Myrmecobius fasciatus* (Myrmecobiidae): history of decline and potential for recovery. Proceedings of the Ecological Society of Australia 16: 369–77.

Fuentes ER, Hajek ER. 1979. Patterns of landscape modification in relation to agricultural practice in central Chile. Environmental Conservation 6: 265–71.

Gaertner M, Den Breeÿen A, Hui C, Richardson DM. 2009. Impacts of alien plant invasions on species richness in Mediterranean-type ecosystems: a meta-analysis. Progress in Physical Geography 33: 319–38.

Gaertner M, Biggs R, Te Beest M, Hui C, Molofsky J, Richardson DM. 2014. Invasive plants as drivers of regime shifts: identifying high priority invaders that alter feedback relationships. Diversity and Distributions 20: 733–44.

Gaertner M, Larson BMH, Irlich UM, Holmes PM, Stafford L, van Wilgen BW, Richardson DM. 2016. Managing invasive species in cities: lessons from Cape Town, South Africa. Landscape and Urban Planning 151: 1–9.

Garkaklis M, Calver M, Wilson B, Hardy GE. 2004. Habitat alteration caused by an introduced plant disease, *Phytophthora cinnamomi*: a potential threat to the conservation of Australian forest fauna. Pages 899–913 in Lunney D, ed. Conservation of Australia's forest fauna, Royal Zoological Society of New South Wales, Mossman, NSW.

Garrett KA, Dendy SP, Frank EE, Rouse MN, Travers SE. 2006. Climate change effects on plant disease: genomes to ecosystems. Annual Review of Phytopathology 44: 489–509.

Garzón MB, de Dios RS, Ollero HS. 2008. Effects of climate change on the distribution of Iberian tree species. Applied Vegetation Science 11: 169–78.

Gibson M, Richardson DM, Marchante E, Marchante H, Rodger JG, Stone GN, Byrne M, Fuentes-Ramírez A, George N, Harris C, Johnson SD, Le Roux JJ, Miller JT, Murphy DJ, Pauw A, Prescott MN, Wandrag EM, Wilson JRU. 2011. Reproductive ecology of Australian acacias: important mediator of invasive success? Diversity and Distributions 17: 911–33.

Greuter W. 1994. Extinction in Mediterranean areas. Philosophical Transactions of the Royal Society London Series B 344: 41–6.

Gritti ES, Smith B, Sykes MT. 2006. Vulnerability of Mediterranean Basin ecosystems to climate change and invasion by exotic plant species. Journal of Biogeography 33: 145–57.

Groves RH, di Castri F. 1991. Biogeography of Mediterranean invasions. Cambridge: Cambridge University Press.

Guiot J, Cramer W. 2016. Climate change: the 2015 Paris Agreement thresholds and Mediterranean Basin ecosystems. Science 354: 465–8.

Gutierrez AG, Barbosa O, Christie DA, Del-Val E, Ewing HA. 2008. Regeneration patterns and persistence of the fog-dependent Fray Jorge forest in semiarid Chile during the past two centuries. Global Change Biology 14: 161–76.

Hammill KA, Bradstock RA, Allaway WG. 1998. Post-fire seed dispersal and species re-establishment in proteaceous heath. Australian Journal of Botany 46: 407–19.

Hannah L, Midgley G, Hughes G, Bomhard B. 2005. The view from the cape. Extinction risk, protected areas, and climate change. Bioscience 55: 231–42.

Hannah L, Midgley G, Andelman S, Araújo M, Hughes G, Martinez-Meyer E, Pearson R, Williams P. 2007. Protected area needs in a changing climate. Frontiers in Ecology and the Environment 5: 131–8.

Hansen EM, Goheen EM. 2000. *Phellinus weirii* and other native root pathogens as determinants of forest structure and process in western North America. Annual Review of Phytopathology 38: 515–39.

Henn BM, Gignoux CR, Jobin M, Granka JM, Macpherson JM, Kidd JM, Rodriguez-Botigue L, Ramachandran S, Hon L, Brisbin A, Lin AA, Underhill PA, Comas D, Kidd KK, Norman PJ, Parham P, Bustamante CD, Mountain J, Feldman MW. 2011. Hunter-gatherer genomic diversity suggests a southern African origin for modern humans. Proceedings of the National Academy of Science 108: 5154–62.

Hewitson BC, Crane RG. 2006. Consensus between GCM climate change projections with empirical downscaling: precipitation downscaling over South Africa. International Journal of Climatology 26: 1315–37.

Hobbs RJ. 2001. Synergisms among habitat fragmentation, livestock grazing, and biotic invasions in southwestern Australia. Conservation Biology 15: 1522–8.

Hoffman MT. 1997. Human impacts on vegetation. Pages 507–34 in Cowling R, Richardson D, Pierce S, eds. Vegetation of southern Africa. Cambridge, Cambridge University Press.

Hoffman MT, Cramer MD, Gillson L, Wallace M. 2011. Pan evaporation and wind run decline in the Cape Floristic Region of South Africa (1974–2005): implications for vegetation responses to climate change. Climatic Change 109: 437–52.

Holmes PM, Rebelo AG, Dorse C, Wood J. 2012. Can Cape Town's unique biodiversity be saved? Balancing conservation imperatives and development needs. Ecology and Society 17: 28. http://dx.doi.org/10.5751/ES-04552-170228.

Holmgren M. 2002. Exotic herbivores as drivers of plant invasion and switch to ecosystem alternative states. Biological Invasions 4: 25–33.

Holmgren M, Avilés R, Sierralta L, Segura AM, Fuentes ER. 2000. Why have European herbs so successfully invaded the Chilean matorral? Effects of herbivory, soil nutrients, and fire. Journal of Arid Environments 44: 197–211.

Huntley B, Collingham YC, Singarayer JS, Valdes PJ, Barnard P, Midgley GF, Altwegg R, Ohlemüller R. 2016. Explaining patterns of avian diversity and endemicity: climate and biomes of southern Africa over the last 140,000 years. Journal of Biogeography 43: 874–86.

IPCC. 2013. Climate change 2013: the physical science basis. 1535 pp. In Stocker TF, Qin D, Plattner G-K, Tignor M, Allen SK, Boschung J, Nauels A, Xia Y, Bex V, Midgley PM. eds. Contribution of Working Group I to the Fifth Assessment Report of the Intergovernmental Panel on Climate Change. Cambridge University Press/IPCC, Cambridge, United Kingdom/New York, NY, USA.

Jacobsen AL, Davis SD, Babritus SL. 2004. Fire frequency impacts non-sprouting chaparral shrubs in the Santa Monica Mountains of southern California. In: Arianoutsou M, Panastasis VP, eds. Ecology and conservation of mediterranean climate ecosystems. Millpress, Rotterdam, The Netherlands.

Jacobsen AL, Roets F, Jacobs SM, Esler KJ, Pratt RB. 2012. Dieback and mortality of South African fynbos shrubs is likely driven by a novel pathogen and pathogen-induced hydraulic failure. Austral Ecology 37: 227–35.

Jacobsen AL, Tobin MF, Toschi HS, Percolla MI, Pratt RB. 2016. Structural determinants of increased susceptibility to dehydration-induced cavitation in post-fire resprouting chaparral shrubs. Plant, Cell & Environment 39: 2473–285.

Jiménez A, Pauchard A, Cavieres LA, Marticorena A, Bustamante RO. 2008. Do climatically similar regions contain similar alien floras? A comparison between the Mediterranean areas of central Chile and California. Journal of Biogeography 35: 614–24.

Jin Y, Goulden ML, Faivre N, Veraverbeke S, Sun F, Hall A, Hand MS, Hook S, Randerson JT. 2015. Identification of two distinct fire regimes in Southern California: implications for economic impact and future change. Environmental Research Letters 10: doi: 10.1088/1748-9326/10/9/094005.

Jump AS, Penuelas J. 2005. Running to stand still: adaptation and the response of plants to rapid climate change. Ecology Letters 8: 1010–20.

Kafle HK, Bruins HJ. 2009. Climatic trends in Israel 1970–2002: warmer and increasing aridity inland. Climatic Change 96: 63–77.

Kaniewski D, Van Campo E, Morhange C, Guiot J, Zviely D, Le Burel S, Otto T, Artzy M. 2014. Vulnerability of Mediterranean ecosystems to long-term changes along the coast of Israel. PLoS One 9: e102090.

Kaniewski D, Giaime M, Marriner N. 2015. First evidence of agro-pastoral farming and anthropogenic impact in the Taman Peninsula, Russia. Quaternary Science Reviews 114: 43–51.

Keeley JE. 2002. Native American impacts on fire regimes of the Californian coastal ranges. Journal of Biogeography 29: 303–20.

Keeley JE, Brennan TJ. 2012. Fire-driven alien invasion in a fire-adapted ecosystem. Oecologia 169: 1043–52.

Keeley JE, Fotheringham CJ. 2003. Impact of past, present, and future fire regimes on North American Mediterranean shrublands. Pages 218–62 in Fire and climatic change in temperate ecosystems of the Western Americas. Springer, New York.

Keeley JE, Syphard AD. 2016. Climate change and future fire regimes: examples from California. Geosciences 6: 37; doi:10.3390/geosciences6030037.

Keeley JE, Fotheringham CJ, Morais, M. 1999. Re-examining fire suppression impacts on brushland fire regimes. Science 284: 1829–32.

Keeley JE, Bond WJ, Bradstock RA, Pausas JG, Rundel PW. 2012. Fire in Mediterranean ecosystems: ecology, evolution and management. Cambridge University Press, Cambridge.

Kershaw AP, Moss PT, van der Kaars S. 1997. Environmental change and the human occupation of Australia. Anthropologie 35: 35–43.

Kershaw P, van der Kaars S, Moss P, Opdyke B, Guichard F. 2006. Environmental change and the arrival of people in the Australian region. Before Farming 1: 1–24.

Klausmeyer KR, Shaw MR. 2009. Climate change, habitat loss, protected areas and the climate adaptation potential of species in Mediterranean ecosystems world-wide. PLoS One 4: e6392.

Klein RG. 1983. Palaeoenvironmental implications of Quaternary large mammals in the Fynbos region. Pages 116–38 in Deacon HJ, Hendey QB, Lambrechts JJN, eds. Fynbos palaeoecology: a preliminary synthesis. South African National Scientific Programmes Report.

Klein RG. 2001. Southern Africa and modern human origins. Journal of Anthropological Research 57: 1–16.

Kraaij T, Cowling RM, Wilgen BW, Schutte-Vlok A. 2013. Proteaceae juvenile periods and post-fire recruitment as indicators of minimum fire return interval in eastern coastal fynbos. Applied Vegetation Science 16: 84–94.

Lambers H, Ahmedi I, Berkowitz O, Dunne C, Finnegan PM, Hardy GESJ, Jost R, Laliberté E, Pearse SJ, Teste FP. 2013. Phosphorus nutrition of phosphorus-sensitive Australian native plants: threats to plant communities in a global biodiversity hotspot. Conservation Physiology 1: p.cot010.

Latimer AM, Silander JA, Cowling RM. 2005. Neutral ecological theory reveals isolation and rapid speciation in a biodiversity hot spot. Science 309: 1722–5.

Le Floc'h E, Houérou HNL, Mathez J. 1990. History and patterns of plant invasion in northern Africa. Pages 105–33 in di Castri F, Hansen AJ, Debussche M. eds. Biological invasions in Europe and the Mediterranean Basin. Springer, Netherlands.

Le Maitre DC, van Wilgen BW, Chapman RA, McKelly DH. 1996. Invasive plants and water resources in the Western Cape Province, South Africa: modelling the consequences of a lack of management. Journal of Applied Ecology 33: 161–72.

Le Maitre DC, Gaertner M, Marchante E, Ens EJ, Holmes PM, Pauchard A, O'Farrell PJ, Rogers AM, Blanchard R, Blignaut J, Richardson DM. 2011. Impacts of invasive Australian acacias: implications for management and restoration. Diversity and Distributions 17: 1015–29.

Lenihan JM, Drapek R, Bachelet D, Neilson RP. 2003. Climate change effects on vegetation distribution, carbon, and fire in California. Ecological Applications 13: 1667–81.

Lippitt CL, Stow DA, O'Leary JF, Franklin J. 2013. Influence of short-interval fire occurrence on post-fire recovery of fire-prone shrublands in California, USA. International Journal of Wildland Fire 22: 184–93.

Lloret F, Medail F, Brundu G, Hulme PE. 2004a. Local and regional abundance of exotic plant species on Mediterranean islands: are species traits important? Global Ecology and Biogeography 13: 37–45.

Lloret F, Siscart D, Dalmases C. 2004b. Canopy recovery after drought dieback in holm-oak Mediterranean forests of Catalonia (NE Spain). Global Change Biology 10: 2092–9.

Loarie SR, Carter BE, Hayhoe K, McMahon S, Moe R, Knight CA, Ackerly DD. 2008. Climate change and the future of California's endemic flora. PLoS One 3: e2502.

Lumbreras EL. 2001. The micro-reserves as a tool for conservation of threatened plants in Europe. Convention on the Conservation of European Wildlife and Natural Habitats. Nature and Environment Vol. 121. Council of Europe Publishing.

Lynch AH, Beringer J, Kershaw P, Marshall A, Mooney S, Tapper N, Turney C, van der Kaars S. 2007. Using the paleorecord to evaluate climate and fire interactions in Australia. Annual Review of Earth and Planetary Sciences 35: 215–39.

MacDougall AS, Turkington R. 2005. Are invasive species the drivers or passengers of change in degraded ecosystems? Ecology 86: 42–55.

Magrin GO, Marengo JA, Boulanger J-P, Buckeridge MS, Castellanos E, Poveda G, Scarano FR, Vicuña S. 2014. Central and South America. Pages 1499–566 in Barros VR, Field CB, Dokken DJ, Mastrandrea MD, Mach KJ, Bilir TE, Chatterjee M, Ebi KL, Estrada YO, Genova RC, Girma B, Kissel ES, Levy AN, MacCracken S, Mastrandrea PR, White LL.. eds. Climate change 2014: impacts, adaptation, and vulnerability. Part B: Regional aspects. Contribution of Working Group II to the Fifth Assessment Report of the Intergovernmental Panel on Climate Change. Cambridge University Press, Cambridge, United Kingdom and New York, NY, USA.

Malak DA, Pausas JG, Pardo-Pascual JE, Ruiz LA. 2015. Fire recurrence and the dynamics of the enhanced vegetation index in a mediterranean ecosystem. International Journal of Applied Geospatial Research 6: 18–35.

Malo J, Suarez F. 1997. Dispersal mechanism and transcontinental naturalization proneness among Mediterranean herbaceous species. Journal of Biogeography 24: 391–4.

Marchese C. 2015. Biodiversity hotspots: a shortcut for a more complicated concept. Global Ecology and Conservation 3: 297–309.

Marean CW. 2010a. Pinnacle Point Cave 13B (Western Cape Province, South Africa) in context: the Cape Floral Kingdom, shellfish, and modern human origins. Journal of Human Evolution 59: 425–43.

Marean CW. 2010b. When the sea saved humanity. Scientific American 303: 54–61.

Marean CW, Cawthra HC, Cowling RM, Esler KJ, Fisher E, Milewski A, Potts AJ, Singels E, De Vynck J. 2014. Stone Age people in a changing South African Greater Cape Floristic Region. Pages 164–99 in Allsopp N, Colville JF, Verboom GA. eds. Ecology and evolution of fynbos: understanding megadiversity. Oxford University Press, Oxford.

Marr SM, Olden JD, Leprieur F, Arismendi I, Ćaleta M, Morgan DL, Nocita A, Šanda R, Tarkan AS, García-Berthou E. 2013. A global assessment of freshwater fish introductions in Mediterranean-climate regions. Hydrobiologia 719: 317–29.

Marull J, Otero I, Stefanescu C, Tello E, Miralles M, Coll F, Pons M, Diana GL. 2015. Exploring the links between forest transition and landscape changes in the Mediterranean. Does forest recovery really lead to better landscape quality? Agroforestry Systems 89: 705–19.

Maskell LC, Bullock JM, Smart SM, Thompson K, Hulme PE. 2006. The distribution and habitat associations of non-native plant species in urban riparian habitats. Journal of Vegetation Science 17: 499–508.

Matusick G, Ruthrof KX, Hardy GJ. 2012. Drought and heat triggers sudden and severe dieback in a dominant Mediterranean-type woodland species. Open Journal of Forestry 2: 183–6.

Matusick G, Ruthrof K, Brouwers N, Dell B, Hardy GJ. 2013. Sudden forest canopy collapse corresponding with extreme drought and heat in a mediterranean-type eucalypt forest in southwestern Australia. European Journal of Forest Research 132: 497–510.

McDowell N, Pockman WT, Allen CD, Breshears DD, Cobb N, Kolb T, Plaut J, Sperry J, West A, Williams DG, Yepez EA. 2008. Mechanisms of plant survival and mortality during drought: why do some plants survive while others succumb to drought? New Phytologist 178: 719–39.

McIntyre PJ, Thorne JH, Dolanc CR, Flint AL, Flint LE, Kelly M, Ackerly DD. 2015. Twentieth-century shifts in forest structure in California: denser forests, smaller trees, and increased dominance of oaks. PNAS 112: 1458–63.

McLaughlin BC, Ackerly DD, Klos PZ, Natali J, Dawson TE, Thompson SE. 2017. Hydrologic refugia, plants, and climate change. Global Change Biology, 23: 2941–61.

Metzger MJ, Rounsevell MDA, Acosta-Michlik L, Leemans R, Schroter D. 2006. The vulnerability of ecosystem services to land use change. Agriculture, Ecosystems & Environment 114: 69–85.

Midgley GF, Hannah L, Millar D, Rutherford MC, Powrie LW. 2002. Assessing the vulnerability of species richness to anthropogenic climate change in a biodiversity hotspot. Global Ecology and Biogeography 11: 445–51.

Millington JD, Perry GL, Romero-Calcerrada R. 2007. Regression techniques for examining land use/cover change: a case study of a Mediterranean landscape. Ecosystems 10: 562–78.

Minnich RA. 1983. Fire mosaics in southern California and northern Baja California. Science 219: 1287–94.

Murray BR, Phillips ML. 2012. Temporal introduction patterns of invasive alien plant species to Australia. NeoBiota 13: 1–14.

Musil C, Milton S, Davis G. 2005. The threat of alien invasive grasses to lowland Cape floral diversity: an empirical appraisal of the effectiveness of practical control strategies. South African Journal of Science 101: 337–44.

Myers N, Mittermeier RA, Mittermeier CG, Da Fonseca GA, Kent J. 2000. Biodiversity hotspots for conservation priorities. Nature 403: 853–8.

Ndlovu J, Richardson DM, Wilson JRU, Le Roux JJ. 2013. Co-invasion of South African ecosystems by an Australian legume and its rhizobial symbionts. Journal of Biogeography 40: 1240–51.

Neary DG, Klopatek CC, DeBano LF, Ffolliott PF. 1999. Fire effects on belowground sustainability: a review and synthesis. Forest Ecology and Management 122: 51–71.

Ness JH, Bronstein JL. 2004. The effects of invasive ants on prospective ant mutualists. Biological Invasions 6: 445–61.

Newbold T, Hudson LN, Arnell AP, Contu S, De Palma A, Ferrier S, Hill SL, Hoskins AJ, Lysenko I, Phillips HR, Burton VJ. 2016. Has land use pushed terrestrial biodiversity beyond the planetary boundary? A global assessment. Science 353: 288–91.

Newton IP, Knight RS. 2005. The use of a 60-year series of aerial photographs to assess local agricultural transformations of West Coast renosterveld, an endangered South African vegetation type. South African Geographical Journal 87: 18–27.

Ovalle C, Aronson J, Del Pozo A, Avendano J. 1990. The espinal: agroforestry systems of the Mediterranean-type climate region of Chile. Agroforestry Systems 10: 213–39.

Paddock III WAS, Davis SD, Pratt RB, Jacobsen AL, Tobin MF, López-Portillo J, Ewers FW. 2013. Factors determining mortality of adult chaparral shrubs in an extreme drought year in California. Aliso 31: 49–57.

Pauchard A, Garcia RA, Pena E, González C, Cavieres LA, Bustamante RO. 2008. Positive feedbacks between plant invasions and fire regimes: *Teline monspessulana* (L.) K. Koch (Fabaceae) in central Chile. Biological Invasions 10: 547–53.

Pausas JG. 2004. Changes in fire and climate in the eastern Iberian Peninsula (Mediterranean basin). Climatic Change 63: 337–50.

Pausas JG, Pratt RB, Keeley JE, Jacobsen AL, Ramirez AR, Vilagrosa A, Paula S, Kaneakua-Pia IN and Davis SD. 2016. Towards understanding resprouting at the global scale. New Phytologist 209: 945–54.

Peñuelas J, Lloret F, Montoya R. 2001. Severe drought effects on Mediterranean woody flora in Spain. Forest Science 47: 214–18.

Petersen LM, Moll EJ, Collins R, Hockings MT. 2012. Development of a compendium of local, wild-harvested species used in the informal economy trade, Cape Town, South Africa. Ecology and Society 17: 26. http://dx.doi.org/10.5751/ES-04537-170226.

Pignatti E, Pignatti S, Ladd PG. 2002. Comparison of ecosystems in the Mediterranean Basin and Western Australia. Plant Ecology 163: 177–86.

Poyatos R, Latron J, Llorens P. 2003. Land use and land cover change after agricultural abandonment: the case of a Mediterranean mountain area (Catalan Pre-Pyrenees). Mountain Research and Development 23: 362–8.

Pratt RB, Jacobsen AL, Ramirez AR, Helms AM, Traugh CA, Tobin MF, Heffner MS, Davis SD. 2014. Mortality of resprouting chaparral shrubs after a fire and during a record drought: physiological mechanisms and demographic consequences. Global Change Biology 20: 893–907.

Prospero JM, Barrett K, Church T, Dentener F, Duce RA, Galloway JN, Levy H, Moody J, Quinn P. 1996. Atmospheric deposition of nutrients to the North Atlantic Basin. Biogeochemistry 35: 27–73.

Pyšek P. 1998. Alien and native species in central European urban floras: a quantitative comparison. Journal of Biogeography 25: 155–63.

Pyšek P, Pergl J, Essl F, Lenzer B, Dawson W, Kreft H, Weigelt P, Winter M, Kartesz J, Nishino M, Antonova LA, Barcelona JF, Cabezas FJ, Cárdenas-Toro J, Castaño N, Chacón E, Chatelain C, Dullinger S, Ebel AL, Figueiredo E, Fuentes N, Genovesi P, Groom QJ, Henderson L, Kupriyanov A, Masciadri S, Maurel N, Meerman J, Morozova O, Moser D, Nickren D, Nowak PM, Pagad S, Patzelt A, Pelser PB, Seebens H, Shu W, Thomas J, Velayos M, Weber E, Wieringa JJ, Baptiste MP, van Kleunen M. 2017. Naturalized alien flora of the world: species diversity, taxonomic and phylogenetic patterns, geographic distribution and global hotspots of plant invasion. Preslia 89: 203–74.

Radeloff VC, Hammer RB, Stewart SI, Fried JS, Holcomb SS, McKeefry JF. 2005. The wildland-urban interface in the United States. Ecological Applications 15: 799–805.

Raimondo D, von Staden L, Foden W, Victor JE, Helme NA, Turner RC, Kamundi DA, Manyama PA. 2009. Red list of South African plants. Strelitzia 25. South African National Biodiversity Institute, Pretoria, South Africa.

Rebelo AG, Holmes PM, Dorse C, Wood J. 2011. Impacts of urbanization in a biodiversity hotspot: conservation challenges in Metropolitan Cape Town. South African Journal of Botany 77: 20–35.

Rector AL, Verrelli BC. 2010. Glacial cycling, large mammal community composition, and trophic adaptations in the Western Cape, South Africa. Journal of Human Evolution 58: 90–102.

Reid TS, Murphy DD. 1995. Providing a regional context for local conservation action. BioScience 45: S84–S90.

Rejmánek M, Randall JM. 1994. Invasive alien plants in California: 1993 summary and comparison with other areas in North America. Madroño 41: 161–77.

Rejmánek M, Richardson DM. 2013. Trees and shrubs as invasive alien species—2013 update of the global database. Diversity and Distributions 19: 1093–4.

Richardson DM, Gaertner M. 2013. Plant invasions as builders and shapers of novel ecosystems. Pages 102–14 in Hobbs RJ, Higgs EC, Hall CM, eds. Novel ecosystems: intervening in the new ecological world order. Wiley-Blackwell, Oxford.

Richardson DM, Pyšek P. 2008. Fifty years of invasion ecology—the legacy of Charles Elton. Diversity and Distributions 14: 161–8.

Richardson DM, van Wilgen BW. 2004. Invasive alien plants in South Africa: how well do we understand the ecological impacts? Working for Water. South African Journal of Science 100: 45–52.

Richardson DM, Williams PA, Hobbs RJ. 1994. Pine invasions in the southern hemisphere: determinants of spread and invadability. Journal of Biogeography 21: 511–27.

Richardson DM, Pyšek P, Rejmanek M, Barbour MG, Panetta FD, West CJ. 2000. Naturalization and invasion of alien plants: concepts and definitions. Diversity and Distributions 6: 93–107.

Richardson DM, Cambray JA, Chapman RA, Dean WR, Griffiths CL, Le Maitre DC, Newton DJ, Winstanley TJ. 2003. Vectors and pathways of biological invasions in South Africa—past, present and future. Pages 292–49 in Rui G, Carlton J. eds. Invasive species: vectors and management strategies. Island Press, Washington, DC.

Richardson DM, Carruthers J, Hui C, Impson FAC, Robertson MP, Rouget M, Le Roux JJ, Wilson JRU. 2011. Human-mediated introductions of Australian acacias—a global experiment in biogeography. Diversity and Distributions 17: 771–87.

Richardson DM, Hui C, Nunez, MA, Pauchard A. 2014. Tree invasions: patterns, processes, challenges and opportunities. Biological Invasions 16: 473–81.

Riggan PJ, Lockwood RN, Lopez EN. 1985. Deposition and processing of airborne nitrogen pollutants in Mediterranean-type ecosystems of southern California. Environmental Science & Technology 19: 781–9.

Riley SP, Sauvajot RM, Fuller TK, York EC, Kamradt DA, Bromley C, Wayne RK. 2003. Effects of urbanization and habitat fragmentation on bobcats and coyotes in southern California. Conservation Biology 17: 566–76.

Robin C, Desprez-Loustau ML, Capron G, Delatour C. 1998. First record of *Phytophthora cinnamomi* on cork and Holm oaks in France and evidence of pathogenicity. Annales des Sciences Forestières 55: 869–83.

Rössler M. 2006. World heritage cultural landscapes: a UNESCO flagship programme 1992–2006. Landscape Research 31: 333–53.

Rouget M, Richardson DM, Cowling RM, Lloyd JW, Lombard AT. 2003. Current patterns of habitat transformation and future threats to biodiversity in terrestrial ecosystems of the Cape Floristic Region, South Africa. Biological Conservation 112: 63–85.

Rouget M, Robertson MP, Wilson JRU, Hui C, Essl F, Renteria JL, Richardson DM. 2016. Invasion debt—quantifying future biological invasions. Diversity and Distributions 22: 445–56.

Rundel PW, Dickie IA, Richardson DM. 2014. Tree invasions into treeless areas: mechanisms and ecosystem processes. Biological Invasions 16: 663–75.

Rundel PW, Arroyo MT, Cowling RM, Keeley JE, Lamont BB, Vargas P. 2016. Mediterranean biomes: evolution of their vegetation, floras, and climate. Annual Review of Ecology, Evolution, and Systematics 47: 383–407.

Ruthrof K, Fontaine J, Matusick G, Breshears DD, Law DJ, Hardy G, Powell S. 2015a. How drought-induced forest die-off alters microclimate and increases fuel loadings and fire potentials. International Journal of Wildland Fire 25: 116–26.

Ruthrof KX, Matusick G, Hardy GESJ. 2015b. Early differential responses of co-dominant canopy species to sudden and severe drought in a Mediterranean-climate type forest. Forests 6: 2082–91.

Sala OE, Chapin FS, Armesto JJ, Berlow E, Bloomfield J, Dirzo R, Huber-Sanwald E, Huenneke LF, Jackson RB, Kinzig A, Leemans R. 2000. Global biodiversity scenarios for the year 2100. Science 287: 1770–4.

Salvati L, Sabbi A. 2011. Exploring long-term land cover changes in an urban region of southern Europe. International Journal of Sustainable Development & World Ecology 18: 273–82.

Salvati L, Serra P, Rugiero S, Sabbi A. 2016. Complexity in action? Fringe agro-forest systems, demography and societal transformations in Mediterranean peri-urban areas. Agriculture and Agricultural Science Procedia 8: 222–7.

Salvati L, De Zuliani E, Sabbi A. 2017. Land-cover changes and sustainable development in a rural cultural landscape of central Italy: classical trends and counter-intuitive results. International Journal of Sustainable Development & World Ecology 24: 27–36.

Sandberg R. 2013. The response of biological communities to natural and anthropogenic habitat fragmentation in South Outeniqua sandstone fynbos, South Africa. MSc Dissertation, Stellenbosch University, Stellenbosch.

Sandberg RN, Allsopp N, Esler KJ. 2016. The use of fynbos fragments by birds: stepping-stone habitats and resource refugia. Koedoe 58: a1321.http://dx.doi.org/10.4102/koedoe.v58i1.1321.

Saruwatari MW, Davis SD. 1989. Tissue water relations of three chaparral shrub species after wildfire. Oecologia 80: 303–8.

Schurr FM, Midgley GF, Rebelo AG, Reeves G, Poschlod P, Higgins SI. 2007. Colonization and persistence ability explain the extent to which plant species fill their potential range. Global Ecology and Biogeography 16: 449–59.

Seager R, Ting M, Held I, Kushnir Y, Lu J, Vecchi G, Huang HP, Harnik N, Leetmaa A, Lau NC, Li C. 2007. Model projections of an imminent transition to a more arid climate in southwestern North America. Science 316: 1181–4.

Serra P, Pons X, Saurí D. 2008. Land-cover and land-use change in a Mediterranean landscape: a spatial analysis of driving forces integrating biophysical and human factors. Applied Geography 28: 189–209.

Seydack AH, Bekker SJ, Marshall AH. 2007. Shrubland fire regime scenarios in the Swartberg Mountain Range, South Africa: implications for fire management. International Journal of Wildland Fire 16: 81–95.

Shane MW, Szota C, Lambers H. 2004. A root trait accounting for the extreme phosphorus sensitivity of *Hakea prostrata* (Proteaceae). Plant, Cell and Environment 27: 991–1004.

Sharma GP, Muhl SA, Esler KJ, Milton SJ. 2010. Competitive interactions between the alien invasive annual grass *Avena fatua* and indigenous herbaceous plants in South African renosterveld: the role of nitrogen enrichment. Biological Invasions 12: 3371–8.

Shearer BL, Crane CE, Cochrane A. 2004. Quantification of the susceptibility of the native flora of the South-West Botanical Province, Western Australia, to *Phytophthora cinnamomi*. Australian Journal of Botany 52: 535–443.

Shearer BL, Crane CE, Barrett S, Cochrane A. 2007. *Phytophthora cinnamomi* invasion, a major threatening process to conservation of flora diversity in the south-west botanical province of Western Australia. Australian Journal of Botany 55: 225–38.

Shearer BL, Crane CE, Cochrane A, Dunne CP. 2013. Variation in susceptibility of threatened flora to *Phytophthora cinnamomi*. Australasian Plant Pathology 42: 491–502.

Simmons MT, Cowling RM. 1996. Why is the Cape Peninsula so rich in plant species? An analysis of the independent diversity components. Biodiversity and Conservation 5: 551–73.

Singels E, Potts AJ, Cowling RM, Marean CW, de Vynk JC, Esler KJ. 2016. Foraging potential of underground storage organs in the southern Cape, South Africa. Journal of Human Evolution 101: 79–89.

Sleeter BM, Wilson TS, Soulard CE, Liu J. 2011. Estimation of late twentieth century land-cover change in California. Environmental Monitoring and Assessment 173: 251–66.

Smith BD, Zeder MA. 2013. The onset of the Anthropocene. Anthropocene 4: 8–13.

Southey D. 2009. Wildfires in the Cape Floristic Region: exploring vegetation and weather as drivers of fire frequency (Unpublished MSc). Department of Botany, University of Cape Town, South Africa.

Suarez AV, Case TJ. 2002. Bottom-up effects on persistence of a specialist predator: ant invasions and horned lizards. Ecological Applications 12: 291–8.

Symeonakis E, Caccetta PA, Wallace JF, Arnau-Rosalen E, Calvo-Cases A, Koukoulas S. 2015. Multi-temporal forest cover change and forest density trend detection in a Mediterranean environment. Land Degradation & Development 28: 1188–98.

Syphard A, Franklin J, Keeley J. 2006. Simulating the effects of frequent fire on southern California coastal shrublands. Ecological Applications 16: 1744–56.

Syphard AD, Radeloff VC, Keeley JE, Hawbaker TJ, Clayton MK, Stewart SI, Hammer RB. 2007. Human influence on California fire regimes. Ecological Applications 17: 1388–402.

Syphard AD, Radeloff VC, Hawbaker TJ, Stewart SI. 2009. Conservation threats due to human-caused increases in fire frequency in Mediterranean climate ecosystems. Conservation Biology 23: 758–69.

Syphard AD, Keeley JE, Massada AB, Brennan TJ, Radeloff VC. 2012. Housing arrangement and location determine the likelihood of housing loss due to wildfire. PLoS One 7(3): e33954.

Tamrazian A, Ladochy S, Willis J, Patzert WC. 2008. Heat waves in southern California: are they becoming more frequent and longer lasting? APCG Year Book 70: 59–69.

Taylor RJ, Hoxley G. 2003. Dryland salinity in Western Australia: managing a changing water cycle. Water Science and Technology 47: 201–7.

Tessler N, Sapir Y, Wittenberg L, Greenbaum N. 2015. Recovery of Mediterranean vegetation after recurrent forest fires: insight from the 2010 forest fire on Mount Carmel, Israel. Land Degradation & Development 27: 1424–31.

Timmermann A, Friedrich T. 2016. Late Pleistocene climate drivers of early human migration. Nature 538: 92–5.

Trabaud L. 1990. Fire and an agent of plant invasion? A case study in the French Mediterranean vegetation. Pages 417–37 in di Castri F, Hansen AJ, Debussche M, eds. Biological invasions in Europe and the Mediterranean Basin. Kluwer Academic Publishers, Dordrecht, The Netherlands.

Underwood EC, Viers JH, Klausmeyer KR, Cox RL, Shaw MR. 2009a. Threats and biodiversity in the mediterranean biome. Diversity and Distributions 15: 188–97.

Underwood EC, Klausmeyer KR, Cox RL, Busby SM, Morrison SA, Shaw MR. 2009b. Expanding the global network of protected areas to save the imperiled Mediterranean biome. Conservation Biology 23: 43–52.

UNESCO. 2012. World heritage. [Accessed Feb 10, 2017]. Available from: http://whc.unesco.org/en/about/

Urban MC. 2015. Accelerating extinction risk from climate change. Science 348: 571–3.

Vallejo VR, Smanis A, Chirino E, Fuentes D, Valdecantos A, Vilagrosa A. 2012. Perspectives in dryland restoration: approaches for climate change adaptation. New Forests 43: 561–79.

van Kleunen M, Dawson W, Essl F, Pergl J, Winter M, Weber E, Kreft H, Weigelt P, Kartesz J, Nishino M, Antonova LA, Barcelona JF, Cabezas FJ, Cárdenas D, Cárdenas-Toro J, Castaño N, Chacón E, Chatelain C, Ebel AL, Figueiredo E, Fuentes N, Groom QJ, Henderson L, Inderjit, Kupriyanov A, Masciadri S, Meerman J, Morozova O, Moser D, Nickrent DL, Patzelt A, Pelser PB, Baptiste MP, Poopath M, Schulze M, Seebens H, Shu W, Thomas J, Velayos M, Wieringa JJ, Pyšek P. 2015. Global exchange and accumulation of non-native plants. Nature 525: 100–3.

van Wilgen BW, Wannenburgh AM. 2016. Co-facilitating invasive species control, water conservation and poverty relief: achievements and challenges in South Africa's Working for Water program. Current Opinion in Environmental Sustainability 19: 7–17.

van Wilgen BW, Carruthers J, Cowling RM, et al. 2016. Ecological research and conservation management in the Cape Floristic Region between 1945 and 2015: history, current understanding and future challenges. Transactions of the Royal Society of South Africa 71: 207–303.

Venturas MD, MacKinnon ED, Dario HL, Jacobsen AL, Pratt RB, Davis SD. 2016. Chaparral shrub hydraulic traits, size, and life history types relate to species mortality during California's historic drought of 2014. PLoS One 11: e0159145.

Viers JH, Williams JN, Nicholas KA, Barbosa O, Kotzé, Spence L, Webb LB, Merenlender A, Reynolds M. 2013. Vinecology: pairing wine with nature. Conservation Letters 6: 287–99.

Von Broembsen S. 1984. Occurrence of *Phytophthora cinnamomi* on indigenous and exotic hosts in South Africa, with special reference to the south-western Cape Province. Phytophylactica 16: 221–6.

Wicht C. 1945. Preservation of the vegetation of the South Western Cape (Special publication of the Royal Society of South Africa). Cape Town: Royal Society of South Africa.

Williams P, Hannah L, Andelman S, Midgley G, Araújo M, Hughes G, Manne L, Martinez-Meyer E, Pearson R. 2005. Planning for climate change: identifying minimum-dispersal corridors for the Cape Proteaceae. Conservation Biology 19: 1063–74.

Williamson TN, Graham RC, Shouse PJ. 2004. Effects of a chaparral-to-grass conversion on soil physical and hydrologic properties after four decades. Geoderma 123: 99–114.

Wilson AM, Latimer AM, Silander JA, Gelfand AE, de Klerk H. 2010. A hierarchical Bayesian model of wildfire in a Mediterranean biodiversity hotspot: implications of weather variability and global circulation. Ecological Modelling 221: 106–12.

Wilson D, Stock WD, Hedderson T. 2009. Historical nitrogen content of bryophyte tissue as an indicator of increased nitrogen deposition in the Cape Metropolitan Area, South Africa. Environmental Pollution 157: 938–45.

Wilson JR, Gaertner M, Griffiths CL, et al. 2014. Biological invasions in the Cape Floristic region: history, current patterns, impacts, and management challenges. Pages 273–98 in Allsopp N, Colville JF, Verboom A. eds. Fynbos: ecology, evolution, and conservation of a megadiverse region. Oxford University Press, Oxford.

Yates CJ, Hobbs RJ. 1997. Temperate eucalypt woodlands: a review of their status, processes threatening their persistence and techniques for restoration. Australian Journal of Botany 45: 949–73.

9 Planning for the future

Abstract

Mediterranean-type climate (MTC) regions are highlighted in several global analyses of conservation risk and priorities. These regions have undergone high levels of habitat conversion and yet of all terrestrial biomes they have the second lowest level of land protection. With transformation pressures set to continue (Chapter 8), planning for a sustainable conservation future in MTC regions is therefore essential. Conservation activities are represented by a variety of philosophies and motives, partially driven by the underlying differences in transformation drivers and sociopolitical contexts across MTC regions. These activities include investment in, and best-practice management of, protected areas (land sparing), an interdisciplinary focus on integrated management of production landscapes (land sharing; stewardship), as well as ecological restoration to increase habitat, improve connectivity, and provide a hedge against the impacts of future climate change. These responses need to be applied in a strategic, synergistic manner to minimize future biodiversity loss.

9.1 Introduction

Against a backdrop of global change, and with intense pressures of urbanization and agriculture set to continue (Chapter 8), planning for a sustainable conservation future in mediterranean-type climate (MTC) regions has never been more critical. These areas are highlighted in global analyses of conservation risk and priorities (Brooks et al. 2006), including the hotspot (Myers et al. 2000; Figure 9.1) and Global 200 (Olson and Dinerstein 2002) analyses, both of which argue that effective conservation in regions with exceptional levels of endemism would help to conserve a significant share of the world's imperilled species. MTC regions have undergone a high level

The Biology of Mediterranean-Type Ecosystems. Karen J. Esler, Anna L. Jacobsen, and R. Brandon Pratt,
Oxford University Press (2018). © Karen J. Esler, Anna L. Jacobsen, and R. Brandon Pratt 2018.
DOI 10.1093/oso/9780198739135.001.0001

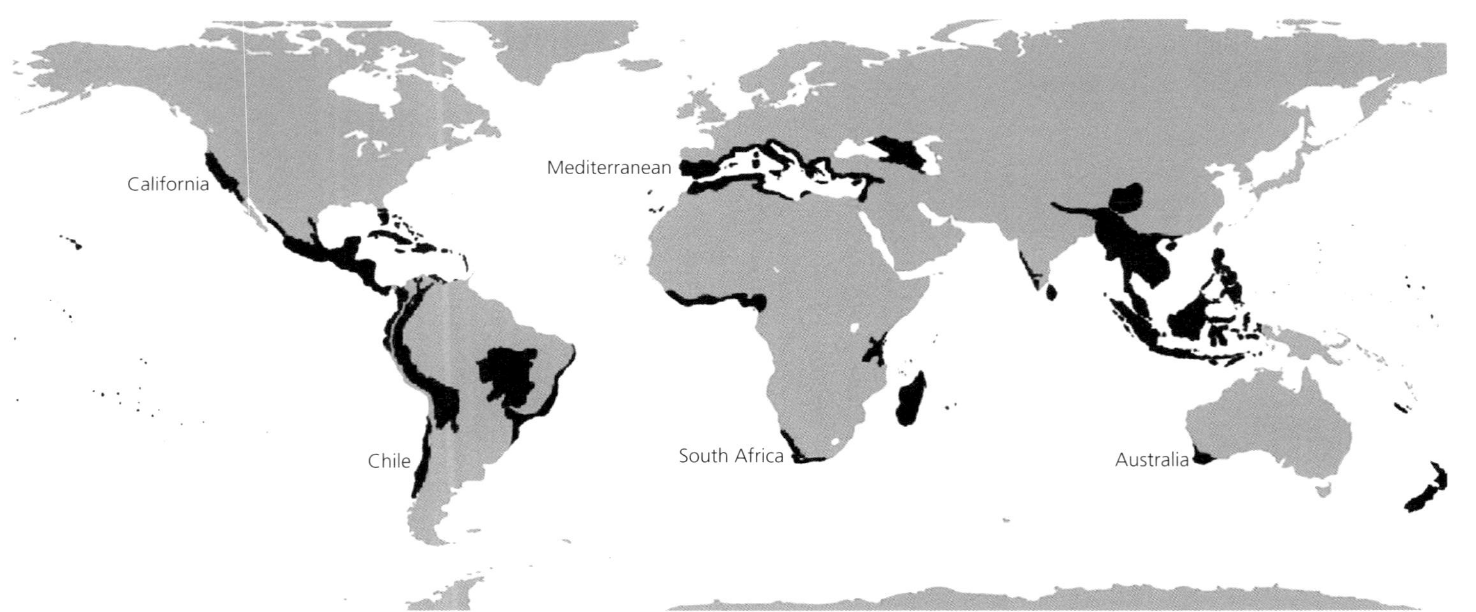

Figure 9.1   **All five mediterranean-type climate regions are featured in the Myers et al. (2000) analysis of hotspots**—areas featuring exceptional concentrations of endemic species and concomitant loss of habitat. Biodiversity hotspots are indicated as regions of black in the figure above. (Modified from Myers et al. 2000.)

Table 9.1 Areas of protected, converted, and impacted land in the five mediterranean-type climate (MTC) regions, according to Cox and Underwood (2011)

MTC region	Australia	California	Chile	Mediterranean	South Africa
Area (km²)[a]	802,523	176,425	148,408	2,077,131	96,953
Area protected (IUCN I–V1; km² and proportion)[a]	86,962 (11%)	34,955 (20%)	1370 (1%)	77,411 (4%)	18,823 (19%)
Area converted to urban/industrial (km² and proportion)[a]	2974 (<1%)	17,092 (10%)	1100 (1%)	30,977 (1%)	1249 (1%)
Area impacted by intensive/cultivated agriculture (km² and proportion)[a]	290,577 (36%)	39,934 (23%)	35,076 (24%)	592,299 (29%)	22,866 (24%)
Area with conservation potential (km² and proportion)[a]	422,010 (53%)	84,444 (48%)	110,861 (75%)	1,376,444 (66%)	54,015 (56%)
Plant species[b]	8000	4300	2400	25,000	8500
Vertebrate species[a]	754	487	198	920	536

IUCN, International Union for Conservation of Nature. Data from [a] Cox and Underwood, 2011; [b] Cowling et al. 1996.

of habitat conversion (41.4 per cent across all mediterranean-type forests, woodlands and scrub; Hoekstra et al. 2005) and yet of all 13 terrestrial biomes they have the second lowest level of land protection (5 per cent of total biome area; Hoekstra et al. 2005) (also see Table 9.1). The extent of habitat conversion in MTC regions exceeds that of habitat protection by a factor greater than eight, second only to temperate grasslands, savannas, and shrublands (Hoekstra et al. 2005). Despite significant projected contractions of MTC areas under climate change over the next century, around 50 per cent of the global mediterranean biome is likely to remain stable (Klausmeyer and Shaw 2009). This implies that investment in, and sound management of, protected areas (**land sparing**) remains a sound conservation strategy. There is also an increasing realization that a focus on protected areas alone is insufficient to stem the tide of loss.

9.2 Conservation approaches

Over the past 50 years, global views of nature and conservation have shifted in focus, from an emphasis on species and a prioritization on wilderness and intact natural habitats (**protected areas, land sparing**) to a more interdisciplinary focus on integrated management of ecosystems that incorporates social and ecological sciences (Mace 2014). These shifts are partially reflected in the scientific focus on conservation activities in MTC regions over time, to a situation today where conservation science and practice is represented by a variety of philosophies and motives (Figure 9.2). For example, attention

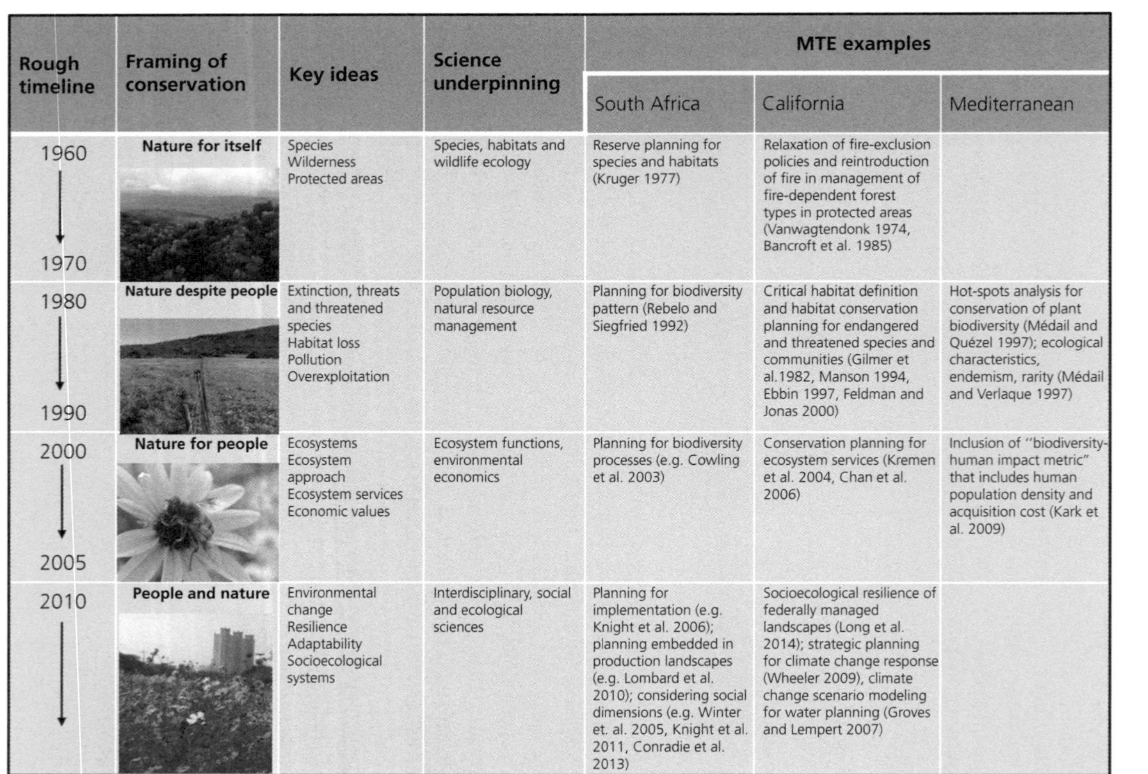

Rough timeline	Framing of conservation	Key ideas	Science underpinning	MTE examples		
				South Africa	California	Mediterranean
1960 → 1970	**Nature for itself**	Species Wilderness Protected areas	Species, habitats and wildlife ecology	Reserve planning for species and habitats (Kruger 1977)	Relaxation of fire-exclusion policies and reintroduction of fire in management of fire-dependent forest types in protected areas (Vanwagtendonk 1974, Bancroft et al. 1985)	
1980 → 1990	**Nature despite people**	Extinction, threats and threatened species Habitat loss Pollution Overexploitation	Population biology, natural resource management	Planning for biodiversity pattern (Rebelo and Siegfried 1992)	Critical habitat definition and habitat conservation planning for endangered and threatened species and communities (Gilmer et al.1982, Manson 1994, Ebbin 1997, Feldman and Jonas 2000)	Hot-spots analysis for conservation of plant biodiversity (Médail and Quézel 1997); ecological characteristics, endemism, rarity (Médail and Verlaque 1997)
2000 → 2005	**Nature for people**	Ecosystems Ecosystem approach Ecosystem services Economic values	Ecosystem functions, environmental economics	Planning for biodiversity processes (e.g. Cowling et al. 2003)	Conservation planning for ecosystem services (Kremen et al. 2004, Chan et al. 2006)	Inclusion of "biodiversity-human impact metric" that includes human population density and acquisition cost (Kark et al. 2009)
2010 →	**People and nature**	Environmental change Resilience Adaptability Socioecological systems	Interdisciplinary, social and ecological sciences	Planning for implementation (e.g. Knight et al. 2006); planning embedded in production landscapes (e.g. Lombard et al. 2010); considering social dimensions (e.g. Winter et. al. 2005, Knight et al. 2011, Conradie et al. 2013)	Socioecological resilience of federally managed landscapes (Long et al. 2014); strategic planning for climate change response (Wheeler 2009), climate change scenario modeling for water planning (Groves and Lempert 2007)	

Figure 9.2 **The changing views of nature and conservation in three mediterranean-type climate regions (adapted from Mace 2014).** Over time, the framing of conservation has evolved, as have key ideas, and the science underpinning these ideas. Current conservation practices include all four framings.

Source: Images from Tessa Oliver.

has increasingly turned to conservation stewardship initiatives in production landscapes (**land sharing**), as well as **ecological restoration** to increase habitat, improve connectivity, and therefore to provide a hedge against the impacts of future climate change. Land sparing and land sharing have been represented in the literature as dichotomous strategies, with land sparing emerging from the view that human settlement is incompatible with biodiversity conservation, in contrast to land sharing, where people are enabled to maintain or even enhance biodiversity through historical processes of land stewardship (Figure 9.3). Some argue that, by intensifying food production in agricultural systems, pressures on natural land can be deflected (Phalan et al. 2011a; 2011b), but others argue that real-world complexities provide multiple opportunities for conservation beyond food production and biodiversity on natural land (Tscharntke et al. 2005; 2012). The reality is that these strategies are not dichotomous; each approach has value depending on context and therefore a variety of conservation strategies exists in MTC regions. This variety is driven partially by the underlying differences in transformation drivers and sociopolitical contexts across MTC regions. For example, the long history of human occupation in the Mediterranean and resultant links between cultural and biological diversity has led to a conservation focus on maintaining biodiversity through maintenance of traditional land

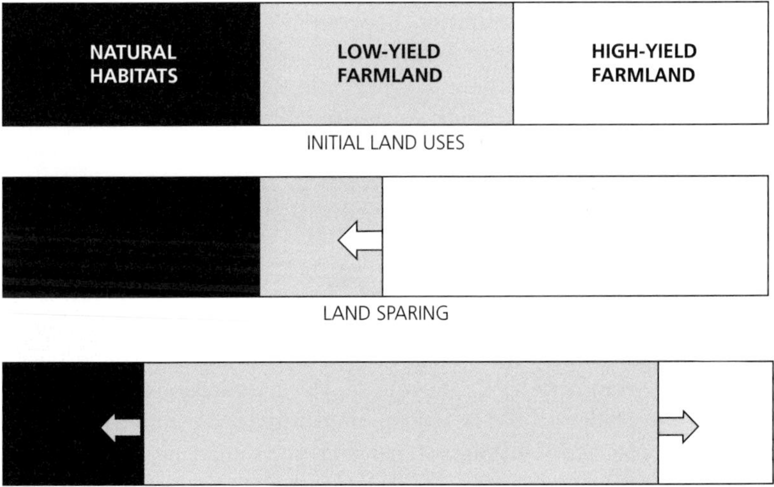

Figure 9.3 **Farming strategies.** Land sparing and biodiversity-friendly farming strategies could still be used to meet increased food demand and achieve conservation objectives. Land sparing (middle) involves increasing farmland yields, while protecting or restoring natural habitats. Biodiversity-friendly farming (bottom) involves increasing the footprint of low-yield farmland at the expense of natural habitats and high-yield farmland. (Adapted from Phalan et al. 2011a.)

uses. In South Africa, major impacts of invasive alien plants have led to a strong ecosystem-service approach to their management and control.

9.2.1 Protected area expansion

Regardless of other initiatives, **protected areas** remain core to MTC region conservation, and momentum to expand the conservation estate has grown, driven partially by global targets. With the exception of the USA (whose regulatory mechanism is the Endangered Species Act of 1973), all MTC region nations are signatories to the Convention on Biological Diversity (CBD 1992) and have agreed to a set of twenty time-bound measurable targets (so called Achi targets, https://www.cbd.int/sp/targets/) that, if translated into strategy and action, will help to slow biodiversity loss. The Rio+20 Sustainable Development goals (http://www.un.org/sustainabledevelopment/) reaffirmed this commitment. At a national scale, biodiversity strategies and action plans (NBSAPs; https://www.cbd.int/nbsap/default.shtml) help to guide sectors whose activities impact diversity. A resulting range of public–private partnerships have enabled the advancement of gap analyses, systematic conservation plans, and acquisition and management of protected areas. As a result, some positive trends exist in MTC region lands being dedicated to conservation of biodiversity. For example, through financial support from local and international donors, several large, multisectoral conservation programmes were initiated in South Africa that included protected area expansion strategies. Between 1990 and 2011, land in the Cape Floristic Region under some form of protection grew by 7 per cent, from approximately 18,000 km^2 (20 per cent; Rebelo 1992) to 23,430 km^2 (27 per cent; Rouget et al. 2014). Advances (although not without challenges; Feldman and Jonas 2000) have also been reported for Southern California, where a regional planning approach to conservation was adopted in the 1990s, departing from past practices that, through the Endangered Species Act, had focused on individual species (Ebbin 1997).

Despite some positive broad-scale trends, MTC regions' protected areas remain patchy and under-representative of certain habitats and biodiversity in general. In an analysis of International Union for Conservation of Nature (IUCN) categories (I–IV) of protection (2006 data) in MTC regions, Underwood et al. (2009b) noted similar protection levels in Australia (11 per cent), South Africa (11 per cent) and California–Baja California (9 per cent), but very low levels of protection in Chile (<1 per cent) and the Mediterranean (<1 per cent). In Chile, a combination of land cover and land-use changes, high land value, and high population densities have compromised the creation of a public system of protected areas (Pauchard and Villarroel 2002). Private protected areas in this MTC region have grown in number but, because they lack formal recognition and public incentives, consistency in their commitment to conservation, expertise, and resources is patchy. There is also no systematic way to account for their number, distribution, or attributes (Tecklin and Sepulveda 2014). In the Mediterranean, fully coordinated conservation planning could increase efficiency and save costs, but the complexities associated with an array of political and socioeconomic contexts in more than 20 countries remains a challenge (Kark et al. 2009). In this region, factors other than biodiversity often determine where conservation priority is focused (Kark et al. 2009).

Within MTC regions, protected areas are also not spatially uniform, with a bias towards montane areas (in South Africa, California–Baja California, Chile, and the Mediterranean) and towards the shrubland vegetation types (11.8 per cent across the five regions, versus <3 per cent each for grassland, scrub, succulent dominated, woodland, and forest vegetation types; Underwood et al. 2009b). As described in Chapter 8, lowlands have been particularly impacted by urbanization and agriculture, and remaining habitat is often in the form of scattered fragments that do not conform well to traditional protected area expansion. This reflects the need for alternative, finer-resolution conservation strategies such as stewardship engagement with private landowners, mainstreaming of biodiversity-related matters into other land-use planning processes, or improving management practices of production sectors (e.g. agriculture, mining).

9.2.2 Conservation stewardship

With the realization that significant levels of MTC region biodiversity reside in the hands of private landowners in production landscapes (particularly in the lowlands), increasing focus has been placed on **stewardship** as a conservation option. The term 'stewardship' refers to the goals, principles, and actions put in place to accomplish the sustainable use of natural resources, while contributing to conservation priorities and stemming environmental degradation. This tactic does, however, introduce complexities, requiring a more interdisciplinary approach and a deeper understanding of social-ecological systems.

The concept of stewardship is applied in a variety of ways at various scales from **earth** or **ecosystem stewardship** at a global scale to more local scales (Figure 9.4). This is described as 'a strategy to respond to and shape social-ecological systems under conditions of uncertainty and change to sustain the supply and opportunities for use of ecosystem services to support human well-being' (Chapin et al. 2010). Some important aspects include regional initiatives such as **market-linked schemes** and local-level initiatives such as **landowner stewardship programmes** and, also playing a part, the establishment of **conservancies,** registered voluntary associations of like-minded communities to share cooperative management of natural resources (Figure 9.4; Barendse et al. 2016).

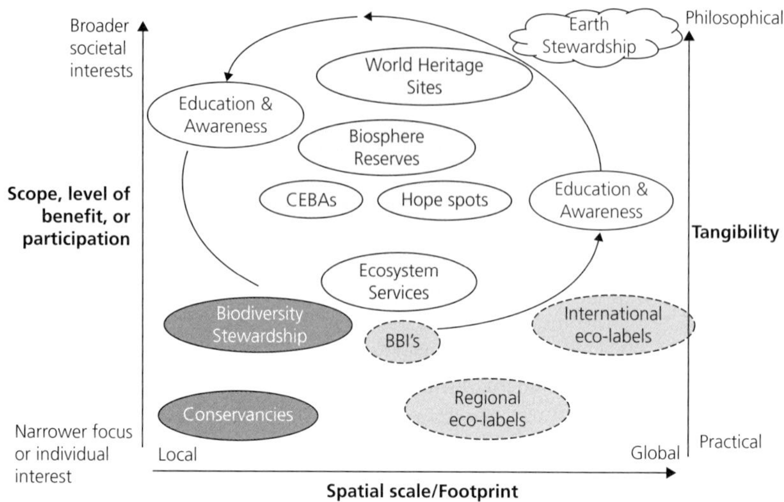

Figure 9.4   **Land stewardship types.** Representation of South African stewardship types in relation to spatial extent and perceived scope, level of benefit, or participation. BBI, Biodiversity and business initiatives; CEBAs, community ecosystems-based adaptation sites. Dotted line ellipses are market-based schemes. Arrows from and to education and awareness schemes indicate their cross-cutting functions. (Adapted from Barendse et al. 2016.)

In the Mediterranean and Australia, stewardship is primarily associated with sustainability in agrienvironmental systems, but with different approaches. The deep agricultural history in the Mediterranean, involving a gradual change from hunter-gatherer societies to agriculture, resulted in a diversification of land-use practices such that high biodiversity, driven by certain land-use practices, has become a focus of conservation concern. Price-support subsidies offer contracts for best-practice management (e.g. creating uncultivated margins around intensive agriculture fields, planting nectar flower mixtures) or for setting aside arable land; however, the effectiveness of the range of measures employed is variable and sometimes contradictory (Abensperg-Traun et al. 2004; Dicks et al. 2013). A useful database providing evidence for the effectiveness of conservation interventions, including farmland conservation, can be found at http://www.conservatio-nevidence.com/. In Western Australia, where land clearing for agriculture was large-scale, rapid, and supported by government land grants and subsidies, no direct subsidies to the farming sector continue today. Here LandCare (https://landcareaustralia.org.au/), a community-driven, grassroots initiative focusing on the adoption of sustainable management of land and catchments, provides the vehicle for stewardship initiatives (Abensperg-Traun et al. 2004). Activities focus predominantly on remnant vegetation (re-vegetation and livestock exclusion) rather than the agricultural matrix itself. It has been suggested that features of the European stewardship system described above could benefit (bird) species in Australian agricultural landscapes (Attwood et al. 2009).

In South Africa, stewardship refers primarily to the protection of biodiversity on privately owned land through 'Biodiversity Stewardship Programmes'. The approach was adopted in the early 2000s, and led to contractual agreements with private landowners to maintain biodiversity on their land. By 2012, a range of legally binding and voluntary agreements had increased the conservation estate on privately owned land by 900 km^2 (Maree and Ralston 2012). Despite being widely heralded as the most cost-effective strategy to conserve (in particular) scattered lowland vegetation, limited human resources and lengthy negotiation times have hampered large-scale roll-out of the programmes (von Hase et al. 2010).

Market-linked schemes, such as 'green' labelling and product certification, which aim to promote sustainability in the production, management, or value chain of specific commodities or services (Barendse et al. 2016) are gaining momentum in MTC regions. There has been substantial focus in particular on the wine industry, where an international collaboration network of researchers and wineries focus on developing conservation initiatives in the two New World MTC regions where there is high potential for vineyard expansion and natural habitat still remains to be conserved (see, for example, Viers et al. 2013). In South Africa, after rapid expansion of the industry footprint in the 1990s, the 'Biodiversity and Wine Initiative' was established in 2004 to promote sustainable production practices and joint planning to avoid areas of high conservation potential (Giliomee 2006; von Hase et al. 2010; http://www.wwf.org.za/?1681/Biodiversity--Wine-Initiative). Similar wine industry-based programmes exist in Chile (Wine, Climate Change and Biodiversity Program; http://www.vccb.cl/english/index.html; Barbosa and Godoy 2014) and in California (California Sustainable Wine Growing Program; Zucca et al. 2009). Market-linked stewardship schemes also exist for biodiversity-based MTC region industries such as the harvest of wildflowers, rooibos tea (*Aspalathus linearis*) and honeybush tea (*Cyclopia* spp.) in South Africa (van Wilgen et al. 2016) and cork oak (*Quercus suber*) in the Mediterranean (Bugalho et al. 2011).

In highlighting the challenges associated with drivers of ecosystem change, the Millennium Ecosystem Assessment (MA 2005) popularized the concept of **ecosystem services**, namely the benefits that humans receive from ecosystems. This concept has gained traction in the conservation arena as a way to persuade policy-makers to invest in the sustained flows of such services, and has a conservation footprint in MTC regions. One of the best known and longest running ecosystem service-driven initiatives is the South African state-funded 'Working for water programme'. Now operating at a national scale, this programme's roots (and indeed much of its earlier activities) started in the Cape, where it was suggested that ongoing, unchecked invasion by woody alien plants would significantly impact stream flows and water supply (van Wilgen et al. 1992; Le Maitre et al. 1996). The Working for Water programme has spent the past 20+ years and significant resources addressing the dual challenges of unemployment and invasive plants. The programme has provided important lessons for the design of resource management approaches worldwide, where the concept of ecosystem services is used to align biological conservation with socioeconomic decisions linked to the environment (e.g. Urgenson et al. 2013; van Wilgen and Wannenburgh 2016).

9.2.3 Biodiversity and spatial planning

Strategies to ensure appropriate management of remnant natural systems beyond protected areas are increasingly gaining attention. In addition to the stewardship approaches described above for agricultural landscapes, focus has also turned to the development sector and land-use planning. Spatial conservation assessments, traditionally confined to the conservation sector, are now being integrated or '**mainstreamed**' into land-use planning at various scales (Figure 9.5). The objective of this process is to integrate biodiversity concerns into development models, policies, and practices. In doing so, biodiversity priority areas (e.g. threatened habitats, freshwater priority areas, areas with key ecological infrastructure) that require appropriate management or conservation actions, are considered in a more ecologically sustainable land use and development framework.

At a global scale, mainstreaming is now implicit in the goals of the Convention of Biological Diversity and the Global Environment Facility, and is gaining momentum in various biomes and sectors, but it is work out of the South African Cape (and neighbouring biomes) that has led the way on how to operationalize such activities (e.g. Brownlie et al. 2005; Pierce et al. 2005; Cowling et al. 2008). In the Cape, mainstreaming has been used to align the National Spatial Biodiversity Assessment with municipal-level Spatial Development Frameworks (development planning tools at a municipal [decision-making] level)(see Rouget et al. 2014) in the development of plans to inform land-use planning at a municipal (decision-making) level (e.g. Holmes et al. 2012) and in efforts to reduce the risks and impacts of natural hazards such as floods and droughts (Nel et al. 2014; Sitas et al. 2014; Reyers et al. 2015). Best-practice guidelines for environmental assessment that adopt a landscape approach to biodiversity assessment in the Western Cape have been published, updated, and are widely referred to (De Villiers et al. 2005; Cadman et al. 2016). Despite these advances, implementation challenges remain, especially in less capacitated municipalities (Wilhelm-Rechmann and Cowling 2011; Wilhelm-Rechmann and Cowling 2013).

It is not clear to what extent biodiversity has been incorporated into spatial planning in other MTC regions, but research out of California is starting to

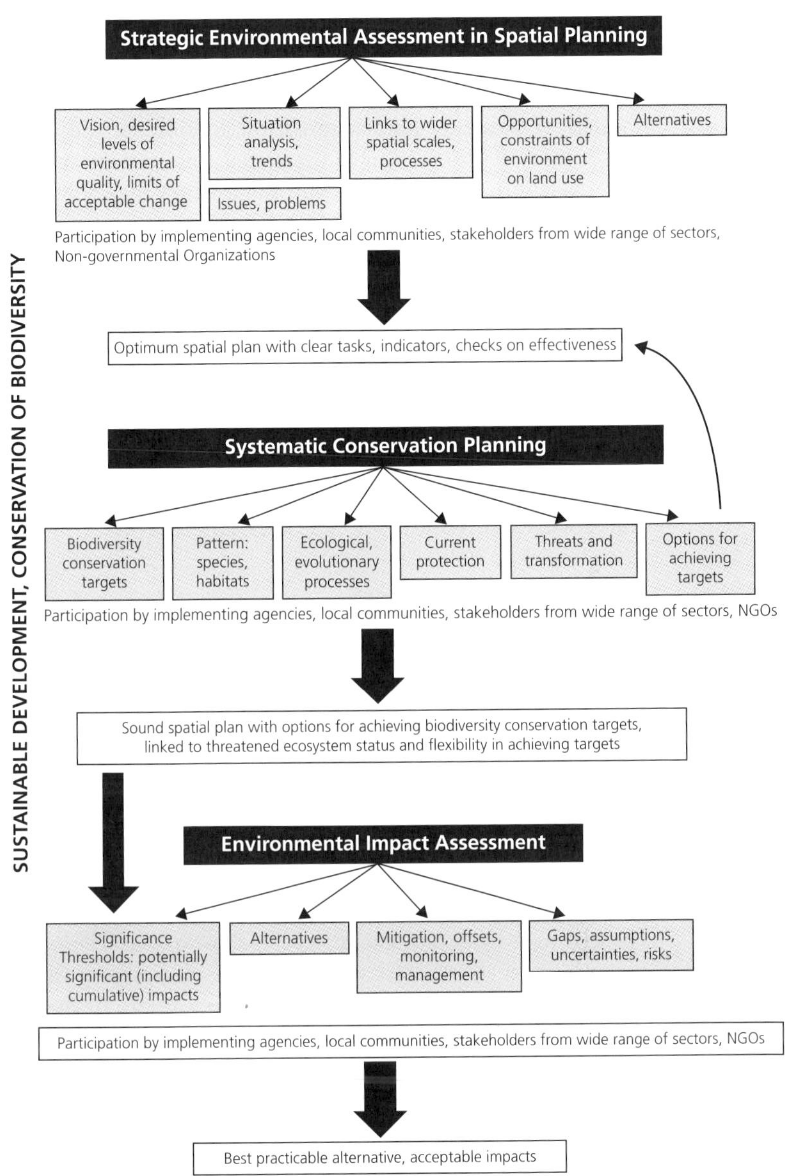

Figure 9.5 **Relationship between environmental impact assessment, systematic conservation planning, and strategic environmental assessment in spatial planning.** Shading of boxes represents comparable steps in each process. NGOs, Non-governmental organizations. (Redrawn from Brownlie et al. 2005.)

focus on the nexus between land-use planning, fire risk, and biodiversity. A regulatory approach to zoning and new construction housing density has been shown to alter projections of fire risk (Syphard et al. 2013), while models of **strategic land acquisition** to simultaneously reduce fire risk and improve conservation indicate that this strategy may well be a more effective approach to dealing with these issues than the traditional vegetation management options (e.g. fuel reduction to reduce fire hazard) (Syphard et al. 2016).

The general consensus is that this type of work requires socially relevant, user-inspired research that is co-produced through the involvement of a range of participants (drawn from research organizations, non-governmental organizations, public and private sectors, parastatals and civil society; Reyers et al. 2015). This type of work can be very time-consuming and challenging, but involving stakeholders in a social process of knowledge co-production results in projects that are deemed democratic and accountable—aspects that are critical in ensuring relevance and legitimacy in decision-making (Cowling et al. 2008).

9.3 Ecological restoration and related activities

Ecological restoration, the 'process of assisting the recovery of an ecosystem that has been degraded, damaged or restored' (SER 2004), falls within a suite of activities intended to initiate or accelerate the repair of ecosystems, including rehabilitation and mitigation (amongst others). Often targeting local indigenous reference conditions, **ecological restoration** generally aims to return former species composition, structure, and function to a site so that it can self-sustain. Efforts to achieve such outcomes are increasingly more challenging as biotic and abiotic thresholds are passed (Figure 9.6), so ecological restoration in its strictest sense may not always be possible. In contrast, **rehabilitation** does not necessarily return a site to its historical trajectory, but rather aims to re-establish ecosystem processes, productivity, and services. **Mitigation**, often used in more transformed situations, aims to reduce or offset impacts on the environment while providing potential for biodiversity conservation. Contracting parties to the Convention on Biological Diversity (CBD 1992) are all obliged to restore or rehabilitate degraded ecosystems, but the extent to which these activities are triggered depends on feasibility, intent, and, to some extent, social and/or legislative context.

Australia has perhaps the best example of an industrial-scale rehabilitation/restoration activity of the MTC regions, underpinned by research and involving earth-moving equipment and technology. Here, efforts to rehabilitate *Eucalyptus marginata* Jarrah forest after bauxite mining started in the 1960s (Koch and Hobbs 2007). At first, this involved replanting eucalypts to restore vegetation cover, but after three decades the focus shifted to include

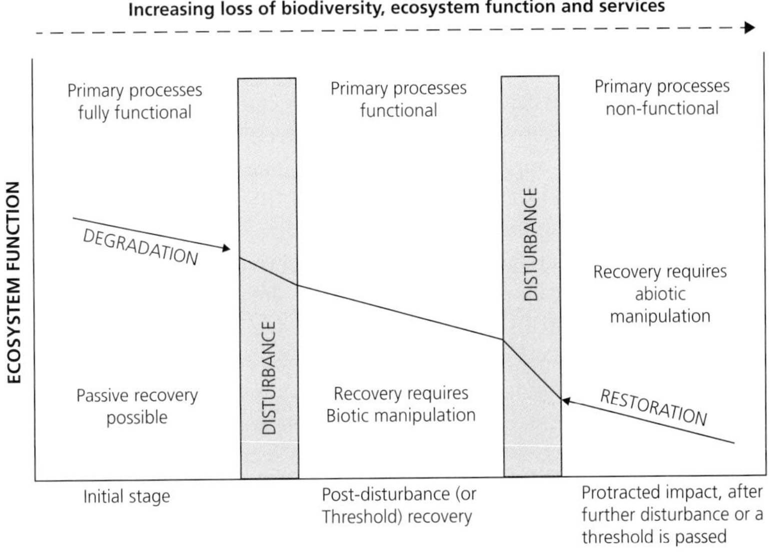

Figure 9.6   **Thresholds in ecological restoration.** Efforts to improve biodiversity, ecosystem function, and services become increasingly more challenging as biotic and abiotic thresholds are passed. (Adapted from Whisenant 2002 and King and Hobbs 2006.)

seeding and planting of a wider range of species, as well as faunal recovery, in an effort to restore a more diverse, functional forest ecosystem (Nichols and Nichols 2003). Restoration activities in this region have since grown to include a wider range of degraded remnant *Eucalyptus* woodlands and shrubland habitats. State-and-transition models have been developed to provide a theoretical framework and a starting framework for these restoration efforts (Yates and Hobbs 1997), but most of the practical experience has emerged from efforts to restore Jarrah forest. While the use of soil seed banks though harvested topsoil is effectively used to re-establish plant communities, a particularly strong feature of restoration-related research in this region has been to seek the best-practice technologies to store and optimally germinate seed for restoration purposes (e.g. Turner et al. 2013). Interestingly, although the groundbreaking discovery that smoke (or smoke-derived products) could be used to stimulate germination was first made by South African botanists (de Lange and Boucher 1990), the Australians were the first to implement this treatment in restoration activities. Today, this seed pre-treatment is used extensively for ecological restoration purposes in MTC regions (Australia: Commander et al. 2009; Rokich and Dixon 2007; Mediterranean: Crosti et al. 2006; South Africa: Holmes 2001; Hall et al. 2017; California: Wilkin et al. 2013).

In South Africa, the primary focus on ecological restoration is associated with the control of invasive alien woody plants, driven, as mentioned in Section 9.2.2, by an ecosystem services approach, and/or the restoration of native biodiversity to threatened habitats (e.g. Waller et al. 2016; Hall et al. 2017). Initially, it was assumed that, if given time, sufficient alien control follow-up and no further actions applied, communities would self-repair after alien control—i.e. **passive restoration**. The realization that invasive species often alter conditions and that these changes may persist as **legacy effects** led to an understanding that restoration trajectories may be inhibited by such legacies, even promoting **secondary invasion** after alien control. For example, nitrogen-fixing legumes such as *Acacia* spp. elevate the soil nitrogen status (Witkowski 1991; Stock et al. 1995). These changed soil conditions may persist up to 10 years after alien control (Nsikani et al. 2017), thereby facilitating the dominance of weedy native species typically associated with disturbance and secondary invaders such as nitrophilous grasses (Yelenik et al. 2004). Similar effects have been reported in Portuguese dune ecosystems (Marchante et al. 2008; 2009; 2011) and in Spanish *Quercus robur* forests (González-Muñoz et al. 2012). In California, legacy effects of Scotch broom (*Cytisus scoparius*) limit the regeneration of the locally dominant Douglas fir (*Pseudotsuga menziesii* var. *menziesii*; Grove et al. 2015). Legacy effects of alien species and ensuing secondary invaders/weedy native species can be self-reinforcing to the extent that restoration to an historical ecosystem is challenging without concerted **active restoration** measures (Gaertner et al. 2014). Active measures, amongst others, include the use of fire to remove alien biomass, and the active reintroduction of key native species via seeding or planting. Active restoration efforts can be justified ecologically and financially (Gaertner et al. 2012), but restoration success is variable and dependent on length of invasion and soil type (Holmes et al. 2000). Restoration activities in South Africa have also involved the removal of invasive fish using piscicides in an effort to restore native fish populations (Impson et al. 2013) as well as repair of riparian ecosystems after alien clearing (see Esler et al. 2008).

A considerable amount of attention has been given to restoration of invaded sites in California, but here the focus has been on the short-lived species that invade these systems. Ironically, it was historical restoration activities that partially resulted in this restoration need today. In the first half of the twentieth century it was thought that post-fire aerial seeding of short-lived non-native species, such as black mustard (*Brassica nigra*), would help with flood control through initiating recovery and soil stabilization (Keeley et al. 2012). Today, such non-native species are a key focus of ecological restoration efforts (e.g. Cione et al. 2002; Cox and Allen 2008). A wide range of restoration approaches have been applied (e.g. mowing, herbicide application, soil disturbance, re-seeding of native species, etc.), but the outcomes are variable and may depend on environmental and community context, making it challenging to translate findings for ecological restoration decision-makers (Stuble et al. 2017). California sage scrub, with its wind-dispersed drought-deciduous species, is less complex to restore than chaparral. Chaparral recruitment is largely fire-dependent, species exhibit lower recruitment, longer establishment times are required, and many have specific associations with mycorrhizal fungi (Horton et al. 1999; Mucina et al. 2017). Dickens et al. (2016) recently addressed the challenges of complexity and context using path analysis to evaluate a range of restoration outcomes in grassland and California sage sites. They concluded that native propagule addition had the desired effect of reducing non-native species and increasing native species, but that the application of herbicide depended on plant community context, and could result in increases of non-targeted, alien species. In Californian forests and woodlands, where livestock grazing, logging, and fire suppression have dramatically altered conditions, a key restoration focus has been on optimal ways to re-establish natural processes to enhance resilience and sustainability. This has involved tree-thinning to reduce tree densities and fuel associated with crown fire risk, protection of large trees, and reintroduction of surface fires to increase herbaceous cover and overall biodiversity (Allen et al. 2002).

California is also one of three regions in the USA where river restoration efforts abound (Bernhardt et al. 2005). There has been some attempt to learn from the wide variety and collective experience of these projects, which include, amongst others, salmonid habitat enhancement, riparian repair, and, more recently, restoration of fluvial process through modification of dam releases. Conventional ideas of stability and equilibrium do not apply to the rivers of this MTC region (and MTC rivers elsewhere), which are highly dynamic, a key consideration when setting restoration goals and strategies (Kondolf et al. 2007). However, as is generally the case with most ecological restoration efforts in MTC regions, a lack of long-term monitoring precludes opportunity to learn from successes and failures (Kondolf et al. 2007). Because environmental statutory law in the USA assumes that no definition is needed for the term 'restoration', these activities in California are widely interpreted to include true ecological restoration and an array of other rehabilitation and mitigation activities, such as the engineering of artificial wetlands through compensatory mitigation projects (Palmer and Ruhl 2015).

In contrast to South Africa and California, a silvicultural approach to restoration (strictly rehabilitation, since soil erosion and runoff are addressed, but not structure, function, and composition of native communities) exists in the Mediterranean and Chile, although a shift in restoration approach has been noted for the Mediterranean (Nunes et al. 2016). In the first half of the twentieth century, most large restoration projects in the Mediterranean were aimed at improving forest productivity, protecting watersheds, and promoting employment in rural areas using the silvicultural approach (planting fast-growing tree species). Towards the latter half of the century in European Union member countries, environmental legislation and policy has driven change towards a more ecosystem-based approach, focusing on a wider variety of species and soil. This is an important development, since environmental heterogeneity and a wide array of land-use histories have resulted in a range of degraded landscapes in the Mediterranean, including overgrazed, invaded, abandoned and cleared sites, and/or sites where processes such as natural fire regimes have been altered (Vallejo et al. 2012). More recent restoration efforts consider a wider range of species, at smaller scales and with less emphasis on plantation techniques. Interestingly, restoration projects in EU countries rely more on native plant species than non-EU countries, possibly because of EU legislation (Nunes et al. 2016). The use of alien species is justified (rightly or wrongly) on the basis that growth rates are faster compared to native species, that species used have greater aesthetic value, and that they act as nurse plants to facilitate establishment of other species (Nunes et al. 2014). Nunes et al. (2016) noted a variety of approaches used, with some being closer to ecological restoration than others. As much as 50 per cent of projects evaluated (n=148) faced unexpected results, largely owing to high mortality and low recovery of biodiversity as a result of drought (Nunes et al. 2016). This is a concern considering that drylands occupy about 67 per cent of the Mediterranean (White and Nackoney 2003), and these systems are slow to recover after degradation (Zdruli 2014).

Low-productivity *Acacia caven*-dominated 'espinales', covering more than 2,000,000 ha of Central Chile, have been the focus of restoration efforts in this MTC region (Aronson et al. 1993; Figures 8.4 and 9.7). Here, pre-European colonization vegetation community structure and composition are largely unknown, and expectations exist that any restoration efforts should lead to improved economic viability of silvopastoral farming activities. The focus has therefore been on the (re)introduction of plants with multipurpose value that improve soil conditions and diversity, including naturalized nitrogen-fixing legumes. Ovalle et al. (1999), reporting on a long-term research and development programme on restoration and rehabilitation of espinales, concluded that approximately 25 per cent of the central Chilean interior drylands lend themselves to active restoration through the reintroduction of native trees and shrubs, but that the remaining 75 per cent are so badly degraded that economic rehabilitation using fast-growing, palatable, non-native legume species (e.g. *Medicago polymorpha*) or complete reallocation to new land-use systems, is needed (Aronson et al. 1993). In sites where remnant vegetation is close by to facilitate regeneration, passive restoration of sclerophyllous forests has yielded some success (Fuentes-Castillo et al. 2012).

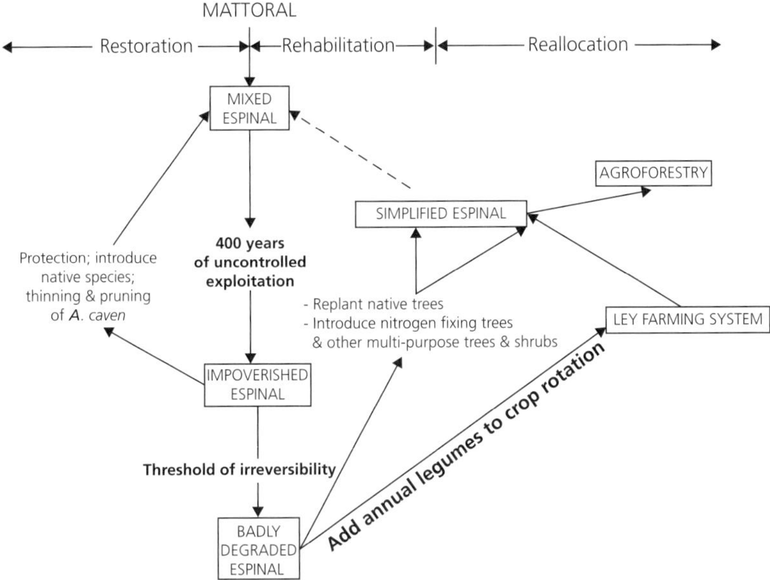

Figure 9.7 **Routes to restoration, rehabilitation, and reallocation of degraded Chilean espinales.** Depending on the state of vegetation, different actions are triggered—restoration, rehabilitation, or reallocation. (Adapted from Aronson et al. 1993.)

It is clear that a wide variety of restoration activities, driven by an array of motivations, have occurred in MTC regions. A major challenge is that these activities are often short term and are site-specific, and the communities targeted for restoration do not always respond predictably. Theoretical models have been used in conceptual attempts to make sense of inconsistencies and unexpected results (e.g. alternative states and positive feedbacks (Suding et al. 2004), including incorporation of state-and-transition models into the structure–function framework (Cortina et al. 2006; Figure 9.6), but this should be coupled with a more pragmatic approach to assess the effectiveness of restoration interventions based on available evidence. The use of path analysis to evaluate a range of restoration outcomes in Californian grassland and coastal sage communities is an example that requires replication in other MTC regions (Dickens et al. 2016).

9.4 Climate change, altered disturbance regimes, and fire management

Chapter 8 outlined the threats to MTC regions associated with changes in fire regime, climate, and non-native species dominance, amongst others.

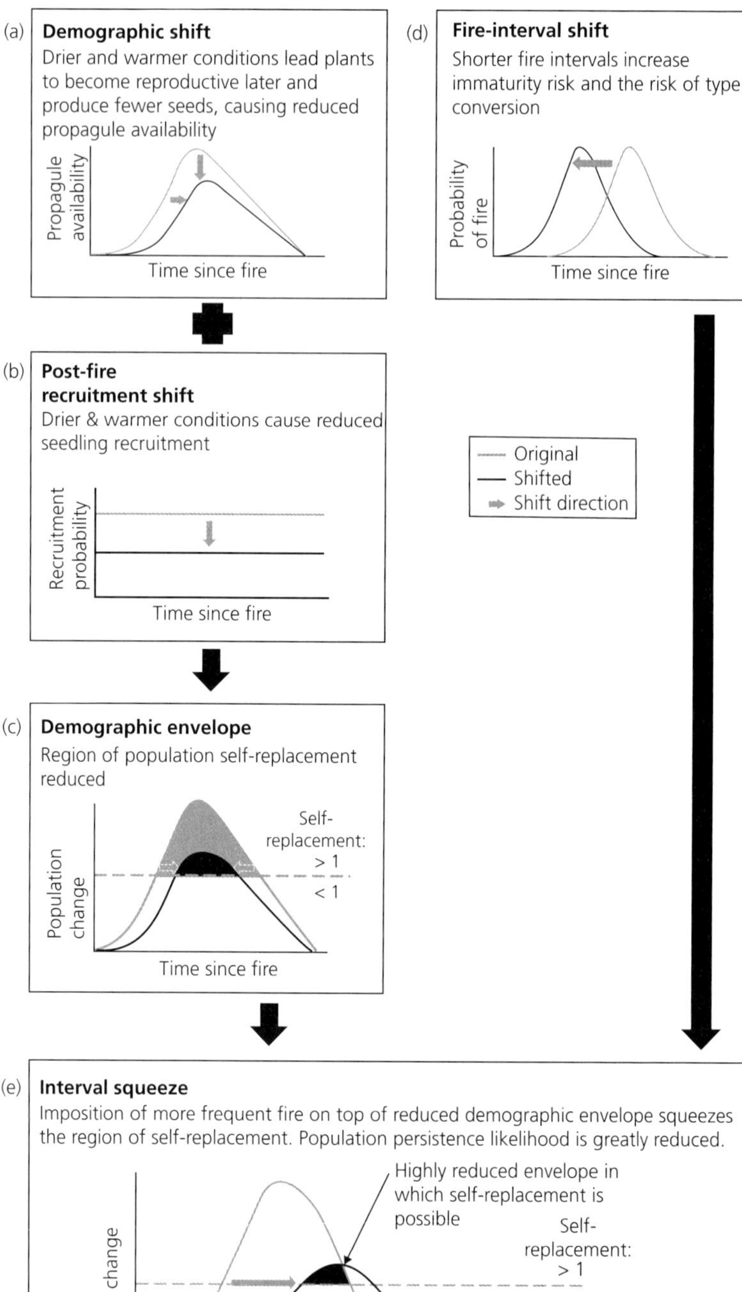
(a) Demographic shift
Drier and warmer conditions lead plants to become reproductive later and produce fewer seeds, causing reduced propagule availability
Propagule availability
Time since fire

(d) Fire-interval shift
Shorter fire intervals increase immaturity risk and the risk of type conversion
Probability of fire
Time since fire

(b) Post-fire recruitment shift
Drier & warmer conditions cause reduced seedling recruitment
Recruitment probability
Time since fire

Original
Shifted
Shift direction

(c) Demographic envelope
Region of population self-replacement reduced
Population change
Self-replacement:
> 1
< 1
Time since fire

(e) Interval squeeze
Imposition of more frequent fire on top of reduced demographic envelope squeezes the region of self-replacement. Population persistence likelihood is greatly reduced.
Highly reduced envelope in which self-replacement is possible
Self-replacement:
> 1
< 1
Population change
Time since fire

These threats are likely to be interactive, resulting in a 'perfect storm' for conservation managers. MTC regions are predicted to become warmer, leading to increases in drought-related mortality and fire frequency, favouring invasive species. Enright et al. (2015) contend that plant species loss in MTC regions is likely to happen more rapidly than currently proposed by separate climate envelope or fire-regime models. Demographic shifts in growth, survival, and fecundity are likely to interact with lower probabilities of recruitment in warmer (possibly drier) post-fire conditions, reducing the range of fire intervals suitable for self-replacement of populations. When combined with the effects of shorter fire return intervals, which lead to insufficient growth and fecundity before fire recurs, these shifts result in what Enright et al. (2015) refer to as an 'interval squeeze'—a further reduction in the range of fire return intervals suitable for population self-replacement. In short, there is the potential for dramatic loss in long-lived woody species, leading to changes in ecosystem composition, structure, and function because of these interactive effects (Figure 9.8). In-situ management responses will be needed to minimize such impacts.

These management responses are likely to be MTC region- and location-specific (Case Study 14). Climate controls on fire occurrence in MTC regions are close to tipping points, where relatively small changes in climatic conditions could lead to dramatic changes in fire regime. While warmer-drier conditions may result in decreased future fire activity, the opposite is predicted under the scenario of a warmer-wetter climate. Findings of geographically variable fire regime changes across MTC regions point to the need for diverse and flexible management strategies, which may include fire suppression, fire abatement, weather forecasting as a decision-support tool for management fires, monitoring of seed banks prior to burning, and potentially even ex-situ seed banking for future ecological restoration.

Figure 9.8   **(Opposite) Shifts that result in an interval squeeze.** A conceptual model developed by and adapted from Enright et al. (2015) demonstrating how, as climate warms and dries, changes in demographic rates, post-fire recruitment rates, and fire return interval can together reduce the fire-interval envelope that is suitable for post-fire population replacement of a perennial seeder. a) The grey curve represents the relationship between propagule availability and stand age prior to climate change, and the black curve demonstrates the demographic shift in age-related propagule availability as climate warms and dries. b) Depicts how post-fire recruitment declines (black line) with an increased incidence of drought after fire. c) Combines the demographic shifts shown in (a) and (b) and indicates how population replacement opportunity is reduced under climate change. d) Shows how the probability of fire shifts to shorter fire intervals under climate change. e) Illustrates the interval squeeze that results when demographic shifts and fire-interval shifts combine and further restrict the opportunity for post-fire stand self-replacement.

Case Study 14 Fire and climate change in mediterranean-type ecosystems

Max A. Moritz, Department of Environmental Science, Policy, and Management, University of California, Berkeley (UC Berkeley), Berkeley, California, USA

Enric Batllori, CEMFOR-CTFC, InForest Joint Research Unit, CSIC-CTFC-CREAF, Solsona, Spain; CREAF, Cerdanyola del Vallès, Spain

To project changes in fire activity for a given region, it is important to understand the processes that are currently limiting fire occurrence. Fire occurrence requires enough accumulating biomass to propagate fire across a landscape, a period when that biomass is dry enough to support combustion, and a coincident ignition; these three factors make up the 'fire regime triangle' controlling fire activity over broad scales of space and time (Moritz 1999; Moritz et al. 2005; Parisien and Moritz 2009; Krawchuk et al. 2009). Through global analyses, fire has been shown to display a unimodal and roughly humpshaped response to a biomass productivity gradient (Krawchuk et al. 2009), a dominant constraint that was first hypothesized by Ryan (1991). Varying constraints thus affect the probabilities of fire along this primary gradient (Krawchuk and Moritz 2011; 2014). Where ignitions are not limiting, fire activity is often seen as being more fuel-limited (i.e. generally low productivity, constrained by decadal and inter-annual pulses in precipitation) or more flammability-limited (i.e. generally more abundant fuel, constrained by occurrence of drought or dry hot and dry winds that reduce fuel moisture and enhance combustion).

Mediterranean-type ecosystems (MTEs) are characterized by substantial seasonal and interannual climatic variability, which favours the regular occurrence of fire. Several regions within MTEs, however, lie close to a threshold that discriminates between fuel-limited and flammability-limited conditions (e.g. Westerling and Bryant 2008; Bradstock 2010; Pausas and Paula 2012; Batllori et al. 2013; Turco et al. 2014). This is important because it could lead to highly divergent alterations in fire activity under future climates, depending on the direction of precipitation changes. Sustained warmer-drier conditions, for example, should push large portions of the current mediterranean-type biome toward being fuel-limited by the end of the twenty-first century. This would represent a shift in the dominant driver of fire activity in many areas, meaning that widespread fire increases reported in MTEs in recent decades could eventually be dampened by biomass availability constraints in the long term. In contrast, warmer-wetter conditions promoting biomass growth could favour fire occurrence over much of the driest portions of the mediterranean-type biome.

MTEs currently encompass a broad enough climate space to examine hypotheses about how spatially varied fire responses will be within different future climatic syndromes. Such hypotheses are now being used routinely in climate change adaptation planning exercises to identify which adaptation measures accommodate multiple future possible outcomes. In the case of MTEs, predictability is facilitated somewhat by a relatively pronounced moisture gradient that spans from roughly 30° latitude at the driest end and extends poleward to 40–45° latitude at the wetter end. Complexity is added, however, by topographic variation (e.g. mountain ranges in California, the Mediterranean, and Chile) and greater longitudinal spans (e.g. the Mediterranean and Australia) for some MTEs.

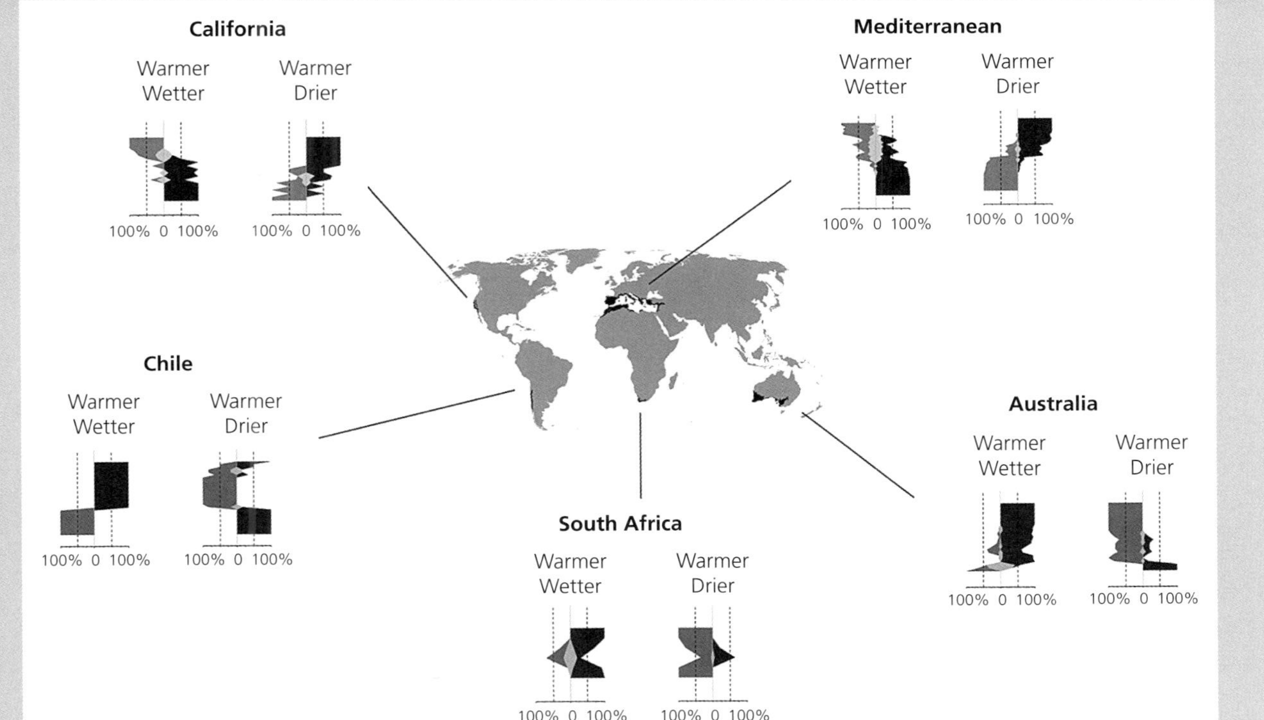

Case Study 14 Figure. Latitudinal patterns over the mediterranean-type biome of predicted changes in climate-driven fire probabilities for 2070–2099, as an estimate of fire frequency. Results are from climate-based statistical fire models driven by ensembles of six global climate models (GCMs) depicting warmer-drier and warmer-wetter climatic syndromes or 'bookends'. Five bioclimatic variables capturing annual temperature and precipitation seasonality and extremes were used to build fire probability models, which were fitted to a baseline 1971–2000 period and then projected to future climatic conditions. Inter-GCM level of agreement in expected fire changes under the two climatic syndromes was assessed using a 66.7% threshold (i.e. where less than four GCMs of six agreed in direction of fire changes were considered to have high uncertainty). Here black shows increasing fire frequencies, dark grey shows fire decreases, and light grey shows areas below the agreement threshold. See Batllori et al. (2013) for a detailed description of the modelling approach.

Continued

Case Study 14 (*Continued*)

Analysing 'bookend' ensembles of warmer-drier and warmer-wetter future scenarios reveals that substantial shifts in fire activity can be expected by the end of the current century, and also that the location and direction of change varies by MTE and by scenario (Case Study 14 Figure). In northern hemisphere MTEs under warmer-drier future conditions, potential climate-driven fire increases would be concentrated at high, relatively productive, latitudes, whereas areas in low, more xeric, latitudes could be subject to climate-induced fuel limitation and thus fire decreases. The opposite patterns would emerge under warmer-wetter conditions. Intermediate latitudes may present higher uncertainty because unique fire drivers are less strong there. Southern hemisphere MTEs are more latitudinally restricted and may exhibit less large-scale spatial variation in fire shifts; the Chilean MTE is an exception, showing similar but reversed fire patterns from northern hemisphere areas. These potential long-term climate-driven alterations in fire-proneness may certainly be over-ridden at finer scales, however.

It should be noted that projections such as those shown here are often based on long-term climate norms at relatively coarse spatial scales and they omit the possible influence of inter-annual variability. Local drivers such as land-use changes, human-induced modifications of landscape flammability and ignition density, or invasive species can all play important roles on any given landscape. Regardless, the climate-space and global arrangement of MTEs places many of these landscapes at or near 'tipping points' (Batllori et al. 2015) in terms of future fire activity. Despite adaptations of many MTE species to fire, alterations in fire frequencies could result in rapid changes to ecosystem composition, threatening biodiversity and the ecosystem services these areas provide. Characterizing fire-controlled critical thresholds of ecosystem resilience is thus crucial for effective adaptive management and conservation strategies.

9.5 Conclusion

A variety of responses exist to address the breadth and magnitude of impacts facing MTC regions, from protected area expansion to the appropriate management of remnant natural systems beyond protected areas to the recovery of degraded systems through ecological restoration. These responses need to be applied in a strategic, synergistic manner to minimize future biodiversity loss. An understanding of MTC systems, and how they function, is core to this ideal, and was a primary motivation for this book. However, future success will also rely on a broader framing of conservation action, to include people's perceptions and practices and a greater understanding of the social context within which these systems are embedded.

Literature cited

Abensperg-Traun M, Wrbka T, Bieringer G, Hobbs R, Deininger F, Main BY, Milasowszky N, Sauberer N, Zulka KP. 2004. Ecological restoration in the slipstream of agricultural policy in the old and new world. Agriculture, Ecosystems & Environment 103: 601–11.

Allen CD, Savage M, Falk DA, Suckling KF, Swetnam TW, Schulke T, Stacey PB, Morgan P, Hoffman M, Klingel JT. 2002. Ecological restoration of southwestern ponderosa pine ecosystems: a broad perspective. Ecological Applications 12: 1418–33.

Aronson J, Ovalle C, Avendano J. 1993. Ecological and economic rehabilitation of degraded 'Espinales' in the subhumid mediterranean-climate region of central Chile. Landscape and Urban Planning 24: 15–21.

Attwood SJ, Park SE, Maron M, Collard SJ, Robinson D, Reardon-Smith KM, Cockfield G. 2009. Declining birds in Australian agricultural landscapes may benefit from aspects of the European agri-environment model. Biological Conservation 142: 1981–91.

Bancroft L, Nichols T, Parsons D, Graber D, Evison B, Van Wagtendonk J. 1985. Evolution of the natural fire management program at Sequoia and Kings Canyon National Parks. Pages 174–80 in JE Lotan, BM Kilgore, WC Fischer, RW Mutch, technical coordinators. Proceedings of the symposium and workshop on wilderness fire. US Forest Service General Technical Report INT-182 1985 Apr.

Barbosa O, Godoy K. 2014. Conservación Biológica en Viñedos: conceptos claves y actividades prácticas. Primera Edición, Universidad Austral de Chile, ISBN 9789569412.02.8.

Barendse J, Roux D, Currie B, Wilson N, Fabricius C. 2016. A broader view of stewardship to achieve conservation and sustainability goals in South Africa. South African Journal of Science 112: 1–15.

Batllori E, Ackerly DD, Moritz MA. 2015. A minimal model of fire-vegetation feedbacks and disturbance stochasticity generates alternative stable states in grassland–shrubland–woodland systems. Environmental Research Letters 10: 034018.

Batllori E, Parisien M, Krawchuk MA, Moritz MA. 2013. Climate change-induced shifts in fire for Mediterranean ecosystems. Global Ecology and Biogeography 22: 1118–29.

Bernhardt ES, Palmer MA, Allan JD, Alexander G, Barnas K, Brooks S, Carr J, Clayton S, Dahm C, Follstad-Shah J, Galat D. 2005. Synthesizing US river restoration efforts. Science 308: 636–7.

Bradstock RA. 2010. A biogeographic model of fire regimes in Australia: current and future implications. Global Ecology and Biogeography 19: 145–58.

Brooks TM, Mittermeier RA, da Fonseca GA, Gerlach J, Hoffmann M, Lamoreux JF, Mittermeier CG, Pilgrim JD, Rodrigues AS. 2006. Global biodiversity conservation priorities. Science 313: 58–61.

Brownlie S, De Villiers C, Driver A, Job N, von Hase A, Maze K. 2005. Systematic conservation planning in the Cape Floristic Region and Succulent Karoo, South Africa: enabling sound spatial planning and improved environmental assessment. Journal of Environmental Assessment Policy and Management 7: 201–28.

Bugalho MN, Caldeira MC, Pereira JS, Aronson J, Pausas JG. 2011. Mediterranean cork oak savannas require human use to sustain biodiversity and ecosystem services. Frontiers in Ecology and the Environment 9: 278–86.

Cadman M, De Villiers C, Holmes P, Rebelo A, Helme N, Euston-Brown D, Clark B, Milton S, Dean WR, Brownlie S, Snaddon K, Day L, Ollis D, Job N, Dorse C, Wood J, Harrison J, Palmer G, Cadman M, Maree K, Manuel J, Holness S, Ralston S, Driver A. 2016. Fynbos forum ecosystem guidelines for environmental assessment in the Western Cape, Edition 2. Fynbos Forum, Fishoek, Cape Town. Available for download on bgis.sanbi.org.

CBD. 1992. Convention on Biological Diversity. http://www.cbd.int/doc/legal/cbd-en.pdf. Accessed 1 June 2017.

Chan KM, Shaw MR, Cameron DR, Underwood EC, Daily GC. 2006. Conservation planning for ecosystem services. PLoS Biol 4(11): p.e379.

Chapin FS, Carpenter SR, Kofinas GP, Folke C, Abel N, Clark WC, Olsson P, Smith DMS, Walker B, Young OR, Berkes F. 2010. Ecosystem stewardship: sustainability strategies for a rapidly changing planet. Trends in Ecology & Evolution 25: 241–9.

Cione NK, Padgett PE, Allen EB. 2002. Restoration of a native shrubland impacted by exotic grasses, frequent fire, and nitrogen deposition in southern California. Restoration Ecology 10: 376–84.

Commander LE, Merritt DJ, Rokich DP, Dixon KW. 2009. Seed biology of Australian arid zone species: germination of 18 species used for rehabilitation. Journal of Arid Environments 73: 617–25.

Conradie B, Treurnicht M, Esler K, Gaertner M. 2013. Conservation begins after breakfast: the relative importance of opportunity cost and identity in shaping private landholder participation in conservation. Biological Conservation 158: 334–41.

Cortina J, Maestre FT, Vallejo R, Baeza MJ, Valdecantos A, Pérez-Devesa M. 2006. Ecosystem structure, function, and restoration success: are they related? Journal for Nature Conservation 14: 152–60.

Cowling RM, Rundel PW, Lamont BB, Arroyo MK, Arianoutsou M. 1996. Plant diversity in mediterranean-climate regions. Trends in Ecology and Evolution 11: 362–6.

Cowling RM, Pressey RL, Rouget M, Lombard AT. 2003. A conservation plan for a global biodiversity hotspot—the Cape Floristic Region, South Africa. Biological Conservation 112: 191–216.

Cowling RM, Egoh B, Knight AT, O'Farrell PJ, Reyers B, Rouget M, Roux DJ, Welz A, Wilhelm-Rechman A. 2008. An operational model for mainstreaming ecosystem services for implementation. Proceedings of the National Academy of Sciences 105: 9483–8.

Cox RD, Allen EB. 2008. Stability of exotic annual grasses following restoration efforts in southern California coastal sage scrub. Journal of Applied Ecology 45: 495–504.

Cox RL, Underwood EC. 2011. The importance of conserving biodiversity outside of protected areas in mediterranean ecosystems. PLOS One 6(1): e14508. https://doi.org/10.1371/journal.pone.0014508.

Crosti R, Ladd PG, Dixon KW, Piotto B. 2006. Post-fire germination: the effect of smoke on seeds of selected species from the central Mediterranean basin. Forest Ecology and Management 221: 306–12.

De Lange JH, Boucher C. 1990. Autecological studies on *Audouinia capitata* (Bruniaceae). I. Plant-derived smoke as a seed germination cue. South African Journal of Botany 56: 700–3.

De Villiers C, Driver A, Clark B, Euston-Brown D, Day L, Job N, Helme N, Holmes P, Brownlie S, Rebelo T. 2005. Fynbos forum ecosystem guidelines for environmental assessment in the Western Cape. Fynbos Forum, Kirstenbosch, Cape Town.

Dickens SJM, Mangla S, Preston KL, Suding KN. 2016. Embracing variability: environmental dependence and plant community context in ecological restoration. Restoration Ecology 24: 119–27.

Dicks LV, Ashpole JE, Dänhardt J, James K, Jönsson A, Randall N, Showler DA, Smith RK, Turpie S, Williams D, Sutherland WJ. 2013. Farmland conservation: evidence for the effects of interventions in northern and western Europe. Exeter, Pelagic Publishing.

Ebbin MJ. 1997. Is the southern California approach to conservation succeeding? Ecology Law Quarterly 24: 695–708.

Enright NJ, Fontaine JB, Bowman DMJS, Bradstock RA, Williams RJ. 2015. Interval squeeze: altered fire regimes and demographic responses interact to threaten woody species persistence as climate changes. Frontiers in Ecology and the Environment 13: 265–72.

Esler KJ, Holmes PM, Richardson DM, Witkowski ETF. 2008. Riparian vegetation management in landscapes invaded by alien plants: insights from South Africa. South African Journal of Botany 74: 397–400.

Feldman TD, Jonas AEG. 2000. Sage scrub revolution? Property rights, political fragmentation, and conservation planning in southern California under the Federal Endangered Species Act. Annals of the Association of American Geographers 90: 256–92.

Fuentes-Castillo T, Miranda A, Rivera-Hutinel A, Smith-Ramírez C, Holmgren M. 2012. Nucleated regeneration of semiarid sclerophyllous forests close to remnant vegetation. Forest Ecology and Management 274: 38–47.

Gaertner M, Nottebrock H, Fourie H, Privett SD, Richardson DM. 2012. Plant invasions, restoration, and economics: perspectives from South African fynbos. Perspectives in Plant Ecology, Evolution and Systematics 14: 341–53.

Gaertner M, Biggs R, Te Beest M, Hui C, Molofsky J, Richardson DM. 2014. Invasive plants as drivers of regime shifts: identifying high-priority invaders that alter feedback relationships. Diversity and Distributions 20: 733–44.

Giliomee JH. 2006. Conserving and increasing biodiversity in the large-scale, intensive farming systems of the Western Cape, South Africa: commentary. South African Journal of Science 102: 375–8.

Gilmer DS, Miller MR, Bauer RD, LeDonne JR. 1982. California's Central Valley wintering waterfowl: concerns and challenges. Pages 441–52 in K. Sabol, ed. Transactions of the 47th North American Wildlife and Natural Resources Conference. US Fish and Wildlife Publications, Paper 41, Washington, DC, USA.

González-Muñoz N, Costa-Tenorio M, Espigares T. 2012. Invasion of alien *Acacia dealbata* on Spanish *Quercus robur* forests: impact on soils and vegetation. Forest Ecology and Management 269: 214–21.

Grove S, Parker IM, Haubensak KA. 2015. Persistence of a soil legacy following removal of a nitrogen-fixing invader. Biological Invasions 17: 2621–31.

Groves DG, Lempert RJ. 2007. A new analytical method for finding policy-relevant scenarios. Global Environmental Change 17: 73–85.

Hall SA, Newton RJ, Holmes PM, Gaertner M, Esler KJ. 2017. Heat and smoke pre-treatment of seeds to improve restoration of an endangered Mediterranean climate vegetation type. Austral Ecology 42: 354–66.

Hoekstra JM, Boucher TM, Ricketts TH, Roberts C. 2005. Confronting a biome crisis: global disparities of habitat loss and protection. Ecology Letters 8: 23–9.

Holmes PM. 2001. Shrubland restoration following woody alien invasion and mining: effects of topsoil depth, seed source, and fertilizer addition. Restoration Ecology 9: 71–84.

Holmes PM, Richardson DM, van Wilgen BW, Gelderblom C. 2000. Recovery of South African fynbos vegetation following alien woody plant clearing and fire: implications for restoration. Austral Ecology 25: 631–9.

Holmes PM, Rebelo A, Dorse C, Wood J. 2012. Can Cape Town's unique biodiversity be saved? Balancing conservation imperatives and development needs. Ecology and Society 17: http://www.ecologyandsociety.org/vol17/iss2/art28/.

Horton TR, Bruns TD, Parker VT. 1999. Ectomycorrhizal fungi associated with *Arctostaphylos* contribute to *Pseudotsuga menziesii* establishment. Canadian Journal of Botany 77: 93–102.

Impson DN, van Wilgen BW, Weyl OL. 2013. Coordinated approaches to rehabilitating a river ecosystem invaded by alien plants and fish. South African Journal of Science 109: 1–4.

Kark S, Levin N, Grantham HS, Possingham HP. 2009. Between-country collaboration and consideration of costs increase conservation planning efficiency in the Mediterranean Basin. Proceedings of the National Academy of Sciences 106: 15368–73.

Keeley JE, Bond WJ, Bradstock RA, Pausas JG, Rundel PW. 2012. Fire in Mediterranean ecosystems: ecology, evolution and management. Cambridge University Press, Cambridge.

King EG, Hobbs RJ. 2006. Identifying linkages among conceptual models of ecosystem degradation and restoration: towards an integrative framework. Restoration Ecology 14: 369–78.

Klausmeyer KR, Shaw MR. 2009. Climate change, habitat loss, protected areas and the climate adaptation potential of species in mediterranean ecosystems worldwide. PLoS One 4(7): p.e6392.

Knight AT, Cowling RM, Boshoff AF, Wilson SL, Pierce SM. 2011. Walking in STEP: lessons for linking spatial prioritisations to implementation strategies. Biological Conservation 144: 202–11.

Koch JM, Hobbs RJ. 2007. Synthesis: is Alcoa successfully restoring a jarrah forest ecosystem after bauxite mining in Western Australia? Restoration Ecology 15: S137–S144.

Kondolf GM, Anderson S, Lave R, Pagano L, Merenlender A, Bernhardt ES. 2007. Two decades of river restoration in California: what can we learn? Restoration Ecology 15: 516–23.

Krawchuk MA, Moritz MA, Parisien MA, Van Dorn J, Hayhoe K. 2009. Global pyrogeography: the current and future distribution of wildfire. PLOS One 4(4): e5102. https://doi.org/10.1371/journal.pone.0005102.

Krawchuk MA, Moritz MA. 2011. Constraints on global fire activity vary across a resource gradient. Ecology 92: 121–32.

Krawchuk MA, Moritz MA. 2014. Burning issues: statistical analyses of global fire data to inform assessments of environmental change. Environmetrics 25: 472–81.

Kremen C, Williams NM, Bugg RL, Fay JP, Thorp RW. 2004. The area requirements of an ecosystem service: crop pollination by native bee communities in California. Ecology Letters 7: 1109–19.

Kruger FJ. 1977. Ecological reserves in Cape-fynbos—toward a strategy for conservation. South African Journal of Science 73: 81–5.

Le Maitre DC, van Wilgen BW, Chapman RA, McKelly DH. 1996. Invasive plants and water resources in the Western Cape Province, South Africa: modelling the consequences of a lack of management. Journal of Applied Ecology 33: 161–72.

Lombard AT, Cowling RM, Vlok JHJ, Fabricius C. 2010. Designing conservation corridors in production landscapes: assessment methods, implementation issues, and lessons learned. Ecology and Society 15: 7.

Long JW, Quinn-Davidson LN, Skinner CN. 2014. Science synthesis to support socioecological resilience in the Sierra Nevada and southern Cascade Range. General Technical Report PSW-GTR-247. US Dept. of Agriculture, Forest Service, Pacific Southwest research Station, Albany, CA.

MA (Millennium Ecosystem Assessment). 2005. Ecosystems and human well-being: synthesis. Washington, DC, Island Press.

Mace GM. 2014. Whose conservation? Science 345: 1558–60.

Manson C. 1994. Natural communities conservation planning: California's new ecosystem approach to biodiversity. Environmental Law Quarterly 24: 603–16.

Marchante E, Kjøller A, Struwe S, Freitas H. 2008. Short- and long-term impacts of *Acacia longifolia* invasion on the belowground processes of a Mediterranean coastal dune ecosystem. Applied Soil Ecology 40: 210–17.

Marchante E, Kjøller A, Struwe S, Freitas H. 2009. Soil recovery after removal of the N2-fixing invasive *Acacia longifolia*: consequences for ecosystem restoration. Biological Invasions 11: 813–25.

Marchante H, Freitas H, Hoffmann JH. 2011. Post-clearing recovery of coastal dunes invaded by *Acacia longifolia*: is duration of invasion relevant for management success? Journal of Applied Ecology 48: 1295–304.

Maree K, Ralston S. 2012. Protected areas and biodiversity mainstreaming. Pages 5–22 in Turner AA, ed. Western Cape province state of biodiversity 2012, CapeNature Scientific Services, Stellenbosch, South Africa.

Medail F, Quezel P. 1997. Hot-spots analysis for conservation of plant biodiversity in the Mediterranean Basin. Annals of the Missouri Botanical Garden 84: 112–27.

Médail F, Verlaque R. 1997. Ecological characteristics and rarity of endemic plants from southeast France and Corsica: implications for biodiversity conservation. Biological Conservation 80: 269–81.

Moritz MA. 1999. Controls on disturbance regime dynamics: fire in Los Padres National Forest. Unpublished PhD dissertation, University of California at Santa Barbara.

Moritz MA, Morais ME, Summerell LA, Carlson JM, Doyle J. 2005. Wildfires, complexity, and highly optimized tolerance. Proceedings of the National Academy of Sciences of the United States of America 102: 17912–17.

Mucina L, Bustamante-Sánchez MA, Duguy B, Holmes P, Keeler-Wolf T, Armesto JJ, Dobrowolski M, Gaertner M, Smith-Ramírez C, Vilagrosa A. 2017. Ecological restoration in mediterranean-type shrublands and woodlands. Pages 173–96 in Allison S, Murphy S, eds. Routledge handbook of ecological and environmental restoration. Taylor & Francis, Abingdon, UK.

Myers N, Mittermeier RA, Mittermeier CG, Da Fonseca GA, Kent J. 2000. Biodiversity hotspots for conservation priorities. Nature 403: 853–8.

Nel JL, Le Maitre DC, Nel DC, Reyers B, Archibald S, van Wilgen BW, Forsyth GG, Theron AK, O'Farrell PJ, Kahinda JMM, Engelbrecht FA. 2014. Natural hazards in a changing world: a case for ecosystem-based management. PloS One 9(5): p. e95942.

Nichols OG, Nichols FM. 2003. Long-term trends in faunal recolonization after bauxite mining in the Jarrah forest of Southwestern Australia. Restoration Ecology 11: 261–72.

Nsikani MM, Novoa A, van Wilgen B, Keet JH, Gaertner M. 2017. *Acacia saligna's* soil legacy effects persist longer than ten years after clearing. South African Journal of Botany 109: 361.

Nunes A, Oliveira G, Cabral MS, Branquinho C, Correia O. 2014. Beneficial effect of pine thinning in mixed plantations through changes in the understory functional composition. Ecological Engineering 70: 387–96.

Nunes A, Oliveira G, Mexia T, Valdecantos A, Zucca C, Costantini EA, Abraham EM, Kyriazopoulos AP, Salah A, Prasse R, Correia O. 2016. Ecological restoration across the Mediterranean Basin as viewed by practitioners. Science of the Total Environment 566: 722–32.

Olson DM, Dinerstein E. 2002. The Global 200: priority ecoregions for global conservation. Annals of the Missouri Botanical Garden 89: 199–224.

Ovalle C, Aronson J, Del Pozo A, Avendaño J. 1999. Restoration and rehabilitation of mixed espinales in central Chile: 10-year report and appraisal. Arid Soil Research and Rehabilitation 13: 369–81.

Palmer MA, Ruhl JB. 2015. Aligning restoration science and the law to sustain ecological infrastructure for the future. Frontiers in Ecology and the Environment 13: 512–19.

Parisien M, Moritz MA. 2009. Environmental controls on the distribution of wildfire at multiple spatial scales. Ecological Monographs 79: 127–54.

Pauchard A, Villarroel P. 2002. Protected areas in Chile: history, current status, and challenges. Natural Areas Journal 22: 318–30.

Pausas JG, Paula S. 2012. Fuel shapes the fire-climate relationship: evidence from Mediterranean ecosystems. Global Ecology & Biogeography 21: 1074–82.

Phalan B, Balmford A, Green RE, Scharlemann JPW. 2011a. Minimising harm to biodiversity of producing more food globally. Food Policy 36: S62–S71.

Phalan B, Onial M, Balmford A, Green RE, 2011b. Reconciling food production and biodiversity conservation: land sharing and land sparing compared. Science 333: 1289–91.

Pierce SM, Cowling RM, Knight AT, Lombard AT, Rouget M, Wolf T. 2005. Systematic conservation planning products for land-use planning: interpretation for implementation. Biological Conservation 125: 441–58.

Rebelo AG. 1992. Preservation of biotic diversity. Pages 309–44 in Cowling RM, ed. The ecology of fynbos: nutrients, fire and biodiversity. Oxford University Press, Cape Town, South Africa.

Rebelo AG, Siegfried WR. 1992. Where should nature reserves be located in the Cape Floristic Region, South Africa? Models for the spatial configuration of a reserve network aimed at maximizing the protection of floral diversity. Conservation Biology 6: 243–52.

Reyers B, Nel JL, O'Farrell PJ, Sitas N, Nel DC. 2015. Navigating complexity through knowledge coproduction: mainstreaming ecosystem services into disaster risk reduction. Proceedings of the National Academy of Sciences 112: 7362–8.

Rokich DP, Dixon KW. 2007. Recent advances in restoration ecology, with a focus on the Banksia woodland and the smoke germination tool. Australian Journal of Botany 55: 375–89.

Rouget M, Barnett M, Cowling RM, Cumming T, Daniels F, Hoffman M, Manuel J, Nel J, Parker A, Raimondo D, Rebelo AG. 2014. Conserving the Cape Floristic Region. Pages 321–36 in Allsopp N, Colville JF, Verboom GA, eds. Fynbos: ecology, evolution and conservation of a megadiverse region. Oxford University Press, Oxford.

Ryan KC. 1991. Wildfires vegetation and wildland fire: implications of global climate change. Environment International 17: 169–78.

SER (Society for Ecological Restoration International Science & Policy Working Group) 2004. The SER international primer on ecological restoration. www.ser.org & Tucson: Society for Ecological Restoration International.

Sitas N, Prozesky HE, Esler KJ, Reyers B. 2014. Exploring the gap between ecosystem service research and management in development planning. Sustainability 6: 3802–24.

Stock WD, Wienand KT, Baker AC. 1995. Impacts of invading N_2-fixing Acacia species on patterns of nutrient cycling in two Cape ecosystems: evidence from soil incubation studies and ^{15}N natural abundance values. Oecologia 101: 375–82.

Stuble KL, Fick SE, Young TP. 2017. Every restoration is unique: testing year effects and site effects as drivers of initial restoration trajectories. Journal of Applied Ecology 54: 1051–7 .

Suding KN, Gross KL, Houseman GR. 2004. Alternative states and positive feedbacks in restoration ecology. Trends in Ecology & Evolution 19: 46–53.

Syphard AD, Bar Massada A, Butsic V, Keeley JE. 2013. Land use planning and wildfire: development policies influence future probability of housing loss. PLoS One 8(8): e71708.

Syphard A, Butsic V, Bar-Massada A, Keeley J, Tracey J, Fisher R. 2016. Setting priorities for private land conservation in fire-prone landscapes: are fire risk reduction and biodiversity conservation competing or compatible objectives? Ecology and Society 21:2 http://dx.doi.org/10.5751/ES-08410-210302.

Tecklin DR, Sepulveda C. 2014. The diverse properties of private land conservation in Chile: growth and barriers to private protected areas in a market-friendly context. Conservation and Society 12: 203–17.

Tscharntke T, Klein AM, Kruess A, Steffan-Dewenter I, Thies C. 2005. Landscape perspectives on agricultural intensification and biodiversity–ecosystem service management. Ecology Letters 8: 857–74.

Tscharntke T, Clough Y, Wanger TC, Jackson L, Motzke I, Perfecto I, Vandermeer J,Whitbread A. 2012. Global food security, biodiversity conservation and the future of agricultural intensification. Biological conservation 151: 53–9.

Turco M, Llasat M, von Hardenberg J, Provenzale A. 2014. Climate change impacts on wildfires in a mediterranean environment. Climatic Change 125: 369–80.

Turner SR, Steadman KJ, Vlahos S, Koch JM, Dixon KW. 2013. Seed treatment optimizes benefits of seed bank storage for restoration-ready seeds: the feasibility of prestorage dormancy alleviation for mine-site revegetation. Restoration Ecology 21: 186–92.

Underwood EC, Klausmeyer KR, Cox RL, Busby SM, Morrison SA, Shaw MR. 2009b. Expanding the global network of protected areas to save the imperiled Mediterranean biome. Conservation Biology 23: 43–52.

Urgenson L, Prozesky H, Esler KJ. 2013. Stakeholder perceptions of an ecosystem services approach to clearing invasive alien plants on private land. Ecology and Society 18:1 http://dx.doi.org/10.5751/ES-05259-180126.

Vallejo VR, Allen EB, Aronson J, Pausas JG, Cortina J, Gutiérrez JR. 2012. Restoration of Mediterranean-type woodlands and shrublands. Pages 130–44 in van Andel J, Aronson J, eds. Restoration ecology: the new frontier, 2nd edn. Blackwell Publishing, Malden, MA.

Van Wagtendonk JW. 1974. Revised burning prescriptions for Yosemite National Park. National Park Service Occasional Paper Number Two. Washington, DC, USA.

van Wilgen BW, Wannenburgh A. 2016. Co-facilitating invasive species control, water conservation and poverty relief: achievements and challenges in South Africa's Working for Water programme. Current Opinion in Environmental Sustainability 19: 7–17.

van Wilgen BW, Bond WJ, Richardson DM. 1992. Ecosystem management. Pages 345–71 in Cowling RM, ed. The ecology of fynbos: nutrients, fire and diversity. Cape Town, Oxford University Press.

van Wilgen BW, Carruthers J, Cowling RM, Esler KJ, Forsyth AT, Gaertner M, Hoffman MT, Kruger FJ, Midgley GF, Palmer G, Pence GQ. 2016. Ecological research and conservation management in the Cape Floristic Region between 1945 and 2015: history, current understanding and future challenges. Transactions of the Royal Society of South Africa 71: 207–303.

Viers JH, Williams JN, Nicholas KA, Barbosa O, Kotzé I, Spence L, Webb LB, Merenlender A, Reynolds M. 2013. Vinecology: pairing wine with nature. Conservation Letters 6: 287–99.

von Hase A, Rouget M, Cowling RM. 2010. Evaluating private land conservation in the Cape Lowlands, South Africa. Conservation Biology 24: 1182–9.

Waller PA, Anderson PML, Holmes PM, Allsopp N. 2016. Seedling recruitment responses to interventions in seed-based ecological restoration of Peninsula Shale Renosterveld, Cape Town. South African Journal of Botany 103: 193–209.

Westerling AL, Bryant BP. 2008. Climate change and wildfire in California. Climatic Change 87: 231–49.

Wheeler S. 2009. California's climate change planning: policy innovation and structural hurdles. Pages 125–35 in Davoudi S, Crawford J, Mehmood A, eds. Planning for climate change. Strategies for mitigation and adaptation for spatial planners. Earthscan, Stirling, VA, USA.

Whisenant SG. 2002. Terrestrial systems. Pages 83–105 in Perrow MR, Davy AJ, eds. Handbook of ecological restoration. Volume 1. Principles of restoration. Cambridge University Press, New York.

White RP, Nackoney J. 2003. Drylands, people, and ecosystem goods and services: a web-based geospatial analysis (PDF version). World Resources Institute (http:// pdf.wri.org/drylands;pdf accessed on 11 April 2017).

Wilhelm-Rechmann A, Cowling RM. 2011. Framing biodiversity conservation for decision makers: insights from four South African municipalities. Conservation Letters 4: 73–80.

Wilhelm-Rechmann A, Cowling RM. 2013. Local land-use planning and the role of conservation: an example analysing opportunities. South African Journal of Science 109: 1–6.

Wilkin KM, Holland VL, Keil D, Schaffner A. 2013. Mimicking fire for successful chaparral restoration. Madroño 60: 165–72.

Winter SJ, Esler KJ, Kidd M. 2005. An index to measure the conservation attitudes of landowners towards Overberg Coastal Renosterveld, a critically endangered vegetation type in the Cape Floral Kingdom, South Africa. Biological Conservation 126: 383–94.

Witkowski ET. 1991. Effects of invasive alien acacias on nutrient cycling in the coastal lowlands of the Cape fynbos. Journal of Applied Ecology 28: 1–15.

Yates CJ, Hobbs RJ. 1997. Woodland restoration in the Western Australian wheatbelt: a conceptual framework using a state and transition model. Restoration Ecology 5: 28–35.

Yelenik SG, Stock WD, Richardson DM. 2004. Ecosystem level impacts of invasive *Acacia saligna* in the South African fynbos. Restoration Ecology 12: 44–51.

Zdruli P. 2014. Land resources of the Mediterranean: status, pressures, trends and impacts on future regional development. Land Degradation & Development 25: 373–84.

Zucca G, Smith DE, Mitry DJ. 2009. Sustainable viticulture and winery practices in California: what is it, and do customers care. International Journal of Wine Research 2: 189–94.

Index

Tables and figures are indicated by an italic *t* and *f* following the page number.

A

Abies
  A. concolor 115*t*, 128
  A. magnifica 128
Acacia 84, 114*t*, 117*t*, 125, 153, 157*t*, 221*t*, 228*t*, 266, 267, 304
  A. acuminata 114*t*
  A. aneura 114*t*
  A. baileyana 248*t*
  A. caven 94*f*, 117*t*, 256, 257*f*, 268, 306, Plate 6b, Plate 15
  A. cyclops 248, 248*t*, 264*f*, Plate 16a
  A. dealbata 248, 248*t*
  A. decurrens 248*t*
  A. denticulosa 94*f*, Plate 6a
  A. elata 248*t*
  A. implexa 248*t*
  A. longifolia 248, 248*t*
  A. mearnsii 248, 248*t*
  A. melanoxylon 248, 248*t*, 249
  A. paradoxa 248*t*
  A. podalyrifolia 248*t*
  A. pycnantha 248*t*
  A. retinoides 248*t*
  A. saligna 248*t*, 249
  A. stricta 248*t*
  invasive Australian acacia species in MTC regions 247–50
  subgenus *Phyllodineae* 247
Acmispon glaber 228*t*, 231
actinomycete bacteria 228
actinorhizal symbionts 228
active restoration 304, 306
adaptations 54, 69, 179
  plant 54, 70, 84, 227–9
  see also organismal adaptations
adaptive breakthrough 89
adaptive radiation 33, 156, 157, 158
adaptive traits 69–70, 85, 91, 199, 209
  in animals 87
  diversity and endemism 89

fire-associated 84
in microorganisms 88
root 77
use of phosphorus 228
Adenanthos 205*t*
Adenostoma 28, 160
  A. fasciculatum 45*f*, 72*f*, 116*t*, 122, 183, 184*f*, 221*t*, 223, 275, Plate 4g, Plate 12c
  A. sparsifolium 116*t*
Adesmia 157*t*
  A. microphylla 124
Aextoxicon 153
  A. punctatum 116*t*
Africa
  diverse genera of 158*t*
  heathland occurrence in 124*f*
Afromontane forests and thickets 131, 133, 154
Agathosma 157*t*
agrosilvopastural savannas 117*t*, 129
Agrostis pourretii 135
Allocasuarina 228*t*
Alyssum 158*t*
ammonium 229
anagenesis 155
Anchusa 158*t*
Andalusia, Spain 158*t*, 231
Andes 33, 43, 50, 131, 153, 201, 203
Angophora 128
animals
  adaptations to MTC environments 70
  adaptive traits 69, 87–8
  behavioural traits 87
  granivores 93
  invasive 267–8
  and plant interactions 92–7
  as pollinators 94–5
  response to fire 88, 200
  as seed dispersers 95–6
annuals 73–4*t*, 76, 110, 114*t*, 120, 133–4, 135, 152, 205, 231, 266
  see also plants

Anthropocene 39, 244
ants
  Argentine 267
  dispersal of seeds by 97, 205
Araucaria 89*f*, 153
  A. araucana 131, 132*f*
Arbutus 28, 152
  A. menziesii 115*t*
  A. unedo 198*t*
Arctostaphylos 12*f*, 72*f*, 116*t*, 157*t*, 158, 159*f*, 163, 185,
    198*t*, 205*t*
  A. glandulosa 159*f*, 184*f*, Plate 11a, Plate 11c,
    Plate 11d, Plate 11f, Plate 12e
  A. glauca 159*f*, 184*f*, Plate 11b, Plate 11e, Plate 11g,
    Plate 12e
  resprouting and non-resprouting 159*f*, Plate 11
Arcto-Tertiary taxa 152
Argentine ant 267
Argillaceous rocks 51*t*
arid shrub communities 126*f*, Plate 8
Arizona 119, 120, 181
Artemisia 72*f*, 152
  A. californica 116*t*, 124
arum lily 264*f*, 267, Plate 16h
Aspalathus 157*t*, 205*t*, 228*t*, 231
  A. hirta 198*t*
  A. linearis 299
Asteraceae 117*t*, 123, 153, 157*t*, 158*t*, 160
Astragalus 157*t*, 158*t*, 228*t*
Australia
  alien annual grasses 265
  areas of protected, converted and impacted
    land in 293*t*
  bird pollinators from 96*f*, Plate 7e
  density of shrubs and trees 122
  development of modern MTC region 32–3
  diversity 168
  edaphic communities in 92*t*
  European colonization of 256
  evergreen sclerophyllous shrublands stand
    structure and productivity 221*t*
  examples of vegetation community types 113–14*t*
  forests 113*t*, 114*t*
  geoflora connections to Gondwana 153
  grasslands 114*t*, 135
  habitat type conversion 275
  heathlands 122, 124*f*
  highest point of 51
  human impact 244*t*
  Kwongon stand structure and productivity 221*t*
  landscape 50
  leaf mass area 182
  Mallee stand structure and productivity 221*t*

  most speciose genera in 157*t*
  MTC region of 119
  population density 244
  post-fire seeders 205*t*
  protection levels 296
  rehabilitation/restoration activity 303
  shrubland 114*t*
  soil and foliar nutrient content 226*t*
  southwestern 5, 17
  thicket stand structure and productivity 221*t*
  woodlands 114*t*, 128
Australian Cricket Ball Hakea 46*f*, Plate 5d
Avena 118*t*, 135, 264*f*, Plate 16e
  A. barbata 116*t*
  A. fatua 116*t*, 265
azonal riparian and wetland vegetation 111

B

Baccharis 137
Bacillus megaterium 82, 83*t*
Banksia 12*f*, 26*f*, 28, 80, 84, 114*t*, 185, 205*t*, 206, 228,
    Plate 3c
  B. attenuata 198*t*
  B. carlinoids 274*f*, Plate 18a
  B. menziesi 198*t*
  B. ornata 221*t*
  fire regime changes and habitat type
    conversion 261
  leaf mass area 179*f*
  resprouting following fire 45*f*, Plate 4e
batha 118*t*, 125
biodiversity
  hotspots 17–19, 291
  intactness 244
biogeographical history 23, 55
biomass 219–20
biomes 4, 224
  mediterranean-type shrubland biome 6
  terrestrial 4, 240, 244, 293
  see also mediterranean-type ecosystems (MTEs)
birds
  dispersal of seeds by 96, 97
  granivorous 93, 94
  as pollinators 95
black mustard 304
Bluff Knoll, Australia 51
Botryosphaeria dothidea 268, 269*f*, Plate 17a
Bradyrhizobium 83, 228
  B. elkanii 82, 83*t*
Brassica 262*f*
  B. nigra 265, 304
  Brome see *Bromus*
  Bromus 116*t*, 117*t*, 118*t*, 262*f*, 264*f*, 265, Plate 16e

B. carinatus 133
B. erectus 135
B. hordeaceus 135
Brunia 198*t*, 270
B. noduliflora 198*t*
Bruniceae 198*t*, 269*f*, Plate 17c
Burkholderia 228
Byblis 229

C

Cabrillo, Juan Rodriguez 44
Cactaceae 6*f*, 7, Plate 2a, Plate 2c
Caesalpinioideae 228
Caladenia 157*t*
calcareous rocks 51*t*
Calceolaria 157*t*
Caldcluvia 85
California 5
  areas of protected, converted and impacted
    land in 293*t*
  chaparral *see* chaparral
  climate-biotic communities study comparing
    Chile and 14
  development of modern MTC region 33
  drought-deciduous soft-leaved shrublands 124
  edaphic communities 92*t*
  fire return interval (FRI) 45
  Floristic Province 17, 85, 120, 152, 157
  forests 115*t*, 128
  generalist or specialist pollinators from 96*f*,
    Plate 7b
  grassland 116*t*
  growing season 224
  habitat fragmentation 258
  highest point of 50
  human impact 244*t*
  invasive Australian acacia species in 248*t*
  invasive plants 265
  land-cover change 246–56
  landscape 50
  legacy effects of Scotch broom 304
  Madrean-Tethyan flora 152
  mediterranean-type vegetation 182
  Mojave Desert 127
  most speciose genera in 157*t*
  post-fire community 138
  post-fire seeders 205*t*
  protection levels 296
  restoration of 305
  shrublands 33, 116*t*
  soil and foliar nutrient content 226*t*
  urban and suburban development in
    southern 247*f*
  vegetation community types 115–16*t*
  wildfires 261
  woodland 115–16*t*
California pitcher plant 230*f*, Plate 13
California poppy 264*f*, Plate 16g
California sage scrub 116*t*, 124
canopy
  closed 111, 118, 119, 121, 122, 220, 224, 232
  open 115*t*, 232
Cape Floristic Region, South Africa 5, 17, 119,
    125, 154, 167, 168, 229, 244, 247, 251, 257,
    260, 263, 296
  flora 157
  Greater Cape Floristic Region 125, 241
  MTV in 119
Cape Town
  comparison of Los Angeles and 251*t*
  land-use changes in 251–6
Carduus pycnocephalus 265
Carex 157*t*
  C. flacca 135
  C. humilis 135
case studies
  Chilean MTC flora 85–6
  early mediterranean ecosystem comparisons 14–16
  fire and climate change in MTEs 310–12
  freezing and plant distribution within MTEs 39–42
  invasion of MTEs by Australian acacias 247–50
  land-use change comparison between the city of
    Cape Town, South Africa and Los Angeles
    County, USA 251–6
  large old (venerable) trees of fire-prone MTC
    regions 129–33
  life histories of MTC region species 203*f*
  linking fire traits and historical fire regimes in the
    mediterranean-type environment 202–3
  Mediterranean heathland and fynbos 160–3
  mediterranean-type floras 27–30
  MTV outside of mediterranean-type climate
    regions 119–20
  origin of the mediterranean biome 167–8
  proteoid root clusters in three mediterranean
    regions 80–3
  role of environmental stability in explaining
    variation in plant diversity in MTEs 34–5
Casuarina 228*t*
cats 268
cattle 254, 256, 266, 268
cavitation 195, 196, 197, 199–200, 275
Ceanothus 30, 84, 116*t*, 122, 157*t*, 158, 185, 205*t*,
    224, 228
  adaptive leaf traits 184*f*, Plate 12
  C. crassifolius 40*f*, Plate 12d

Ceanothus (*cont.*)
  C. megacarpus 40*f*, 46*f*, 47*f*, 184*f*, 221*t*, Plate 5e,
    Plate 12f
  C. oliganthus 40*f*
  C. spinosus 40*f*, 94*f*, Plate 6e
  C. vestitus 188*f*
  re-establishing through seeding following fire 46*f*
Cenophytic 24*f*, 25
Cenozoic 24*f*, 25, 27, 31, 32, 34, 35, 161
Centaurea 158*t*
  C. solstitialis 265
Cephalotus 229
  C. follicularis 230*f*, Plate 13e
Cercocarpus betuloides 228*t*
Chamaemelum
  C. fuscatum 135
  C. mixtum 135
chamise 28, 116*t*, 183, 184*f*, 223, Plate 12c
chaparral 5*f*, 119–20, 122, 221*t*, Plate 1b
  aboveground NPP 223
  mortality during droughts 274*f*, Plate 18b
  restoration of 305
chenopod 114*t*, 125
Chenopodiaceae 152, 153
Chile 5, 17
  alien annual grasses 265
  alien grasslands 266
  areas of protected, converted and impacted land
    in 293*t*
  climate-biotic communities study comparing
    California and 14–16
  degraded Chilean landscape 257*f*, Plate 15
  density of shrubs and trees 122
  development of modern MTC region 33
  diversity 168
  drought-deciduous soft-leaved shrublands 124
  effects of invasive animals in 268
  espinales 129, 256, 306, 307*f*
  establishment of MTC 33
  European colonization of 256
  examples of vegetation community types 116–17*t*
  fire return interval 45
  forests 116*t*
  generalist or specialist pollinators from 96*f*,
    Plate 7a
  geofloras 153
  grasslands 117*t*, 135
  growing season 224
  highest point of 50
  human impact 244*t*
  invasive Australian acacia species in 248*t*
  landscape 50
  matorral stand structure and productivity 222*t*

  mediterranean-type vegetation 182
  most speciose genera in 157*t*
  population density 243–4
  Protea family (Proteaceae) 26*f*
  protection levels 296
  restoration of 305
  shrublands 117*t*
  soil and foliar nutrient content 226*t*
  temperate taxa 131
  wine industry-based programmes 299
  woodlands 117*t*, 128
Christmas tree 229
Cistus 125, 158, 185, 198*t*, 205*t*
cladogenesis 155
Cliffortia 94*f*, 157*t*, 205*t*, Plate 6f
climate 4, 9, 35–43, 71
climate change 271–5, 307–12
climate diagrams ('climadiagrams') 36
  from different cities within the USA 36*f*
  for selected cities from each of the global
    MTC regions 37*f*
coastal sage scrub 111, 124, 275, 305
see also sage scrub
co-evolution 93, 94, 208
cohesion-tension model 180
Colliguaja 153
  C. odorifera 117*t*
common dormouse 87
community assembly processes 85, 110, 149–50
conservancies 297
conservation approaches 293–302
  biodiversity and spatial planning 300–2
  conservation stewardship 297–300
  protected area expansion 296–7
Convention on Biological Diversity 296, 300, 302
convergent evolution 6–7, 28, 53–6, 160–4, 180
  among MTE communities 164*f*
  between heathland communities 162*f*
  of a succulent plant form in different deserts 6*f*,
    Plate 2
see also evolution
Cordillera de la Costa, Valparaíso Region, Chile 5*f*,
    Plate 1c
Coriaria 137
  C. myrtifolia 228*t*
cork oak 26*f*, 128, 246*f*, 268, 299, Plate 14b
Corsica, France 5*f*, Plate 1d
Corymbia 128
  C. calophylla 40*f*
Crepis capillaris 135
crown fires 9, 43, 44, 56, 167, 201, 202, 203, 252,
    263, 305
see also fire

Cryptantha 157t
Cupressaceae 130, 131, 156
cyanobacteria 228
cycads 228
Cypressus sempervirons 131
Cystisus 228t
Cytisus scoparius 304

D

Darlingtonia californica 230f, 231f, Plate 13
Darwin, Charles 4
Daviesia 228t
decomposition 178, 231
deforestation 129
dehydration
   avoidance 196–7, 198t
   tolerance 88, 196, 197, 198t, 199–200, 209
demography, role of fire in population
     dynamics 200–10
deposition, nitrogen 246, 270–1
desert shrublands along arid margins 111, 125–7
di Castri, Francesco 14, 15, 16
Disa 158
disturbance regimes 239, 244, 256, 307–12
divergence 34, 155–9
diversity
   and community structure 109–46
   drivers of 166–70
   and endemism 89–91
   indices and topographic and climate stability
     across MTEs 35f
   in MTC regions 165–6
DNA barcoding 268
Dombey's beech 128
Douglas fir 130, 304
downy oak 129
Drimys winteri 116t, 156, 156f, Plate 10a
Drosera 229, 230f, Plate 13b, Plate 13c, Plate 13d
Drosophyllum lusitanicum 230f, 231f, Plate 13f
drought
   cavitation resistance and water status 199f
   global change-type (hot drought) 196
   -induced dieback and mortality 273, 274f, Plate 18
   regimes 196
   response of shrubs to 196–200
drought-deciduous
   soft-leaved shrubs 11, 13, 71, 110, 111, 124–5
   taxa 180
Dryandra 80, 157t

E

early Cenozoic 25, 167
earth stewardship 297

Echinops 117t, 135
Echium plantagineum 135
ecological and evolutionary adaptive traits 91–7
   animal and plant interactions 92–7
   edaphic communities 92
ecological restoration 257, 295, 312
   and related activities 302–7
   thresholds in 303f
ecological sorting 28, 54
ecosystem drivers 4, 8t, 9, 23, 33
   climate 35–43
   fire 43–7
   and processes 35–53
   soils and mineral nutrition 51–3
   topography and geology 47–51
ecosystem processes 119, 219–36, 275
ecosystems 3–4
   novel 250f
   structure and primary productivity 219–25
ecosystem services 3, 245, 263, 297, 299–300,
    304, 312
ecosystem stewardship 297
edaphic communities 92
edaphic effects 77
Elton, Charles 263
Elymus glaucus 133
endemic taxa 89–91, 273
endemism 160, 161, 166, 239, 252, 273, 291
   in desert shrublands along arid margins 127
   diversity and 89–91
   evergreen sclerophyll shrublands and 118, 119
   in heathlands 123
   in MTC regions 17, 34
Engelmann oak 128
environment 4
environmental impacts 252, 267
environment impacts 301f
Erica 118t, 137, 155, 157t, 160, 205t, 270
Ericaceae 270
Eriogonum 72f, 157t
Eschscholzia californica 264f, Plate 16g
espinales 129, 256, 306, 307f
eucalypts 40, 128, 130, 303
Eucalyptus 41, 111, 128, 129, 135, 157t, 206, 247, 264f,
    266, 303, Plate 16d
   E. diversicolor 45f, 113t, 129, 130, 132f, Plate 4c
   E. incrassata 221t
   E. jacksonii 129, 130, 132f
   E. marginata 40f, 41, 113t, 130, 132f, 302
   E. patens 40f
   E. wandoo 40f
Euclea 154
   E. racemosa 114t

Euphorbia acanthothamnos 94*f*, Plate 6c
Euphorbiaceae 6*f*, 7, Plate 2b, Plate 2d
Europe, heathland occurrence in 124*f*
evergreen California live oak 128
evolution 149–76
  co-evolution 93, 94, 208
  convergence 160–4
  divergence 155–9
  diversity 165–6
  diversity drivers 166–70
  extinction 165
  macroevolutionary patterns 155–65
  origins 149–55
  parallel 29, 30, 53
  see also convergent evolution
evolutionary convergence *see* convergent evolution
evolutionary radiation 33, 156, 157, 158
exaptations 28, 29, 30, 54, 70, 183
extinction 30, 149, 165, 166, 168, 243, 257, 258, 268
  debt 258, 259
  risk 273

F
facultative seeders 207, 208
farming 254, 295*f*, 298, 306
feedback loops 249
  in Acacia invasion dynamics 250*f*
Felis catus 268
fennel 265
Festuca christiani-bernardii 135
filters 110, 150*f*
  abiotic 110
  biotic 110, 121
  climatic 110
  edaphic 110
  environmental 110, 121
fire 43–7, 84–5, 125, 160
  biomass and 220
  causes of 43–4
  and climate change in MTEs 310–12
  endemics 138
  and evolutionary radiation 158
  evolution of traits in response to 206–10
  fire-dependent reproduction 138
  fire-stimulated flowering 138
  and nutrient cycling 53, 229
  post-fire flowering bulb from South Africa 45*f*
  post-fire resprouting 45*f*, Plate 4
  post-fire seeding 46*f*, 202–3, 205*t*, 273, Plate 5
  response of animals to 88, 200
  role in population dynamics 200–10
  shrub response to 204–6
fire management 307–12

fire regimes 9, 44, 200–3, 243
  and habitat type conversion 259–63
fire-return intervals (FRI) 9, 44, 122, 161, 201, 246,
    260, 261
  impact on species persistence 47*f*
  Santa Monica Mountains, California 262*f*
fish 267
Fitzroya 131
  F. cupressoides 130
flowering plants (angiosperms) 25
  fire stimulated flowering patterns 84
  flowering patterns 77, 84
  generalist or specialist pollinators 96*f*, Plate 7
  proportion of species flowering by month 78*f*
  see also plants
fluoroacetate 93
Foeniculum vulgare 265
food-producers 241
foraging 225, 241, 257, 260
forbs 134*f*, 135, 265, 266
forests 111, 128–33
  Australia 113*t*
  California 115*t*, 128
  Chile 116*t*
  Jarrah 41, 269*f*, 303
  Mediterranean 117*t*
  sclerophyllous 111, 306
  South Africa 114*t*, 128
  Spain 304
  and venerable trees in fire-prone MTC
    regions 132*f*
fountain grass 264*f*, Plate 16f
foxes 268
France 13, 16, 43, 45, 129, 135
Frankia 228
freezing 195, 199
  as a driver of plant distribution within MTEs 39–42
  -induced cavitation 195
French broom 266
FRI *see* fire-return intervals (FRI)
frost 40, 41, 42
fungal root disease 268
fungi 84, 227, 305
fynbos 5*f*, 122, 138, 161, 163, 222*t*, 252, 270, Plate 1e
  aboveground NPP 224
  habitat fragmentation 259

G
galactolipids 228
Galium 158*t*
garigue 111
garrigue 111, 118, 161, 222*t*
Gastrolobium 93, 228*t*

Genista 155, 205*t*, 228*t*, 266
  G. acanthoclada 222*t*
geofloras 149, 150, 151, 152, 153, 154
geological timescale 24*f*
geophytes 71–6, 127, 167, 231, 241, 260, 267
Gevuina 85
Giant sequoia 91
Gladiolus 158
Global 200, 291
Global Naturalized Alien Flora database 263
Gompholobium 228*t*
Gondwana 25, 27, 85, 86*f*, 110, 131, 153, 154
gorse 266
granivores 93, 225
grasses, yellow 264*f*, Plate 16f
grasslands 111, 133–5
  Australia 114*t*, 135
  California 116*t*
  Chile 117*t*, 135, 266
  Mediterranean 118*t*
  in MTC region 134*f*, Plate 9
  South Africa 115*t*, 135
Greater Cape Floristic Region, South Africa *see* Cape
  Floristic Region, South Africa
Greece 16, 34, 50, 125, 223, 224, 271
  diverse genera of 158*t*
  Mount Olympus 50
Grevillea 26*f*, 28, 157*t*, 184*f*, 205*t*, Plate 3c, Plate 12b
growth
  reproductive 193
  seasonality of 193
  vegetative 193

H

habitat
  conversion 239, 257, 276, 291, 293
  effect of fire on 88
  fire regime changes and habitat type
    conversion 259–63
  fragmentation 258–9, 260, 267, 270
  protection 293
  type-conversion 134, 135, 246, 259–63, 265, 275
Hajek, Ernesto 14
Hakea 26*f*, 80, 93, 157*t*, 179, 185, 198*t*, 205*t*, 206, 247,
  266, Plate 3c
  H. laurina 82
  H. platysperma 46*f*, Plate 5d
  H. polyanthema 198*t*
  H. prostrata 80
Haplopappus 157*t*
haustorium 229
heathlands 119, 122, 123–4, 161
hemiparasite 229

herbivores 87, 193, 219, 266, 268
  plant–herbivore interactions 92, 93–4
herriza 123, 161, 163
Hesperocyparis 206
Heteromeles 160, 198*t*
  H. arbutifolia 182
Hieracium pilosella 135
Hoary-leaf *Ceanothus* 184*f*, Plate 12d
Holarctic group 151–2
holm oak 128, 129, 268–70
holoparasite 229
Homo
  H. erectus 242*f*
  H. sapiens 241, 242*f*
Hordeum 135
  H. bulbosum 135
horned lizard 267
hotspots 17–19, 291
Humboldt, Alexander von 4, 14
Huntley, Brian 16
hydraulic redistribution 232
Hydrobiologica (journal) 137
hydrology 231–2
  and fire 232
  and vegetation 232
Hydrophyllaceae 138

I

Ilex 137
  I. aquifolium 198*t*, 199*f*
immaturity risk 46
Indonesian archipelago 243
insects 87, 95, 96, 229
International Biological Program (IBP) 13, 14, 17
International Society for Mediterranean Ecology
  (ISOMED) 13, 17
introduction dynamics 263
invasive species 247–50, 263–70
  invasive animals 267–8
  invasiveness of 263
  invasive plants 263–7, Plate 16
  legacy effects of 304
  pathogens 268–70
invertebrates 87, 88
Irano-Turanian taxa 152
ISOMED 13, 17
Israel 13, 125, 241, 261
Italy 129, 240*t*, 245
  Mont Blanc 50

J

Jarrah forest/woodland 41, 269*f*, 303, Plate 17
Joshua tree 95, 116*t*

Jubaea 89*f*
  J. chilensis 91
Juglans 152
Juniperus 117*t*, 131
  J. californica 116*t*
  J. thurifera 131
  J. thurifera 132*f*

K

karoo *see* succulent karoo, South Africa
kermes oak 128, 129
key adaptations 89
Köppen classification system 38
kwongan 28, 44, 111, 118, 119, 123, 161, 243

L

Lampranthus 157*t*
LandCare 298
landowner stewardship programmes 297
land sharing 295
land sparing 293, 295*f*
land stewardship 295, 298*f*
land-use and land-cover change (LULC) 244–58
  in Cape Town, South Africa and Los Angeles
    County, USA 251–6
land use, traditional 246*f*, Plate 14
Last Glacial Maximum (LGM) 33
late Cretaceous 25, 31
late Palaeogene 31
Laurasia 25, 27, 151, 152
Laurus 152
  L. nobilis 198*t*
Lavandula 155
leaching 226, 229
leaves
  broad sclerophyllous 160
  ericoid 123, 183, 184*f*, 185, Plate 12
  evergreen 77, 125, 154, 180, 186, 200, 227
  leaf area index 224–5
  leaf mass area 178, 179*f*, 224
  leaf size 183
  leaf traits 30, 179, 182–5, Plate 12
  longevity and sclerophylly 181*f*
  needle-like 160, 184*f*, Plate 12
  sclerophyllous 13, 28, 177–82
  seasonally dimorphic 71, 124, 125
  shedding of 197
  specific leaf area 178
  structure by region 179*f*
legacy effects, of alien species 304
leghaemoglobin 228
Leucadendron 205*t*, 206, 270
  L. salicifolium 81*f*, 83, 198*t*

Leucopogon 157*t*
  L. conostephioides 198*t*
Leucospermum 26*f*, 205*t*
life histories 122, 202, 203*f*, 208, 209, 210
lignotubers 43, 45*f*, 130, 204, 210
Limonium 158*t*
lineage sorting 155
Linepithema humile 267
Lithraea 12*f*, 85, 160
  L. caustica 85, 86*f*, 117*t*, 198*t*
litter, decomposition of 231
local determinism 170
lodgepole pine 128
Lomatia 26*f*, 85
Los Angeles 252
Los Angeles Basin 247*f*, 252, 254, 271
Los Angeles County 251
  comparison of Cape Town and 251*t*
  demographic and land-use changes 253*f*
  population growth in 252
Lotus corniculatus 135, 245, 257, 260, 273
lowlands 297
LULC *see* land-use and land-cover change (LULC)
Lupinus 157*t*
  L. albus 83, 83*t*

M

macroevolutionary patterns 155–65
  convergence 160–4
  divergence 155–9
  extinction 165
Macrozamia 228
Madrean-Tethyan flora 152
Madro-Tertiary taxa 152
mainstreaming 300
Malajzuk, Nick 82
mallee 5*f*, 114*t*, 118, 119, 221*t*, 258, Plate 1a
Malosma laurina 40, 45*f*, 198*t*, 268, 269*f*, Plate 4f,
  Plate 17a
maquis 5*f*, 111, 118, Plate 1d
maritime pine 246*f*, Plate 14a
market-linked schemes 297, 299
matorral 5*f*, 110, 111, 118, 123, 161, 225, Plate 1c
  coastal 111, 117*t*, 124
Maytenus 137
Medicago polymorpha 135, 306
Mediterranean
  agro-pastoral use of woodlands and
    shrublands 245
  areas of protected, converted and impacted land
    in 293*t*
  arid shrubland 127
  development of modern MTC region 32

diverse genera of the 158*t*
edaphic communities 92*t*
forest 117*t*
Garrigue stand structure and productivity 222*t*
generalist or specialist pollinators from 96f,
    Plate 7d
grassland 118*t*
highest point of 50
human impact 244*t*
invasive plants 265
land area 50
Laurasian geoflora 151–2
matorral vegetation 123
mediterranean-type vegetation 182
Phrygana stand structure and productivity 222*t*
post-fire seeders 205*t*
restoration of 305
shrubland 118*t*
soil and foliar nutrient content 226*t*
traditional land use 246f
vegetation community types 117–18*t*
woodland 117*t*
Mediterranean Basin 5, 17
    heathlands 161
    invasive Australian acacia species in 248*t*
Mediterranean land snails 87
Mediterranean lizards 89
mediterranean-type climate (MTC) 27
    aspects that affect mediterranean-type vegetation
        species 186f
mediterranean-type climate (MTC) regions 6, 9
    areas of protected, converted and impacted land
        in 293*t*
    cities and population centres 240*t*
    determinants of wildfire in 259f
    differences in ecosystem drivers 9
    early comparative research on 11–17
    forests and venerable trees in 132f
    future of 291–321
    global map 10f
    hotspots 292f
    human interactions with 241–4
    nature and conservation views of 295f
    pathogen impacts on plants in 269f, Plate 17
    plant community types in 113f
    population density 240
    predicted changes due to climate change in 272f
    river characteristics 137f
    threats to 239–40
    vegetation community types 113–18*t*
    vegetation profiles 112f
mediterranean-type ecosystems (MTEs) 5–7
    adaptive traits 69–70

characteristics of 23–65
defining 7–11
ecological convergence of 34
fire and climate change in (case study) 310–12
global location 7*t*
interactions between species and their
    environment 8*t*
landscape 5f, 48f, Plate 1
most speciose genera 157*t*
nutrient status of plants 53
origin and distribution of plant taxa 8*t*
origins 149–55
soil mineral nutrition of 52f
species from each MTE region representing
    sclerophyllous genera 12f
species pools and filters 150f
species structure, physiology and other organismal
    features 8*t*
systems for identifying 7–8*t*
temperature, precipitation regimes, and rainfall
    reliability 7–8*t*
terrain, drainage, and soil types within a broad
    climate region 8*t*
mediterranean-type vegetation (MTV) 6, 11, 25
    in mediterranean-type and non-mediterranean-
        type climate regions 120f
    origins 150
    outside of mediterranean-type climate regions 119
Melaleuca 157*t*
    M. scabra 198*t*
    M. uncinata 221*t*
Mesembryanthemaceae (Aizoaceae) 115*t*, 127
Mesorhizobium 228
meta-analyses 54, 249, 263
Mexico 120
    Tehuacán Valley 181
microbes 82, 83, 88, 228
Micropterus dolomieu 267
Millennium Ecosystem Assessment 299
Mimulus 157*t*
mineral nutrients 92, 168, 190, 192, 225–31, 270
    fire, decomposition and 229–31
    plant adaptations to low nutrients 227–9
    soils and 51–3
mitigation 302, 305
Mojave Desert, California 126f, Plate 8b
Molineriella minuta 135
Monterey pine 264f, Plate 16b
Moraea 157*t*
    M. flacida 267
Mount Whitney, California 50
MTC regions *see* mediterranean-type climate (MTC)
    regions

MTEs *see* mediterranean-type ecosystems (MTEs)
MTV *see* mediterranean-type vegetation (MTV)
Muehlenbeckia 137
mustard 265, 304
mycorrhizal fungi 84, 227, 305
Myrceugenia 86, 153
Myrica 228*t*
myrmecochory 97, 205, 267
Myrtaceae 114*t*, 153, 157*t*, 198*t*, 268

N

Nassella
    N. cernua 133
    N. pulchra 116*t*, 133
National Science Foundation, USA 13, 14
Native Americans 243, 260
Neogene 31, 34, 43
neotropical taxa 85, 153
Nerium 152
    N. oleander 185
net ecosystem exchange 220
net primary productivity (NPP) 223
    aboveground 223–4
neutral theory 170
Nicotiana glauca 264*f*, Plate 16c
nitrate 229, 267
nitrogen 226, 227, 229, 270
    foliar nitrogen content of MTV species 191*f*
    soil content 52
nitrogen-fixing bacteria 84, 93, 227, 228, 228*t*
nitrogen-fixing legumes 304, 306
Nolana 157*t*
North America 118, 152, 153, 243
northern hemisphere 25, 26, 43, 201, 210, 312
    common taxa in MTEs 26*f*, Plate 3
Nothofagus 116*t*, 120, 128, 131, 153
    N. alessandri 85, 116*t*
    N. dombeyi 116*t*, 128
    N. glauca 116*t*
    N. obliqua 116*t*
Notholithocarpus densiflorus 270
novel ecosystems 250*f*
nutrient enrichment 270–1
Nuytsia floribunda 229, 230*f*, Plate 13a

O

oak 26*f*, 224, 246, 246*f*, 252, 268, 270, 299,
        Plate 3a
    in California 128
    in Mediterranean countries 129
    woodland 115–16*t*, 117*t*
oat 264*f*, Plate 16e
obligate seeders 43, 207, 208, 209, 210, 273

'old, climatically buffered, infertile landscapes'
        (OCBILs) 50, 168
Olea 137, 152, 154
    O. europaea 117*t*, 131, 132*f*
    O. oleaster 118*t*
Oncorhynchus mykiss 267
Ononis 158*t*
Ophrys 158, 158*t*
Orchidaceae 157*t*, 158*t*
orchids, pollination of 95
organismal adaptations 69–88
    animals 87–8
    microbes 88
    plants 70–86
organisms and their interactions 69–108
    diversity and endemism 89–91
    ecological and evolutionary 91–7
    organismal adaptations 69–88
outgroups 53, 55
overgrazing 134, 246, 258, 306
Oxalis 157*t*
oxygen 200, 228
    isotope levels, from the deep sea 31*f*
Oxylobium 93

P

paleoclimates 23
Palmeria 85
Papua New Guinea 243
parallel evolution 29, 30, 53
Paranomus 205*t*
parasitic plants 229, 230*f*, Plate 13
parasitism 229
parent materials 51, 226
passive restoration 304, 306
pathogens 268–70
    impacts on MTC region plants 269*f*, Plate 17
patterns in species richness 169
Pelargonium 157*t*
Pennisetum setaceum 264*f*, Plate 16f
Penstemon 157*t*
perennial grasses 114*t*, 133, 135
Pérez-Fernández, María 82
Peumus 85
    P. boldus 85, 86*f*, 116*t*
Phacelia 157*t*
Phellinus weirii 268
phenology 54, 97, 195
    of plants 70–7
    reproductive 71
    of some California MTE species from different
        communities 72*f*
Phillyrea 185

P. latifolia 198*t*
phospholipids 228
phosphorus 77, 79*f*, 83, 226, 227, 228, 270
  soil content 52
photosynthesis 219, 224
  and foliar nitrogen content of MTV
    species 192*f*
  and growth 186–96
  rates of MTV species 190*f*
  stomatal and photosynthetic response to
    available 188*f*
  and temperature 187*f*
  transpiration compromise 189
  and water 187
phrygana 125, 222*t*, 223, 224
Phrynosoma coronatum 267
Phylica 157*t*
  P. cephalantha 222*t*
Phyllodoce 184*f*, 247, Plate 12a
physiognomy 129
  plants 13
Phytophthora 269*f*, 270, Plate 17
  P. cinnamomi 82, 268, 269*f*, 270, Plate 17b
  P. ramorum 270
pine trees 274*f*, Plate 18c
Pinus 117*t*, 131, 152, 154, 206, 247, 266
  P. contorta 128
  P. coulteri 46*f*, Plate 5b
  P. edulus 116*t*
  P. jeffreyii 115*t*
  P. lambertiana 115*t*
  P. monticola 128
  P. pinaster 246*f*, Plate 14a
  P. ponderosa 115*t*, 128
  P. radiata 264*f*, Plate 16b
  P. sabiniana 115*t*
Pistacia 12*f*, 94, 117*t*, 154
  P. lentiscus 118*t*, 198*t*
pitcher plant 230*f*, Plate 13e
plant communities 109–18
  classification of 111
plants
  adaptations 54, 70, 84, 227–9
  adaptations to low nutrients 227–9
  carnivorous 229, 230*f*, 231*f*, Plate 13
  diversity and endemism in MTEs 90*t*
  evergreen sclerophyllous 39
  and fire 84–5
  flammability of 207
  green 219
  and herbivores 93–4
  interactions between pollinators and 94–5
  phenology and life form 70–7

physical defence structures related to
  herbivory 94*f*, Plate 6
plant drought response 154
resprouting following fire 45*f*
roots, soils, and symbioses 77–84
and seed dispersers 95–7
seeding following fire 46*f*
species richness of vascular 18*f*, 19*f*
see also annuals; flowering plants (angiosperms)
Platanus 29, 116*t*, 152
Poa
  P. badensis 135
  P. secunda 133
Poaceae 116*t*
Podalyria 205*t*
Podocarpaceae 114*t*, 133, 156
Podocarpus 133, 137
  P. drouynianus 130
  P. latifolius 156, 156*f*, Plate 10c
  P. saligna 116*t*
Polemoniaceae 138
pollinators, interactions between plants and 94–5
Populus 116*t*, 137
Portugal 240*t*, 271
  dune ecosystems 304
  Serra de Estrela 50
Portuguese oak 128
Potentilla neumanniana 135
pre-adaptations *see* exaptations
precipitation 37
  mean annual precipitation (MAP) 36*f*, 37, 38, 77
primary consumers 219
primary producers 219
primary production
  gross 220
  net 220
productivity 223
  aboveground NPP 223–4
  belowground 225
Prosopis 153
  P. chilensis 256
Protea 46*f*, 121, 160, 198*t*, 205*t*, 206, Plate 3d, Plate 5
  P. amplexicaulis 96*f*, Plate 7f
  P. nitida 45*f*, Plate 4a
  P. repens 187*f*, 222*t*
  transplant experiments 121
Proteaceae 26*f*, 27, 153, 270
protected areas 131, 255, 293, 296
  expansion of 296–7
Prunus 152
Pseudomonas 83
  P. putida 83*t*
Pseudotsuga menziesii var. menziesii 130

Psoralea 228*t*
Purnell, Helen 80
Pyrenean oak 128
pyroendemics (fire endemics) 138
Pythium 270

Q

Quercus 26, 26*f*, 45*f*, 115*t*, 117*t*, 152, 154, 160, 198*t*,
   220, Plate 4d
  Q. agrifolia 26*f*, 128, Plate 3a
  Q. berberidifolia 198*t*
  Q. chrysolepis 115*t*
  Q. coccifera 118*t*, 121, 128, 129, 222*t*
  Q. douglasii 128
  Q. dumosa 116*t*
  Q. engelmannii 128
  Q. faginea 128
  Q. ilex 128, 129, 131, 268–70
  Q. lobata 128
  Q. pubescens 117*t*, 129
  Q. pyrenaica 128
  Q. robur 304
  Q. suber 26*f*, 128, 131, 246*f*, 268, 299, Plate 3b,
    Plate 14b
Quillaja 85, 153
  Q. saponaria 85, 117*t*, 198*t*

R

rabbits 268
  see also herbivores
rainbow trout 267
Ranunculus 158*t*
Raunkiaer life forms 73–5*t*
red-eyed wattle 264*f*, Plate 16a
Red fir 128
redwood 129
refugia 40, 128, 137, 166, 204, 273
rehabilitation 302, 306, 307*f*
relict species 155, 156*f*, 166
  Jubaea and *Araucaria* 89*f*
  limited distributions 156*f*, Plate 10
renosterveld 118, 122, 123, 127, 252, 258
reproduction 76, 138, 204, 206, 208, 209, 225
respiration 189, 220, 228
resprouting 44, 45*f*, 55, 84, 202, 204, 210
  Adenostoma fasciculatum 275
  Arctostaphylos 159*f*
  and non-resprouting species from
    Arctostaphylos 159*f*, Plate 11
Restoniaceae 153
restoration 257, 295, 302–7, 312
Retama 228*t*
Rhamnaceae 28, 30, 55, 93, 198*t*, 205*t*, 228*t*

Rhamnus 152
  R. oleoides 94*f*, Plate 6d
Rhizobium 228
Rhododendron 152
Rhus 152, 154
  R. ovata 40*f*, 187*f*
Rio+20 Sustainable Development goals 296
riparian communities 136–8
riparian plants 111
Rodríguez-Sánchez, Jesus 82
Romulea rosea 267
roots 223
  adaptations to nutrient-poor soils 79*f*
  adaptive traits 77
  cluster 77, 79*f*, 80, 81*f*, 83*t*, 227–8, 270
  nodules 77, 84
Rytidosperma 135

S

sage scrub 72*f*, 111, 116*t*, 124, 246, 265, 275, 305
  see also coastal sage scrub
Sahelian taxa 152
Salix 116*t*, 137
Salsola 152
Salvia 116*t*, 158, 158*t*
  S. mellifera 124
Sambucus 153
sandstone 119, 226, 257
Santa Monica Mountains, southern California 40
Santa Ynez Mountains, California 5*f*, Plate 1b
Sardinia, Italy 158*t*
sclerophyllous forests 111, 306
sclerophyllous leaves 6, 28, 87, 177–82, 227
Scotch broom 304
Searsia 154
  S. glauca 114*t*
  S. undulata 198*t*
seasonality 35–6, 37, 38
secondary invasion 267, 304
seed banks 202, 204, 205, 206, 207, 208, 249, 265,
   303, 309
seed dormancy 206
seeds, germination of fire-cued 205
Senecio 157*t*
senescence 46, 130, 181
Sequoiadendron giganteum 91, 130, 132*f*, 156, 156*f*,
   Plate 10b
Sequoia sempervirens 111, 129, 130
serotinous species 206
serotiny 43, 84, 202, 205, 206
Seweweekspoortpiek, South Africa 50
shrublands 111, 177–218
  evergreen sclerophyll 110–11, 118–23, 177–82, 221–2*t*

shrubs
    evergreen sclerophyllous 76, 77, 91, 115*t*
    flowering times of 194*f*
    response to fire 204–6
Silene 158*t*
Siliceous rocks 51*t*
small mouth bass 267
smoke, to stimulate germination 303
soil
    and foliar nutrient content 226*t*
    formation 226
    infertility 55
    and mineral nutrition 51–3
    nitrogen soil content 52
    nutrient-poor 225
    phosphorus soil content 52
Solanum 157*t*
Sophora macrocarpa 228*t*
Sorbus 152
South Africa
    areas of protected, converted and impacted
      land in 293*t*
    'Biodiversity and Wine Initiative' 299
    bird pollinators from 96*f*, Plate 7c
    Cape Region of *see* Cape Floristic Region,
      South Africa
    common Proteaceae genera in 26*f*
    conservation status of plant species in 257
    development of modern MTC region 33
    diversity 168
    ecological restoration 304
    edaphic communities 92*t*
    edaphic filters 110
    evergreen sclerophyll shrublands 122
    forests 114*t*, 128
    fynbos stand structure and productivity 222*t*
    geofloras 154
    grasslands 115*t*, 135
    Greater Cape Floristic Region *see* Cape Floristic
      Region, South Africa
    habitat fragmentation 258–9
    highest point of 50
    human impact 244*t*
    invasive Australian acacia species in 248*t*
    invasive plants 266
    landscape 50
    land-use, land-cover change in 257
    mainstreaming 300
    most speciose genera in 157*t*
    nitrogen deposition rates 271
    post-fire flowering bulb from 45*f*
    post-fire seeders 205*t*
    Protea species 46*f*, Plate 5a
    protection levels 296
    rodent-pollinated flower from 96*f*, Plate 7f
    shrublands 114–15*t*
    soil and foliar nutrient content 226*t*
    stewardship 299
    trees 131
    use of fire by early South African Cape
      dwellers 260
    vegetation community types 114–15*t*
    woodland (thicket) 114*t*
    woody invasive taxa in 266
    'Working for water programme' 300
South America 14, 31, 33, 85, 130, 152, 153, 243, 256
southern hemisphere 6, 25, 27, 97, 153, 195, 312
    common taxa in MTEs 26*f*, Plate 3
Spain 34, 240*t*, 271
    arid shrubland 127
    diverse genera of Eastern Andalusia, Spain 158*t*
    forests and woodland 129
    FRI 45, 261
    habitat fragmentation 258
    Mulhacén 50
    plant drought response 154
    Poniente wind 43
    Quercus robur forests 304
    Sierra Nevada range 50
    tomillares 125
Spanish cork oak 26*f*, Plate 3b
spatial planning, biodiversity and 300–2
species distributions 3, 4, 39, 41, 121
specific leaf area 178
Spergularia rubra 135
Spinifex grassland 135
Stachys 158*t*
stand structure 220, 224
    and productivity 221–2*t*
stewardship, conservation 295, 297–300
Stipa
    S. laevissima 117*t*
    S. pennata 135
Stirlingia latifolia 198*t*
Stirling, James 44
Stirling Range, Western Australia, Australia 5*f*, 51,
    Plate 1a
stomatal crypts 184*f*, 185, Plate 12
strandveld 111
strategic land acquisition 302
Stroebe 160
Stylidium 157*t*
substrata, soil 51, 56
succulence 7
succulent karoo, South Africa 17, 38, 125, 126*f*, 127,
    154, Plate 8a

sudden oak death 270
Sugar Palm 91
sulfolipids 228
sundews 229, 230*f*, Plate 13b, Plate 13c
symbioses 79, 84
systematic conservation planning 296, 301*f*

T
tanoak 270
Tegeticula 95
Teline (Genista) monspessulana 266
temperatures
    affect on phenology and biogeography of
      species 195
    definitions for MTC regions 38
Tetraclinus 117*t*
    T. articulata 131
Teucrium 158*t*
thistles 118*t*, 265
thorn scrub, Chile 126*f*, Plate 8c
tomillares 111, 125
topography and geology 47–51
traits, sharing of traits by species in
      MTC regions 29, 29*f*
transformer species 266
transplant experiments 121
tree parasite 229
trees, venerable 129–33
tree tobacco 264*f*, Plate 16c
Trifolium
    T. glomeratum 135
    T. subterraneum 135
Triticum 118*t*, 135
trophic structure 4
Tupungato Mountain, Chile 50
Turkey 158*t*, 271
    diverse genera of 158*t*
    Mount Demirkazik 50
    Taurus Range 131

U
Ulex 205*t*, 228*t*
   U. europaeus 266
Ultramafic rocks 51*t*
Utricularia 229

V
Valeriana 157*t*
valley oak 128
vascular plants, species richness of 18*f*, 19*f*
vegetation
    dynamics 138–9

evergreen shrub-dominated 5–6
    mediterranean-type *see* mediterranean-type
      vegetation (MTV)
venerable trees 129–33
Verbascum 158*t*
vernal pool communities 133, 258
vertebrates 87, 88, 207, 267, 268
    diversity and endemism of 90*t*
    seed dispersal by 96
Verticordia 157*t*
Vicia 158*t*
Viola 157*t*
volcanic activity 43, 47, 49*f*, 131
von Humboldt, Alexander 4, 14
Vulpes vulpes 268
Vulpia geniculata 135

W
wallaby grasses 135
water
    limitations 208–9
    potential 188*f*, 197, 198*t*, 199*f*, 232
    responding to limited 196–200
weathering 226
Western Cape, South Africa 5*f*, 50, 254, 300,
    Plate 1e
western white pine 128
wild oats 116*t*, 265
winds
    Berg 42–3
    foehn 42
    katabatic 42
    Santa Ana 43, 252, 261
    seasonal 42
wine industry 299
winter-deciduous
    blue oak 128
    forests and woodlands 128
    riparian taxa 138
    shrubs 71
winter-rainfall desert shrublands 125–7
woodlands 111, 128–33
    see also forests

Y
'young, often disturbed, fertile landscapes'
    (YODFELs) 50, 168–9
Yucca brevifolia 95, 116*t*
yucca moths 95

Z
Zantedeschia aethiopica 264*f*, 267, Plate 16h